TABLE

OF

MOLECULAR WEIGHTS

A COMPANION VOLUME TO

THE MERCK INDEX

NINTH EDITION

Editors
Martha Windholz
Susan Budavari
Margaret Noether Fertig
Georg Albers-Schönberg

Published by

MERCK & CO., INC.

RAHWAY, N.J., U.S.A.

1978

Merck & Co., Inc.
Rahway, New Jersey, U.S.A.

MERCK SHARP & DOHME
West Point, Pa.

MERCK SHARP & DOHME INTERNATIONAL
Rahway, N. J.

MERCK ANIMAL HEALTH DIVISION
Rahway, N. J.

MERCK CHEMICAL MANUFACTURING DIVISION
Rahway, N. J.

MERCK CHEMICAL DIVISION
Rahway, N. J.

MERCK SHARP & DOHME RESEARCH LABORATORIES
Rahway, N. J. / West Point, Pa.

BALTIMORE AIRCOIL COMPANY, INC.
Baltimore, Md.

CALGON CORPORATION
Pittsburgh, Pa.

HUBBARD FARMS, INC.
Walpole, N. H.

KELCO DIVISION
San Diego, Calif.

1st Edition—1978

Library of Congress Catalog
Card Number 77-95432
ISBN Number 911910-73-5

Printed in the U. S. A.

EDITORS' PREFACE

In the early 1970s, soon after publication of *The Merck Index, Eighth Edition,* it came to our attention that Drs. Henry M. Fales and Tung Sun of the Laboratory of Chemistry, National Heart, Lung and Blood Institute, had compiled a computer listing from *The Merck Index* which was very useful to mass spectroscopists. It contained the molecular weights of selected compounds described in *The Merck Index,* listed in ascending order along with the corresponding empirical formulae. At one time we considered including such a list in the ninth edition of *The Merck Index,* but decided against it because of space limitations.

Shortly after publication of *The Merck Index, Ninth Edition,* several readers and, most particularly, Dr. R. L. Wolen of the Lilly Laboratory for Clinical Research, inquired whether, in light of the fact that the monographs were stored in computer memory, we planned to prepare a list, similar to Dr. Fales', of compounds from the ninth edition.

Prompted by these outside stimuli, we discussed the possibility of such a project with Dr. Georg Albers-Schönberg, Senior Investigator, Merck Sharp & Dohme Research Laboratories and were very fortunate to enlist his help as co-editor and consultant in mass spectrometry. Under his guidance the selection of compounds was made and the format for presentation was chosen.

We are now pleased to present this information in the form of a book entitled *Table of Molecular Weights, A Companion Volume to The Merck Index, Ninth Edition.* The book contains high-resolution molecular weights, arranged in ascending order associated with empirical formulae, compound names, and monograph numbers under which the specific compounds are listed in *The Merck Index.* We hope that this volume used with *The Merck Index* will serve as an important tool in making the identification of chemical compounds easier.

It was our aim to produce this volume as rapidly as possible utilizing much of the computer programming developed for the production of *The Merck Index, Ninth Edition* and without extensive editorial refinement. We shall welcome any comments from our readers, be they corrections of errors or suggestions for improving the book in the next edition.

We wish to thank the American Society for Mass Spectrometry for providing us with a forum in which to determine the extent of interest in this publication, and individual members of the Society, in particular, Drs. J. A. McCloskey, C. Fenselau, J. H. Futrell, and most especially, Dr. Wm. J. A. VandenHeuvel, for their help and useful suggestions. The editors are also indebted to Dr. A. H. Wapstra of the Instituut voor Kernphysisch Onderzoeck, Amsterdam, Holland, for providing the reference to his list of high resolution atomic masses which has been used in the calculation of the molecular weights.

Publication of this book would not have been possible without the outstanding contribution of the Automation & Control Department of Merck & Co.,

Inc. We express our gratitude to Mr. M. W. Cline for design of the system, to Mr. R. J. Cimato for coordinating and managing the project, to Miss B. J. Wasowski and Mr. J. C. Weber for their programming efforts, and to Mr. G. Murchake for keeping the text-editing equipment in good running order. We acknowledge with thanks the technical assistance provided by Miss J. R. Gallipeau, Miss E. V. Gannon and Mrs. M. Nunez.

Finally, we wish to extend thanks to representatives of mass spectrometer manufacturers who provided us with useful information and encouraged us to undertake this endeavor.

The Editors

EXPLANATORY NOTES

The Table of Molecular Weights is a listing of over nine thousand compounds selected from *The Merck Index, Ninth Edition,* arranged in order of increasing molecular weight and paired with empirical formulae. The list provides mass spectroscopists and other scientists with a link between high resolution molecular weights of compounds and their detailed descriptions in *The Merck Index.*

For the purpose of this book, the following data were extracted by computer from *The Merck Index* monographs: monograph number, monograph title, empirical formula, derivative of title compound and derivative empirical formula. (*See* Explanatory Notes, page ix, *The Merck Index* for a detailed explanation of the format of the monographs.) Programs were written to calculate high resolution molecular weights from the empirical formulae, to group compounds with the same molecular weight, to alphabetize within groups of compounds, and to arrange groups in order of increasing molecular weight.

This computer-generated list was then selectively edited to increase its utility for mass spectroscopists. The rules observed during editing were: a) all salts of organic acids and bases were deleted in favor of the corresponding free acids and bases; b) the anionic portions of quaternary salts were removed and only the cationic portions listed; c) the formulae and names of the individual components of mixtures and complexes were given, the empirical formulae and names of the combinations deleted. In accordance with these rules, compounds and empirical formulae not already in the data base were added and their molecular weights calculated. Therefore, a small number of compounds in the table will not be found in the identical form in *The Merck Index.*

Except for the changes detailed above, no attempt was made to limit the list to those compounds for which mass spectra can be or have been run or for which molecular ion peaks can be expected. In particular, the applicability of field-desorption techniques to the analysis of some inorganic compounds has led us to retain, at least for the present edition, these compounds in the list. See, for example, the discussion by H.-R. Schulten in *Methods of Biochemical Analysis,* D. Glick, Ed. (John Wiley-Interscience, New York, 1977) pp 340-342.

The information is arranged in three columns. A running head is provided which indicates the integer portion of the molecular weight of the first entry on even-numbered pages and of the last entry on odd-numbered pages.

Molecular Weight. The molecular weights listed in the first column were calculated to six decimal places from the atomic masses of the most abundant isotope of each element. The values used were those reported by A. H. Wapstra and K. Bos in *Atomic Data and Nuclear Data Tables* **19**(3), 177-214 (1977). A list of atomic masses appears immediately following these explanatory notes. In addition, a complete table of the atomic masses and abundances of naturally occurring isotopes, which was compiled using the 1975 IUPAC list published in *Pure and Applied Chemistry* **47**(1), 77-95 (1976) and the Wapstra, Bos publication mentioned above, is included in the appendix.

For the most part, a molecular weight appears once in the table. Exceptions to this rule occur; for example, an uncharged compound and a positively charged quaternary compound can have the same composition and, therefore, the molecular weight is entered twice.

Empirical Formula. The empirical formulae from which the molecular weights were calculated are listed in the second column. Empirical formulae of multiply-charged ions appear twice in the table; once next to the molecular weight divided by the charge (m/e) and again at the molecular weight. (The order of elements within formulae conforms to Chemical Abstracts' rules.)

Compound Name, Monograph Number. The names of the compounds corresponding to the empirical formulae are listed in the third column. One or more numbers follow the name, with the first being the number of the principal monograph describing the compound.

In most cases, the compound name is the monograph title alone or in combination with the derivative type. Occasionally, the monograph title was modified to represent a different form of the title compound, or was deleted when the derivative type alone served to name the compound. For the few cases where it was not possible to derive the compound name from the monograph title, trivial names or simple chemical names were used.

On occasion, one entry, such as Aminophenols, 470-472, served to describe several isomers, with the relevant monograph numbers grouped. When a particular compound appeared more than once in the data base, it was listed once and all the relevant monograph numbers were given. (A notable exception was made for common acids such as maleic acid where only the number of the principal monograph was listed.) If a number of monographs treated various forms of a particular compound, the name of the parent compound was given, as was its principal monograph number and the range of the monograph numbers of its salts, e.g. Quinine, 7853, 7854-7882.

Since this table was generated well over a year after the first printing of the ninth edition of *The Merck Index,* there has been an opportunity to correct errors that were detected in empirical formulae appearing in the original printing. This accounts for discrepancies between the empirical formulae found there and in this volume.

It is intended that this volume serve as a workbook. The number of entries per page, therefore, has been limited to allow space for the user to annotate the text and insert additional molecular weights and compound names. In this way, the individual scientist may further adapt the volume to his/her specialized needs.

Publication of this book represents a first effort; comments and suggestions regarding possible improvements for future editions will be welcomed.

Atomic Masses of Most Abundant Isotopes[a]

Symbol	Element	Atomic Mass	Symbol	Element	Atomic Mass
Ac	Actinium	227.02775091	Mn	Manganese	54.9380463
Ag	Silver	106.905095	Mo	Molybdenum	97.9054050
Al	Aluminum	26.9815413	N	Nitrogen	14.003074008
Am	Americium	243.061374114	Na	Sodium	22.9897697
Ar	Argon	39.9623831	Nb	Niobium	92.9063780
As	Arsenic	74.9215955	Nd	Neodymium	141.907731
At	Astatine	209.987143	Ne	Neon	19.9924391
Au	Gold	196.966560	Ni	Nickel	57.9353471
B	Boron	11.0093053	No	Nobelium	255.09326
Ba	Barium	137.905236	Np	Neptunium	237.048168800
Be	Beryllium	9.0121825	O	Oxygen	15.99491464
Bi	Bismuth	208.980388	Os	Osmium	191.961487
Bk	Berkelium	247.070300	P	Phosphorus	30.9737634
Br	Bromine	78.9183361	Pa	Protactinium	231.035880917
C	Carbon	12.000000000	Pb	Lead	207.976641
Ca	Calcium	39.9625907	Pd	Palladium	105.903475
Cd	Cadmium	113.9033607	Pm	Promethium	144.912754
Ce	Cerium	139.905442	Po	Polonium	208.982422
Cf	Californium	251.079581	Pr	Praseodymium	140.907657
Cl	Chlorine	34.968852729	Pt	Platinum	194.964785
Cm	Curium	247.070349	Pu	Plutonium	244.064200
Co	Cobalt	58.9331978	Ra	Radium	226.025405993
Cr	Chromium	51.9405097	Rb	Rubidium	84.9117996
Cs	Cesium	132.905243	Re	Rhenium	186.955765
Cu	Copper	62.9295992	Rh	Rhodium	102.905503
Dy	Dysprosium	163.929183	Rn	Radon	222.017573772
Er	Erbium	165.930305	Ru	Ruthenium	101.9043475
Es	Einsteinium	254.088021	S	Sulfur	31.9720718
Eu	Europium	152.921243	Sb	Antimony	120.9038237
F	Fluorine	18.99840325	Sc	Scandium	44.9559136
Fe	Iron	55.9349393	Se	Selenium	79.9165205
Fm	Fermium	257.095103	Si	Silicon	27.9769284
Fr	Francium	223.019734	Sm	Samarium	151.919741
Ga	Gallium	68.9255809	Sn	Tin	119.9021990
Gd	Gadolinium	157.924111	Sr	Strontium	87.9056249
Ge	Germanium	73.9211788	Ta	Tantalum	180.948014
H	Hydrogen	1.007825037	Tb	Terbium	158.925350
He	Helium	4.00260325	Tc	Technetium	96.906362
Hf	Hafnium	179.946561	Te	Tellurium	129.906229
Hg	Mercury	201.970632	Th	Thorium	232.038053805
Ho	Holmium	164.930332	Ti	Titanium	47.9479467
I	Iodine	126.904477	Tl	Thallium	204.974410
In	Indium	114.903875	Tm	Thulium	168.934225
Ir	Iridium	192.962942	U	Uranium	238.050785782
K	Potassium	38.9637079	V	Vanadium	50.9439625
Kr	Krypton	83.9115064	W	Tungsten	183.950953
La	Lanthanum	138.906355	Xe	Xenon	131.904148
Li	Lithium	7.0160045	Y	Yttrium	88.9058560
Lr	Lawrencium	260.10536	Yb	Ytterbium	173.938873
Lu	Lutetium	174.940785	Zn	Zinc	63.9291454
Md	Mendelevium	258.09857	Zr	Zirconium	89.9047080
Mg	Magnesium	23.9850450			

[a] From data in A. H. Wapstra, K. Bos, *Atomic Data and Nuclear Data Tables* **19**(3), 177-214 (1977).

TABLE
OF
MOLECULAR WEIGHTS

Molecular Weight	Empirical Formula	Compound Name, Monograph Number
8.023830	HLi	Lithium Hydride, 5370
11.027833	BeH_2	Beryllium Hydride, 1194
16.031300	CH_4	Methane, 5809
17.026549	H_3N	Ammonia, 506
18.010565	H_2O	Water, 9701
20.006228	FH	Hydrogen Fluoride, 4689
20.023118	D_2O	Deuterium Oxide, 2897
22.056610	BH_4Li	Lithium Borohydride, 5361
23.034729	H_2LiN	Lithium Amide, 5357
23.997595	HNa	Sodium Hydride, 8377
24.018744	HLiO	Lithium Hydroxide, 5371
25.007097	BeO	Beryllium Oxide, 1199
25.012379	BN	Boron Nitride, 1360
26.000695	H_2Mg	Magnesium Hydride, 5492
26.014408	FLi	Lithium Fluoride, 5368
26.015650	C_2H_2	Acetylene, 82
27.010899	CHN	Hydrogen Cyanide, 4688
27.994915	CO	Carbon Monoxide, 1819
28.031300	C_2H_4	Ethylene, 3728
28.065561	B_2H_6	Diborane(6), 2981
29.997989	NO	Nitric Oxide, 6396
30.005016	AlH_3	Aluminum Hydride, 340
30.010565	CH_2O	Formaldehyde, Gas, 4093
30.024365	CBe_2	Beryllium Carbide, 1190
30.026924	Li_2O	Lithium Oxide, 5376
30.046950	C_2H_6	Ethane, 3649
31.042199	CH_5N	Methylamine, 5889
32.008229	H_4Si	Silane, 8230
32.026215	CH_4O	Methanol, 5814
32.037448	H_4N_2	Hydrazine, 4653, 3147, 9362
33.021464	H_3NO	Hydroxylamine, 4730

33.

Molecular Weight	Empirical Formula	Compound Name, Monograph Number
33.987722	H_2S	Hydrogen Sulfide, 4695
33.997239	H_3P	Phosphine, 7149
34.005479	H_2O_2	Hydrogen Peroxide, 4691
34.021878	CH_3F	Fluoromethane, 4060
35.976678	ClH	Hydrogen Chloride, 4687
37.032777	FH_4N	Ammonium Fluoride, 545
38.028846	AlH_4Li	Aluminum Lithium Hydride, 347
38.030375	BH_4Na	Sodium Borohydride, 8336
39.008494	H_2NNa	Sodium Amide, 8315
39.059133	C_2H_6Be	Dimethylberyllium, 3227
39.093393	B_2BeH_8	Beryllium Borohydride, 1188
39.976928	CSi	Silicon Carbide, 8234
39.979960	MgO	Magnesium Oxide, 5500
39.992509	$HNaO$	Sodium Hydroxide, 8380
40.984615	AlN	Aluminum Nitride, 352
41.026549	C_2H_3N	Acetonitrile, 56
41.978241	CaH_2	Calcium Hydride, 1669
41.984857	$ClLi$	Lithium Chloride, 5364
41.988173	FNa	Sodium Fluoride, 8368
42.010565	C_2H_2O	Ketene, 5148
42.021798	CH_2N_2	Cyanamide, 2693 Diazomethane, 2965
42.046950	C_3H_6	Cyclopropane, 2753 Propylene, 7638
43.005814	$CHNO$	Cyanic Acid, 2694 Isocyanic Acid, 5021
43.017047	HN_3	Hydrogen Azide, 4685
43.017662	BeH_2O_2	Beryllium Hydroxide, 1195
43.042199	C_2H_5N	Ethylenimine, 3739
43.971843	OSi	Silicon Monoxide, 8238
43.989829	CO_2	Carbon Dioxide, 1815
44.001063	N_2O	Nitrous Oxide, 6473

Molecular Weight	Empirical Formula	Compound Name, Monograph Number
44.026215	C_2H_4O	Acetaldehyde, 27 Ethylene Oxide, 3738
44.062600	C_3H_8	Propane, 7591
45.021464	CH_3NO	Formamide, 4096
45.057849	C_2H_7N	Dimethylamine, 3215 Ethylamine, 3690
45.992903	NO_2	Nitrogen Dioxide, 6425
46.005479	CH_2O_2	Formic Acid, 4098
46.041865	C_2H_6O	Alcohol, Anhydrous, 211 Methyl Ether, 5940
46.053098	CH_6N_2	Methylhydrazine, 5954
47.000728	HNO_2	Nitrous Acid, 6472
47.008989	BeF_2	Beryllium Fluoride, 1192
47.037114	CH_5NO	Methoxyamine, 5862
48.001143	$CHFO$	Formyl Fluoride, 4106
48.003372	CH_4S	Methanethiol, 5813
48.992844	$CNNa$	Sodium Cyanide, 8354
48.996392	FNO	Nitrosyl Fluoride, 6467
49.963597	H_2Ti	Titanium Hydride, 9183
49.992328	CH_3Cl	Methyl Chloride, 5916
51.014271	H_5NS	Ammonium Bisulfide, 518
51.971592	$ClHO$	Hypochlorous Acid, 4786
52.006148	C_2N_2	Cyanogen, 2698
53.003227	ClH_4N	Ammonium Chloride, 528
53.026549	C_3H_3N	Acrylonitrile, 127
53.967256	ClF	Chlorine Monofluoride, 2068
53.991721	F_2O	Fluorine Monoxide, 4048
54.004313	BH_4K	Potassium Borohydride, 7392
54.046950	C_4H_6	1,3-Butadiene, 1491
54.115472	B_4H_{10}	Tetraborane(10), 8900
55.042199	C_3H_5N	Propionitrile, 7617
.042696	Be_3N_2	Beryllium Nitride, 1198

Molecular Weight	Empirical Formula	Compound Name, Monograph Number
55.957505	CaO	Calcium Oxide, 1686
55.966448	HKO	Potassium Hydroxide, 7423
55.969667	HNaS	Sodium Bisulfide, 8331
55.974874	MgO_2	Magnesium Peroxide, 5504
56.022493	H_4MgN_2	Magnesium Amide, 5477
56.026215	C_3H_4O	Acrolein, 123 Propargyl Alcohol, 7599
56.037221	CB_4	Boron Carbide, 1358
56.037448	$C_2H_4N_2$	Aminoacetonitrile, 412
56.062600	C_4H_8	1-Butene, 1502 2-Butene, 1503 Cyclobutane, 2719 Isobutylene, 4995
57.039006	H_5F_2N	Ammonium Bifluoride, 514
57.057849	C_3H_7N	Allylamine, 278
57.955305	AlP	Aluminum Phosphide, 359
57.958622	ClNa	Sodium Chloride, 8343
57.962111	FK	Potassium Fluoride, 7411
57.990524	H_2MgO_2	Magnesium Hydroxide, 5493
58.005479	$C_2H_2O_2$	Glyoxal, 4346
58.041865	C_3H_6O	Acetone, 52 Allyl Alcohol, 277 Propionaldehyde, 7613 Propylene Oxide, 7645 Trimethylene Oxide, 9385
58.053098	$C_2H_6N_2$	Acetamidine, 32
58.078250	C_4H_{10}	Butane, 1497
58.982971	CHNS	Thiocyanic Acid, 9062
59.037114	C_2H_5NO	Acetaldoxime, 30 Acetamide, 31
59.048347	CH_5N_3	Guanidine, 4414
59.073499	C_3H_9N	Isopropylamine, 5070 Propylamine, 7631 Trimethylamine, 9379
59.966758	O_2Si	Silicon Dioxide, 8235
60.021129	$C_2H_4O_2$	Acetic Acid Glacial, 43 Methyl Formate, 5942

Use in conjunction with The Merck Index, Ninth Edition

Molecular Weight	Empirical Formula	Compound Name, Monograph Number
60.032363	CH_4N_2O	Urea, 9525, 166, 9526, 9530, 9532
60.057515	C_3H_8O	Ethyl Methyl Ether, 3766 Isopropyl Alcohol, 5069 n-Propyl Alcohol, 7630
60.068748	$C_2H_8N_2$	1,1-Dimethylhydrazine, 3237 1,2-Dimethylhydrazine, 3238 Ethylenediamine, 3731, 476
60.971927	$CClN$	Cyanogen Chloride, 2701
60.998621	CH_3NS	Thioformamide, 9066
61.016378	CH_3NO_2	Nitromethane, 6433
61.052764	C_2H_7NO	Acetaldehyde Ammonia, 28 Ethanolamine, 3654
61.974454	Na_2O	Sodium Oxide, 8411
61.981852	F_2Mg	Magnesium Fluoride, 5488
61.992328	C_2H_3Cl	Vinyl Chloride, 9645
61.994401	F_2HNa	Sodium Bifluoride, 8328
62.000394	CH_2O_3	Performic Acid, 6953
62.000807	H_6Si_2	Disilane, 3373
62.011627	$H_2N_2O_2$	Nitramide, 6388
62.017524	BH_3O_3	Boric Acid, 1348
62.019022	C_2H_6S	Ethanethiol, 3653 Methyl Sulfide, 5996
62.036780	$C_2H_6O_2$	Ethylene Glycol, 3735
62.995643	HNO_3	Nitric Acid, 6393
63.961901	O_2S	Sulfur Dioxide, 8766
63.962591	C_2Ca	Calcium Carbide, 1649
64.007978	C_2H_5Cl	Ethyl Chloride, 3713
64.116952	B_5H_9	Pentaborane(9), 6896
64.966782	CKN	Potassium Cyanide, 7405
64.966841	$ClNO$	Nitrosyl Chloride, 6466
64.987758	$CNNaO$	Sodium Cyanate, 8353
64.991150	$CLiNS$	Lithium Thiocyanate, 5386
64.991307	FNO_2	Nitryl Fluoride, 6479
64.998992	N_3Na	Sodium Azide, 8325

Molecular Weight	Empirical Formula	Compound Name, Monograph Number
65.987068	H_3O_2P	Hypophosphorous Acid, 4790
65.988904	$BNaO_2$	Sodium Metaborate, 8397
65.991721	CF_2O	Carbonyl Fluoride, 1825
66.046950	C_5H_6	Cyclopentadiene, 2743
66.132602	B_5H_{11}	Pentaborane(11), 6897
66.958682	ClO_2	Chlorine Dioxide, 2066
67.042199	C_4H_5N	Allyl Cyanide, 282 Methacrylonitrile, 5797 Pyrrole, 7801
67.090433	$C_4H_{10}Be$	Diethylberyllium, 3093
67.989829	C_3O_2	Carbon Suboxide, 1820
68.004515	BF_3	Boron Trifluoride, 1363
68.012296	CN_4	Cyanogen Azide, 2699
68.026215	C_4H_4O	Furan, 4146
68.037448	$C_3H_4N_2$	Imidazole, 4807 Methyleneaminoacetonitrile, 5927 Pyrazole, 7744
68.040820	H_8N_2S	Ammonium Sulfide, 587
68.062600	C_5H_8	Isoprene, 5061
68.982673	$NNaO_2$	Sodium Nitrite, 8407
68.998142	ClH_4NO	Hydroxylamine Hydrochloride, 4731
69.003822	$LiNO_3$	Lithium Nitrate, 5374
69.032697	$C_2H_3N_3$	1H-1,2,4-Triazole, 9285
69.057849	C_4H_7N	Butyronitrile, 1589 3-Pyrroline, 7805
69.986636	F_2O_2	Fluorine Dioxide, 4047
70.003035	CHF_3	Fluoroform, 4058
70.003355	B_2O_3	Boric Anhydride, 1349
70.005479	$C_3H_2O_2$	Propiolic Acid, 7611
70.016713	$C_2H_2N_2O$	Furazan, 4150 1,3,4-Oxadiazole, 6739
70.041865	C_4H_6O	Crotonaldehyde, 2590 Methyl Vinyl Ketone, 6007 Vinyl Ether, 9646

Use in conjunction with The Merck Index, Ninth Edition

Molecular Weight	Empirical Formula	Compound Name, Monograph Number
70.053098	$C_3H_6N_2$	3-Aminopropionitrile, 480 2-Pyrazoline, 7745
70.078250	C_5H_{10}	Amylene, 645 Cyclopentane, 2745 1-Pentene, 6922 2-Pentene, 6923
70.998284	F_3N	Nitrogen Fluoride, 6426
71.037114	C_3H_5NO	Acrylamide, 125 Hydracrylonitrile, 4644
71.073499	C_4H_9N	Pyrrolidine, 7802
71.929854	FeO	Ferrous Oxide, 3977
71.934663	CaS	Calcium Sulfide, 1710
71.943605	HKS	Potassium Bisulfide, 7390
71.952420	CaO_2	Calcium Peroxide, 1690
72.021129	$C_3H_4O_2$	Acrylic Acid, 126 β-Propiolactone, 7610 Pyruvaldehyde, 7807
72.057515	C_4H_8O	Butyraldehyde, 1583 Crotyl Alcohol, 2595 Cyclopropyl Methyl Ether, 2754 Isobutyraldehyde, 5009 Methyl Ethyl Ketone, 5941 Tetrahydrofuran, 8929
72.093900	C_5H_{12}	Neopentane, 6281 Pentane, 6913
72.103358	AlB_3H_{12}	Aluminum Borohydride, 325
72.998621	C_2H_3NS	Methyl Isothiocyanate, 5957 Methyl Thiocyanate, 6001
73.052764	C_3H_7NO	Acetoxime, 66 N,N-Dimethylformamide, 3233 Propionaldehyde, Oxime, 7613 Propionamide, 7614
73.063997	$C_2H_7N_3$	Methylguanidine, 5950
73.089149	$C_4H_{11}N$	Butylamines, 1527-1529 Diethylamine, 3081 Isobutylamine, 4985
73.930262	NiO	Nickel Monoxide, 6326
73.932561	ClK	Potassium Chloride, 7399
73.953537	ClNaO	Sodium Hypochlorite, 8381
73.968070	CaH_2O_2	Calcium Hydroxide, 1670

Molecular Weight	Empirical Formula	Compound Name, Monograph Number
73.992328	C_3H_3Cl	Propargyl Chloride, 7600
74.000394	$C_2H_2O_3$	Glyoxylic Acid, 4348
74.016753	CLi_2O_3	Lithium Carbonate, 5363
74.036780	$C_3H_6O_2$	Ethyl Formate, 3743 Glycidol, 4324 Methyl Acetate, 5884 Propionic Acid, 7615
74.048013	$C_2H_6N_2O$	N-Nitrosodimethylamine, 6458
74.059246	CH_6N_4	Aminoguanidine, 447
74.073165	$C_4H_{10}O$	Butyl Alcohols, 1524-1526 Ethyl Ether, 3742 Isobutyl Alcohol, 4984 Methyl Propyl Ether, 5983
74.084398	$C_3H_{10}N_2$	Propylenediamine, 7641
74.096974	$[C_4H_{12}N]^+$	Tetramethylammonium, 8939, 8940
74.928112	CoO	Cobaltous Oxide, 2398
75.014271	C_2H_5NS	Thioacetamide, 9048
75.032028	$C_2H_5NO_2$	Ethyl Nitrite, 3770 Glycine, 4325, 8925 Methyl Carbamate, 5910 Nitroethane, 6420
75.043262	CH_5N_3O	Hydrazinecarboxamide, 8191
75.068414	C_3H_9NO	2-Aminopropanol, 479 2-Methylaminoethanol, 5890 Trimethylamine, Oxide, 9379
75.944144	CS_2	Carbon Disulfide, 1817
75.947018	Mg_2Si	Magnesium Silicide, 5515
75.998287	C_2H_4OS	Thioacetic Acid, 9049
76.007978	C_3H_5Cl	Allyl Chloride, 281
76.009520	CH_4N_2S	Ammonium Thiocyanate, 594 Thiourea, 9107
76.016044	$C_2H_4O_3$	Glycolic Acid, 4335 Peracetic Acid, 6945
76.027277	$CH_4N_2O_2$	Hydroxyurea, 4770
76.052430	$C_3H_8O_2$	Methylal, 5888 Methyl Cellosolve®, 5913 Propylene Glycol, 7644 Trimethylene Glycol, 9384
76.063663	$C_2H_8N_2O$	2-Hydrazinoethanol, 4659

Use in conjunction with The Merck Index, Ninth Edition

Molecular Weight	Empirical Formula	Compound Name, Monograph Number
76.134082	B_6H_{10}	Hexaborane(10), 4542
77.011293	CH_3NO_3	Methyl Nitrate, 5967
77.027692	C_2H_4FNO	Fluoroacetamide, 4051
77.029921	C_2H_7NS	Cysteamine, 2777
77.945071	AsH_3	Arsine, 838
77.951611	Na_2S	Sodium Sulfide, 8450
77.952479	GeH_4	Germane, 4239
77.959397	CaF_2	Calcium Fluoride, 1663
77.968339	F_2HK	Potassium Bifluoride, 7385
77.969369	Na_2O_2	Sodium Peroxide, 8416
77.987242	C_2H_3ClO	Acetyl Chloride, 77 Chloroacetaldehyde, 2078
77.989760	AlH_3O_3	Aluminum Hydroxide, 341
78.011708	$C_2H_3FO_2$	Fluoroacetic Acid, 4052
78.013937	C_2H_6OS	Dimethyl Sulfoxide, 3249 2-Mercaptoethanol, 5699
78.023628	C_3H_7Cl	Isopropyl Chloride, 5072 Propyl Chloride, 7635
78.042928	$CH_6N_2O_2$	Ammonium Carbamate, 524
78.046950	C_6H_6	Benzene, 1069
78.924514	CuO	Cupric Oxide, 2648
78.949888	$BeCl_2$	Beryllium Chloride, 1191
78.982491	CH_2ClNO	Carbamyl Chloride, 1784
79.026943	CH_5NO_3	Ammonium Bicarbonate, 513
79.042199	C_5H_5N	Pyridine, 7752, 7754, 7756, 7757
79.924060	OZn	Zinc Oxide, 9812
79.926161	BrH	Hydrogen Bromide, 4686
79.937776	O_2Ti	Titanium Dioxide, 9182
79.956816	O_3S	Sulfur Trioxide, 8775
79.968739	$CCaN_2$	Calcium Cyanamide, 1655
80.002893	C_2H_5ClO	Ethylene Chlorohydrin, 3730
80.022192	$H_4N_2O_3$	Ammonium Nitrate, 563

Molecular Weight	Empirical Formula	Compound Name, Monograph Number
80.037448	$C_4H_4N_2$	Pyrazine, 7741 Pyridazine, 7751 Pyrimidine, 7769 Succinonitrile, 8672
80.961697	CKNO	Potassium Cyanate, 7404
80.961756	$ClNO_2$	Nitryl Chloride, 6478
80.964916	CNNaS	Sodium Thiocyanate, 8465
80.986221	FNO_3	Fluorine Nitrate, 4049
81.084088	$B_3H_6N_3$	s-Triazaborane, 9283
81.932171	H_2Se	Hydrogen Selenide, 4694
81.961140	$AlNaO_2$	Sodium Aluminate, 8313
81.981982	H_3O_3P	Phosphorous Acid, 7155
81.983819	$BNaO_3$	Sodium Perborate, 8413
82.063295	$C_4H_{10}Mg$	Diethylmagnesium, 3103
82.078250	C_6H_{10}	Cyclohexene, 2732 2,3-Dimethyl-1,3-butadiene, 3228
83.013617	H_6NO_2P	Ammonium Hypophosphite, 552
83.073499	C_5H_9N	N-Methylpyrroline, 5988
83.930339	CrO_2	Chromium Dioxide, 2231
83.953356	CH_2Cl_2	Methylene Chloride, 5932
83.961422	$ClHO_3$	Chloric Acid, 2061
83.976751	AlF_3	Aluminum Fluoride, 337
83.982339	$CHNaO_3$	Sodium Bicarbonate, 8327
84.003372	C_4H_4S	Thiophene, 9090
84.021129	$C_4H_4O_2$	Diketene, 5148
84.032363	$C_3H_4N_2O$	Cyanoacetamide, 2696
84.043596	$C_2H_4N_4$	Aminotriazole, 496 Dicyanodiamide, 3068
84.057515	C_5H_8O	Cyclopentanone, 2747 2-Methyl-3-butyn-2-ol, 5908 Senecialdehyde, 8195
84.068748	$C_4H_8N_2$	Lysidine, 5454
84.093900	C_6H_{12}	Cyclohexane, 2728
84.956611	KNO_2	Potassium Nitrite, 7433

Use in conjunction with The Merck Index, Ninth Edition

Molecular Weight	Empirical Formula	Compound Name, Monograph Number
84.977588	$NNaO_3$	Sodium Nitrate, 8406
84.998621	C_3H_3NS	Thiazole, 9039
85.016378	$C_3H_3NO_2$	Cyanoacetic Acid, 2697
85.052764	C_4H_7NO	Acetone Cyanohydrin, 53 2-Pyrrolidone, 7804
85.089149	$C_5H_{11}N$	Piperidine, 7261
85.932620	Cl_2O	Chlorine Monoxide, 2069
85.934341	$BrLi$	Lithium Bromide, 5362
85.963793	F_2OS	Thionyl Fluoride, 9086
86.036780	$C_4H_6O_2$	Butyrolactone, 1588 Crotonic Acid, 2591 Diacetyl, 2923 Erythritol Anhydride, 3597 Isocrotonic Acid, 5019 Methacrylic Acid, 5796 Methyl Acrylate, 5887 Vinyl Acetate, 9644
86.048013	$C_3H_6N_2O$	2-Imidazolidinone, 4810
86.073165	$C_5H_{10}O$	Allyl Ethyl Ether, 286 Cyclopentanol, 2746 Diethyl Ketone, 3102 Isovaleraldehyde, 5093 Methyl Propyl Ketone, 5984 Tetrahydropyran, 8933 n-Valeraldehyde, 9566
86.084398	$C_4H_{10}N_2$	Piperazine, 7254, 7255-7257, 7259
86.096974	$[C_5H_{12}N]^+$	Neurine, Cation, 6302
86.109551	C_6H_{14}	n-Hexane, 4563
86.910118	MnS	Manganese Sulfide, 5571
86.927876	MnO_2	Manganese Dioxide, 5558
87.014271	C_3H_5NS	Ethyl Isothiocyanate, 3755 Ethyl Thiocyanate, 3806
87.032028	$C_3H_5NO_2$	Isonitrosoacetone, 5049
87.068414	C_4H_9NO	n-Butyramide, 1584 N,N-Dimethylacetamide, 3214 Morpholine, 6116, 6117
87.104799	$C_5H_{13}N$	n-Amylamine, 631 Isoamylamine, 4959
87.907011	FeS	Ferrous Sulfide, 3983
87.968973	F_3P	Phosphorus Trifluoride, 7169

87.

Molecular Weight	Empirical Formula	Compound Name, Monograph Number
87.969012	H_2NaO_2P	Sodium Hypophosphite, 8385
87.993613	CF_4	Carbon Tetrafluoride, 1822
88.010743	BF_4H	Fluoboric Acid, 4029
88.016044	$C_3H_4O_3$	Pyruvic Acid, 7809
88.027277	$C_2H_4N_2O_2$	Oxamide, 6747
88.052430	$C_4H_8O_2$	Acetoin, 49 Aldol, 218 Butyric Acid, 1585 Dioxane, 3300 Ethyl Acetate, 3685 Isobutyric Acid, 5010 Methyl Propionate, 5982 Propyl Formate, 7647
88.070829	$C_4H_{12}Si$	Diethylsilane, 3112
88.088815	$C_5H_{12}O$	Isopentyl Alcohol, 5055 2-Methyl-1-butanol, 5905 3-Methyl-2-butanol, 5906 Neopentyl Alcohol, 6282 1-Pentanol, 6916 2-Pentanol, 6917 3-Pentanol, 6918 tert-Pentyl Alcohol, 6934
88.094432	$[C_9H_{24}N_2O]^{2+}$	Prolonium (half mass), 7576
88.932594	$FeHO_2$	Ferric Hydroxide, 3942
88.932673	$CCuN$	Cuprous Cyanide, 2666
89.003227	C_3H_4ClN	β-Chloropropionitrile, 2143
89.011293	$C_2H_3NO_3$	Oxamic Acid, 6746
89.047679	$C_3H_7NO_2$	DL-Alanine, 192 L-Alanine, 193 β-Alanine, 194 Formicin, 4099 Isopropyl Nitrite, 5077 1-Nitropropane, 6449 2-Nitropropane, 6450 n-Propyl Nitrite, 7654 Sarcosine, 8126 Urethan, 9535, 9536
89.084064	$C_4H_{11}NO$	2-Amino-1-butanol, 432 2-Amino-2-methyl-1-propanol, 456, 391 Deanol, 2824-2826
89.937316	CrF_2	Chromous Fluoride, 2240
89.940419	FeH_2O_2	Ferrous Hydroxide, 3973
89.948452	$ClNaO_2$	Sodium Chlorite, 8344

Use in conjunction with The Merck Index, Ninth Edition

Molecular Weight	Empirical Formula	Compound Name, Monograph Number
89.993681	Li_2O_3Si	Lithium Silicate, 5382
89.995309	$C_2H_2O_4$	Oxalic Acid, 6743
90.023628	C_4H_7Cl	1-Chloro-2-butene, 2104 3-Chloro-1-butene, 2105 1-Chloro-2-methylpropene, 2124 3-Chloro-2-methylpropene, 2125
90.029569	$B_2H_4O_4$	Diboron Tetrahydroxide, 2983
90.031694	$C_3H_6O_3$	Dihydroxyacetone, 3163 Glyceraldehyde, 4316 Hydracrylic Acid, 4643 D-Lactic Acid, 5184 DL-Lactic Acid, 5185 L-Lactic Acid, 5186 Methyl Carbonate, 5912 s-Trioxane, 9400
90.050322	$C_4H_{10}S$	n-Butyl Mercaptan, 1565 sec-Butyl Mercaptan, 1566 tert-Butyl Mercaptan, 1567 Ethyl Sulfide, 3801 Isobutyl Mercaptan, 5001
90.054161	CH_6N_4O	Carbohydrazide, 1809
90.068080	$C_4H_{10}O_2$	1,3-Butylene Glycol, 1555 2,3-Butylene Glycol, 1556 tert-Butyl Hydroperoxide, 1559 1,2-Dimethoxyethane, 3211 Dimethylacetal, 3213 2-Ethoxyethanol, 3678
90.905270	CoS	Cobaltous Sulfide, 2402
91.020419	CH_5N_3S	Thiosemicarbazide, 9100
91.054775	$[C_7H_7]^+$	Tropylium, 9449
91.921072	S_2Si	Silicon Disulfide, 8236
91.930852	$CoHO_2$	Cobaltic Oxide Monohydrate, 2383
91.940826	H_2NiO_2	Nickel Hydroxide, 6324
91.964062	ClF_3	Chlorine Trifluoride, 2070
91.968739	C_2CaN_2	Calcium Cyanide, 1657
91.993201	$C_2H_4O_2S$	Thioglycolic Acid, 9070
91.993385	H_8Si_3	Trisilane, 9420
92.002893	C_3H_5ClO	Chloroacetone, 2085 Epichlorohydrin, 3536 Propionyl Chloride, 7618
92.027358	$C_3H_5FO_2$	Fluoroacetic Acid, Methyl ester, 4052

Molecular Weight	Empirical Formula	Compound Name, Monograph Number
92.039278	C_4H_9Cl	n-Butyl Chloride, 1546 sec-Butyl Chloride, 1547 tert-Butyl Chloride, 1548 Isobutyl Chloride, 4993
92.047344	$C_3H_8O_3$	Glycerol, 4319
92.050024	$[C_{12}H_{12}N_2]^{2+}$	Diquat, Cation (half mass), 3370
92.062600	C_7H_8	Toluene, 9225
92.934853	F_2Mn	Manganese Difluoride, 5557
92.938677	CoH_2O_2	Cobaltous Hydroxide, 2394
92.998142	C_2H_4ClNO	Chloroacetamide, 2079
93.057849	C_6H_7N	Aniline, 692 α-Picoline, 7203 β-Picoline, 7204 γ-Picoline, 7205
93.922750	Cl_2Mg	Magnesium Chloride, 5485
93.931746	F_2Fe	Ferrous Fluoride, 3970
93.941811	CH_3Br	Methyl Bromide, 5904
93.976096	C_2H_6Zn	Dimethylzinc, 3254
93.982157	$C_2H_3ClO_2$	Chloroacetic Acid, 2083 Methyl Chlorocarbonate, 5918
93.991094	$C_2H_6S_2$	1,2-Ethanedithiol, 3652
94.008851	$C_2H_6O_2S$	Dimethyl Sulfone, 3248
94.018543	C_3H_7ClO	Propylene Chlorohydrin, 7639 sec-Propylene Chlorohydrin, 7640
94.041865	C_6H_6O	Phenol, 7038
94.053098	$C_5H_6N_2$	α-Aminopyridine, 485 β-Aminopyridine, 486 Glutaronitrile, 4306
94.112625	$[C_{11}H_{28}N_2]^{2+}$	Pentamethonium (half mass), 6911
94.901671	CuS	Cupric Sulfide, 2659
95.037114	C_5H_5NO	Pyridine 1-Oxide, 7755
95.060923	$[C_5H_7N_2]^+$	Methylpyrimidinium, 7769
95.901217	SZn	Zinc Sulfide, 9828
95.918975	O_2Zn	Zinc Peroxide, 9815
95.932154	F_2Ni	Nickel Fluoride, 6322

Use in conjunction with The Merck Index, Ninth Edition

Molecular Weight	Empirical Formula	Compound Name, Monograph Number
95.953356	C$_2$H$_2$Cl$_2$	Acetylene Dichloride, 85 Vinylidene Chloride, 9647
95.964133	H$_3$O$_3$Sc	Scandium Hydroxide, 8145
95.988116	CH$_4$O$_3$S	Methanesulfonic Acid, 5811
95.999349	H$_4$N$_2$O$_2$S	Sulfamide, 8713
96.021129	C$_5$H$_4$O$_2$	Furfural, 4155 Protoanemonin, 7677
96.037528	C$_6$H$_5$F	Fluorobenzene, 4055
96.057515	C$_6$H$_8$O	1-Pentol, 6929
96.093900	C$_7$H$_{12}$	Norcarane, 6498
96.930004	CoF$_2$	Cobaltous Fluoride, 2392
96.935079	CuH$_2$O$_2$	Cupric Hydroxide, 2644
96.938854	CKNS	Potassium Thiocyanate, 7485
96.952710	BrH$_4$N	Ammonium Bromide, 522
96.983365	H$_3$NO$_3$S	Sulfamic Acid, 8712
97.089149	C$_6$H$_{11}$N	Diallylamine, 2930 Isoamyl Cyanide, 4965
97.898452	ClCu	Cuprous Chloride, 2665
97.932620	CCl$_2$O	Phosgene, 7146
97.967380	H$_2$O$_4$S	Sulfuric Acid, 8769
97.969006	C$_2$H$_4$Cl$_2$	Ethylene Dichloride, 3733 Ethylidene Chloride, 3750
97.976897	H$_3$O$_4$P	Phosphoric Acid, 7153
98.000394	C$_4$H$_2$O$_3$	Maleic Anhydride, 5532
98.036780	C$_5$H$_6$O$_2$	Angelica Lactone, 682 Furfuryl Alcohol, 4156 α-Methylene Butyrolactone, 5931
98.073165	C$_6$H$_{10}$O	Allyl Ether, 285 Cyclohexanone, 2731 Meparfynol, 5671 Mesityl Oxide, 5755 Sorbic Alcohol, 8496
98.999015	H$_5$NO$_3$S	Ammonium Bisulfite, 519
99.014271	C$_4$H$_5$NS	Allyl Isothiocyanate, 289 Parathiazine, 6837
99.032028	C$_4$H$_5$NO$_2$	Succinimide, 8671

Molecular Weight	Empirical Formula	Compound Name, Monograph Number
99.043262	$C_3H_5N_3O$	Cyacetacide, 2690
99.068414	C_5H_9NO	Angelic Acid, Amide, 681 Cyclopentanone, Oxime, 2747 Tiglic Acid, Amide, 9161
99.104799	$C_6H_{13}N$	Cyclohexylamine, 2734
99.899344	GaP	Gallium Phosphide, 4193
99.925254	CrO_3	Chromium Trioxide, 2233
99.946717	MgO_3Si	Magnesium Metasilicate, 5514
99.947335	$CCaO_3$	Calcium Carbonate, 1650
99.956277	$CHKO_3$	Potassium Bicarbonate, 7384
99.956336	$ClHO_4$	Perchloric Acid, 6948
99.963044	FHO_3S	Fluosulfonic Acid, 4069
100.009520	$C_3H_4N_2S$	2-Aminothiazole, 494
100.016044	$C_4H_4O_3$	Succinic Anhydride, 8670
100.027277	$C_3H_4N_2O_2$	Hydantoin, 4637
100.028895	C_6H_5Na	Benzene, Sodium deriv, 1069
100.052430	$C_5H_8O_2$	Acetylacetone, 73 Angelic Acid, 681 Crotonic Acid, Methyl ester, 2591 Ethyl Acrylate, 3688 Isocrotonic Acid, Methyl ester, 5019 Isopropenyl Acetate, 5064 Tiglic Acid, 9161
100.063663	$C_4H_8N_2O$	Allylurea, 295
100.088815	$C_6H_{12}O$	Caproic Aldehyde, 1760 Cyclohexanol, 2730 3-Hexen-1-ol, 4570 Isopropylacetone, 5068 Methyl Butyl Ketone, 5907 Pinacolone, 7239
100.125201	C_7H_{16}	n-Heptane, 4521
100.926406	CuF_2	Cupric Fluoride, 2639
100.951526	KNO_3	Potassium Nitrate, 7432
101.022526	$C_2H_3N_3O_2$	Urazole, 9524
101.047679	$C_4H_7NO_2$	2-Azetidinecarboxylic Acid, 919
101.070145	$C_2H_7N_5$	Biguanide, 1234
101.084064	$C_5H_{11}NO$	Isovaleramide, 5094 n-Valeraldehyde, Oxime, 9566

Use in conjunction with The Merck Index, Ninth Edition

Molecular Weight	Empirical Formula	Compound Name, Monograph Number
101.120450	$C_6H_{15}N$	Diisopropylamine, 3180, 3181, 3183 n-Dipropylamine, 3363 Triethylamine, 9342, 9343
101.120450	$[C_{12}H_{30}N_2]^{2+}$	Hexamethonium (half mass), 4557, 4558, 4580
101.908106	BrNa	Sodium Bromide, 8338
101.914539	HORb	Rubidium Hydroxide, 8044
101.925952	F_2Zn	Zinc Fluoride, 9795
101.947827	Al_2O_3	Aluminum Oxide, 355
101.952000	$ClFO_3$	Perchloryl Fluoride, 6949
101.958708	F_2O_2S	Sulfuryl Fluoride, 8777
101.988274	Al_2CaH_8	Aluminum Calcium Hydride, 330
102.025170	$C_3H_6N_2S$	4,5-Dihydro-2-thiazolamine, 1450 2-Imidazolidinethione, 4809
102.031694	$C_4H_6O_3$	Acetic Anhydride, 44 Acetoacetic Acid, 46 Pyruvic Acid, Methyl ester, 7809
102.042928	$C_3H_6N_2O_2$	Cycloserine, 2757
102.046950	C_8H_6	Ethynylbenzene, 3814
102.068080	$C_5H_{10}O_2$	Ethyl Propionate, 3790 Isobutyl Formate, 4997 Isopropyl Acetate, 5066 Isovaleric Acid, 5095 Methyl Butyrate, 5909 Methyl Isobutyrate, 5956 Pivalic Acid, 7293 Propyl Acetate, 7629 Tetrahydrofurfuryl Alcohol, 8931 Valeric Acid, Normal, 9568
102.079313	$C_4H_{10}N_2O$	N-Nitrosodiethylamine, 6457
102.104465	$C_6H_{14}O$	1-Hexanol, 4565 Isopropyl Ether, 5073 Propyl Ether, 7646
102.115699	$C_5H_{14}N_2$	1,5-Pentanediamine, 6914
102.948729	CrH_3O_3	Chromic Hydroxide, 2222
103.038176	$C_2H_5N_3O_2$	Biuret, 1329 Semioxamazide, 8193
103.042199	C_7H_5N	Benzonitrile, 1108
103.045571	C_4H_9NS	Thiamorpholine, 9033

Molecular Weight	Empirical Formula	Compound Name, Monograph Number
103.063329	C$_4$H$_9$NO$_2$	α-Aminobutyric Acid, 434 β-Aminobutyric Acid, 435 γ-Aminobutyric Acid, 436 α-Aminoisobutyric Acid, 453 n-Butyl Nitrite, 1570 tert-Butyl Nitrite, 1571 Isobutyl Nitrite, 5003 Ethylglycocoll, 3745
103.099714	C$_5$H$_{13}$NO	1-(Dimethylamino)-2-propanol, 4853
103.900540	OSr	Strontium Oxide, 8639
103.901566	MgSe	Magnesium Selenide, 5512
103.941861	MgO$_3$S	Magnesium Sulfite, 5519
103.942951	H$_2$KO$_2$P	Potassium Hypophosphite, 7424
103.954410	HNaO$_3$S	Sodium Bisulfite, 8332
103.970541	F$_4$Si	Silicon Tetrafluoride, 8242
104.010959	C$_3$H$_4$O$_4$	Malonic Acid, 5538
104.033425	CH$_4$N$_4$O$_2$	Nitroguanidine, 6430
104.040820	C$_3$H$_8$N$_2$S	sym-Dimethylthiourea, 3252
104.043345	C$_6$H$_5$BO	Benzeneboronic Anhydride, 1071
104.046002	C$_2$H$_6$N$_3$O$_2$	Guanidine-1-carboxamide, 3069
104.047344	C$_4$H$_8$O$_3$	Ethylene Glycol Monoacetate, 3737 β-Hydroxybutyric Acid, 4716 Methyl Lactate, 5959
104.058578	C$_3$H$_8$N$_2$O$_2$	2,3-Diaminopropionic Acid, 2941
104.062600	C$_8$H$_8$	Cubane, 2606 Styrene, 8657
104.064475	C$_3$H$_9$BO$_3$	Trimethyl Borate, 9381
104.065972	C$_5$H$_{12}$S	n-Amyl Mercaptan, 649 Isoamyl Mercaptan, 4970
104.083730	C$_5$H$_{12}$O$_2$	Neopentyl Glycol, 6283 1,5-Pentanediol, 6915
104.107539	[C$_5$H$_{14}$NO]$^+$	Choline, Cation, 2198-2204, 2206-2208
104.921410	CBrN	Cyanogen Bromide, 2700
104.963913	BeO$_4$S	Beryllium Sulfate, 1205
105.006208	C$_2$H$_3$NO$_4$	Acetyl Nitrate, 94
105.017441	CH$_3$N$_3$O$_3$	Nitrourea, 6471

Use in conjunction with The Merck Index, Ninth Edition

Molecular Weight	Empirical Formula	Compound Name, Monograph Number
105.042593	$C_3H_7NO_3$	n-Propyl Nitrate, 7653 Serine, 8207
105.045273	$C_6H_5N_2$	o-Phenylenediamine, 7088
105.078979	$C_4H_{11}NO_2$	2-Amino-2-methyl-1,3-propanediol, 455 Diethanolamine, 3078
105.911008	GeO_2	Germanium Dioxide, 4242
105.943366	$ClNaO_3$	Sodium Chlorate, 8342
105.964283	CNa_2O_3	Sodium Carbonate, 8340
105.964516	$ClLiO_4$	Lithium Perchlorate, 5377
106.008851	$C_3H_6O_2S$	Thiolactic Acid, 9074
106.018543	C_4H_7ClO	n-Butyryl Chloride, 1590 2-Chloroethyl Vinyl Ether, 2119
106.026609	$C_3H_6O_4$	Glyceric Acid, 4318
106.037842	$C_2H_6N_2O_3$	Itramin, 5102
106.041865	C_7H_6O	Benzaldehyde, 1057
106.043008	$C_4H_7FO_2$	Fluoroacetic Acid, Ethyl ester, 4052
106.045237	$C_4H_{10}OS$	2-(Ethylthio)ethanol, 3807
106.054928	$C_5H_{11}Cl$	Amyl Chloride, 643 Isoamyl Chloride, 4964
106.062994	$C_4H_{10}O_3$	Diethylene Glycol, 3100 Orthoformic Acid, Trimethyl ester, 6720
106.078250	C_8H_{10}	Ethylbenzene, 3695 Xylene, 9743
106.986343	$C_2H_5NS_2$	Metham, 5806
107.038257	$C_3H_6FNO_2$	3-Fluoro-D-alanine, 4053
107.073499	C_7H_9N	Benzylamine, 1139 3-Ethylpyridine, 3791 4-Ethylpyridine, 3792 2,6-Lutidine, 5430 Methylaniline, 5892 Toluidines, 9232-9234
107.939172	F_3V	Vanadium Trifluoride, 9580
107.957461	C_2H_5Br	Ethyl Bromide, 3702
107.965685	F_4S	Sulfur Tetrafluoride, 8774
107.980721	N_2O_5	Nitrogen Pentoxide, 6427
107.995565	BeF_4Na	Beryllium Sodium Fluoride, 1204

Use in conjunction with The Merck Index, Ninth Edition

Molecular Weight	Empirical Formula	Compound Name, Monograph Number
107.997807	$C_3H_5ClO_2$	β-Chloropropionic Acid, 2142 Ethyl Chloroformate, 3715 Methyl Chloroacetate, 5917
108.006744	$C_3H_8S_2$	Bis[methylthio]methane, 1278 1,3-Propanedithiol, 7593
108.021129	$C_6H_4O_2$	Quinone, 7898, 7848
108.024501	$C_3H_8O_2S$	Thioglycerol, 9069
108.034193	C_4H_9ClO	tert-Butyl Hypochlorite, 1560
108.057515	C_7H_8O	Anisole, 704 Benzyl Alcohol, 1138 Cresols, 2569-2572
108.068748	$C_6H_8N_2$	2-Amino-4-picoline, 477 m-Phenylenediamine, 7087 p-Phenylenediamine, 7089 Phenylhydrazine, 7098, 7099
108.935719	CrF_3	Chromic Fluoride, 2220
109.052764	C_6H_7NO	Aminophenols, 470-472 Nicotinyl Alcohol, 6347 Phenylhydroxylamine, 7100
109.899488	K_2S	Potassium Sulfide, 7464
109.900296	$CaCl_2$	Calcium Chloride, 1652
109.931865	CH_2S_3	Trithiocarbonic Acid, 9423
109.932271	O_3P_2	Phosphorus Trioxide, 7170
109.941495	C_2N_2Ni	Nickel Cyanide, 6320
109.969006	$C_3H_4Cl_2$	1,3-Dichloropropene, 3051
109.971010	C_2H_6OZn	Methylzinc Methylate, 3254
109.983314	B_2MgO_4	Magnesium Borate, 5480
109.983739	Li_2O_4S	Lithium Sulfate, 5383
109.992688	BF_4Na	Sodium Fluoborate, 8367
110.003766	$C_2H_6O_3S$	Methanesulfonic Acid, Methyl ester, 5811
110.013457	$C_3H_7ClO_2$	3-Chloro-1,2-propanediol, 2141
110.019022	C_6H_6S	Thiophenol, 9093
110.036780	$C_6H_6O_2$	Hydroquinone, 4705 Pyrocatechol, 7780 Resorcinol, 7951
110.053179	C_7H_7F	Fluorotoluene, 4066
110.939346	C_2CoN_2	Cobaltous Cyanide, 2391

Use in conjunction with The Merck Index, Ninth Edition

Molecular Weight	Empirical Formula	Compound Name, Monograph Number
111.032028	$C_5H_5NO_2$	Mecrylate, 5610
111.043262	$C_4H_5N_3O$	Cytosine, 2792
111.048427	C_6H_6FN	p-Fluoroaniline, 4054
111.079647	$C_5H_9N_3$	Betazole, 1213 Histamine, 4595
111.104799	$C_7H_{13}N$	Quinuclidine, 7904
111.104799	$[C_{14}H_{26}N_2]^{2+}$	N,N'-Dimethyltremorinium (half mass), 9270
111.906350	O_2Se	Selenium Oxide, 8183
111.933256	F_3Mn	Manganese Trifluoride, 5572
111.948270	$C_2H_2Cl_2O$	Chloracetyl Chloride, 2024
111.983031	CH_4O_4S	Methyl Sulfate, 5995
111.984656	$C_3H_6Cl_2$	Propylene Dichloride, 7643 Propylidene Chloride, 7650
112.007978	C_6H_5Cl	Chlorobenzene, 2095
112.016044	$C_5H_4O_3$	2-Furoic Acid, 4160
112.027277	$C_4H_4N_2O_2$	Acetylenedicarboxamide, 84 Maleic Hydrazide, 5533 Uracil, 9506
112.052430	$C_6H_8O_2$	Crotonic Acid, Vinyl ester, 2591 Dihydroresorcinol, 3156 Parasorbic Acid, 6836 Sorbic Acid, 8495
112.088815	$C_7H_{12}O$	Cycloheptanone, 2727
112.100048	$C_6H_{12}N_2$	Triethylenediamine, 9344
112.125201	C_8H_{16}	Caprylene, 1763
112.930149	F_3Fe	Ferric Fluoride, 3937
113.022526	$C_3H_3N_3O_2$	Azomycin, 928
113.029921	C_5H_7NS	2,4-Dimethylthiazole, 3251
113.047679	$C_5H_7NO_2$	Ethyl Cyanoacetate, 3722
113.058912	$C_4H_7N_3O$	Acrolein, Semicarbazone, 123 Creatinine, 2557
113.084064	$C_6H_{11}NO$	Caprolactam, 1761 Tetramin, 8948
113.954229	CH_3ClO_2S	Methanesulfonyl Chloride, 5812
113.962295	H_2O_5S	Caro's Acid, 1851

113.

Molecular Weight	Empirical Formula	Compound Name, Monograph Number
113.963920	$C_2H_4Cl_2O$	sym-Dichloromethyl Ether, 3046
114.006542	$C_3H_2N_2O_3$	Parabanic Acid, 6827
114.009914	$H_6N_2O_3S$	Ammonium Sulfamate, 585
114.025170	$C_4H_6N_2S$	2-Amino-4-methylthiazole, 458 2-Amino-5-methylthiazole, 459 Methimazole, 5841
114.028107	$C_6H_4F_2$	p-Difluorobenzene, 3125
114.031694	$C_5H_6O_3$	Reductic Acid, 7925
114.042928	$C_4H_6N_2O_2$	Diazoacetic Ester, 2962 Muscimol, 6137 2,5-Piperazinedione, 7258
114.050322	$C_6H_{10}S$	Allyl Sulfide, 294
114.068080	$C_6H_{10}O_2$	Acetonylacetone, 57 Angelic Acid, Methyl ester, 681 Crotonic Acid, Ethyl ester, 2591 Tiglic Acid, Methyl ester, 9161
114.104465	$C_7H_{14}O$	Cyclohexylcarbinol, 2736 Dipropyl Ketone, 3364 Heptanal, 4519 2-Heptanone, 4525
114.115699	$C_6H_{14}N_2$	Lupetazine, 5434
114.140851	C_8H_{18}	Isooctane, 5051 Octane, 6561
114.922790	$CMnO_3$	Manganese Carbonate, 5554
114.993930	H_5NO_4S	Ammonium Bisulfate, 517
115.003446	H_6NO_4P	Ammonium Phosphate, Monobasic, 573
115.026943	$C_4H_5NO_3$	Maleamic Acid, 5529
115.045571	C_5H_9NS	Isobutyl Thiocyanate, 5007
115.063329	$C_5H_9NO_2$	Proline, 7574
115.074562	$C_4H_9N_3O$	Acetone semicarbazone, 8191
115.099714	$C_6H_{13}NO$	Diacetonamine, 2919 Pinacolone, Oxime, 7239
115.136100	$C_7H_{17}N$	Methylhexaneamine, 5952 Tuaminoheptane, 9461
115.633724	$[C_{13}H_{33}N_3]^{2+}$	Azamethonium (half mass), 911
115.915863	BCl_3	Boron Trichloride, 1362
115.928408	CoF_3	Cobaltic Fluoride, 2382

Use in conjunction with The Merck Index, Ninth Edition

Molecular Weight	Empirical Formula	Compound Name, Monograph Number
115.933493	ClHO$_3$S	Chlorosulfonic Acid, 2152
115.935293	C$_2$N$_2$Zn	Zinc Cyanide, 9792
116.001435	Li$_3$O$_4$P	Lithium Phosphate, 5378
116.010959	C$_4$H$_4$O$_4$	Fumaric Acid, 4137 Maleic Acid, 5531
116.025564	H$_8$N$_2$O$_3$S	Ammonium Sulfite, 590
116.035081	H$_9$N$_2$O$_3$P	Ammonium Phosphite, 574
116.040820	C$_4$H$_8$N$_2$S	Thiosinamine, 9101
116.047344	C$_5$H$_8$O$_3$	Levulinic Acid, 5316 Methyl Acetoacetate, 5885 Pyruvic Acid, Ethyl ester, 7809
116.058578	C$_4$H$_8$N$_2$O$_2$	Dimethylglyoxime, 3235 Succinamide, 8665
116.062600	C$_9$H$_8$	Indene, 4824
116.083730	C$_6$H$_{12}$O$_2$	Butyl Acetates, 1519-1521 tert-Butylacetic Acid, 1522 n-Caproic Acid, 1759 Diacetone Alcohol, 2921 Diethylacetic Acid, 3080, 1661 Ethyl Butyrate, 3706 Ethyl Isobutyrate, 3754 Isoamyl Formate, 4967 Isobutyl Acetate, 4983 Methyl Isovalerate, 5958 Propyl Propionate, 7656
116.094963	C$_5$H$_{12}$N$_2$O	Tetramethylurea, 8947
116.120115	C$_7$H$_{16}$O	1-Heptanol, 4523 2-Heptanol, 4524
116.125732	[C$_{13}$H$_{32}$N$_2$O]$^{2+}$	Plegatil, Cation (half mass), 7312
116.131349	C$_6$H$_{16}$N$_2$	1,6-Hexanediamine, 4564
116.963081	H$_4$NO$_3$V	Ammonium Vanadate(V), 603
116.982885	ClH$_4$NO$_4$	Ammonium Perchlorate, 570
117.000907	BF$_4$NO	Nitrosyl Tetrafluoroborate, 6469
117.024836	C$_4$H$_7$NOS	Homocysteine, Thiolactone, 4615
117.042593	C$_4$H$_7$NO$_3$	Aceturic Acid, 72, 3258
117.053827	C$_3$H$_7$N$_3$O$_2$	Glycocyamine, 4331
117.057849	C$_8$H$_7$N	Benzyl Cyanide, 1146 Indole, 4840 Tolunitriles, 9235, 9236

117.

Molecular Weight	Empirical Formula	Compound Name, Monograph Number
117.078979	$C_5H_{11}NO_2$	β-Aminobutyric Acid, Methyl ester, 435 Betaine, 1207-1210 Isobutyl Carbamate, 4991 Isopentyl Nitrite, 5056 Isovaline, 5099 3-Nitropentane, 6438 Norvaline, 6527 Valine, 9571
117.090212	$C_4H_{11}N_3O$	N,N-Dimethylglycine Hydrazide, 3234
117.115364	$C_6H_{15}NO$	2-Diethylaminoethanol, 3083
117.882044	BrK	Potassium Bromide, 7395
117.885652	Cl_2Ti	Titanium Dichloride, 9181
117.904692	Cl_2OS	Thionyl Chloride, 9085
117.914383	$CHCl_3$	Chloroform, 2120
117.946915	$ClFO_4$	Fluorine Perchlorate, 4050
117.990223	$C_3H_2O_5$	Mesoxalic Acid, 5759
118.008851	$C_4H_6O_2S$	3-Sulfolene, 8753
118.026609	$C_4H_6O_4$	Methyl Oxalate, 5971 Succinic Acid, 8668
118.041865	C_8H_6O	Benzofuran, 1097
118.053098	$C_7H_6N_2$	Benzimidazole, 1088 1H-Indazole, 4823
118.054928	$C_6H_{11}Cl$	Cyclohexyl Chloride, 2737
118.062994	$C_5H_{10}O_3$	Ethyl Carbonate, 3710 Ethyl Lactate, 3757 Methyl Cellosolve® Acetate, 5914
118.074228	$C_4H_{10}N_2O_2$	2,3-Diaminopropionic Acid, Methyl ester, 2941
118.078250	C_9H_{10}	Indan, 4820
118.081622	$C_6H_{14}S$	Propyl Sulfide, 7657
118.091139	$C_6H_{15}P$	Triethyl Phosphine, 9350
118.099380	$C_6H_{14}O_2$	Acetal, 26 Butyl Cellosolve®, 1543 Hexamethylene Glycol, 4560 Hexylene Glycol, 4585 Pinacol, 7238
118.110613	$C_5H_{14}N_2O$	1-[(2-Aminoethyl)amino]-2-propanol, 444
118.909632	Cl_3N	Nitrogen Chloride, 6424
118.917942	$CCoO_3$	Cobaltous Carbonate, 2388

Use in conjunction with The Merck Index, Ninth Edition

Molecular Weight	Empirical Formula	Compound Name, Monograph Number
118.977406	$C_3H_2ClNO_2$	Chloroacetyl Isocyanate, 2088
119.021858	$C_3H_5NO_4$	Hadacidin, 4441
119.037114	C_7H_5NO	2-Furanacrylonitrile, 4148 Phenyl Isocyanate, 7101
119.048347	$C_6H_5N_3$	1H-Benzotriazole, 1119
119.051719	$C_3H_9N_3S$	AET, Free base, 165
119.058243	$C_4H_9NO_3$	3-Amino-4-hydroxybutyric Acid, 450 4-Amino-3-hydroxybutyric Acid, 451 Homoserine, 4622 Isobutyl Nitrate, 5002 Threonine, 9124
119.094629	$C_5H_{13}NO_2$	2-Amino-2-ethyl-1,3-propanediol, 445 Ammonium Valerate, 602
119.877697	SSr	Strontium Sulfide, 8644
119.879111	CaSe	Calcium Selenide, 1704
119.880652	ClRb	Rubidium Chloride, 8043
119.895454	O_2Sr	Strontium Peroxide, 8640
119.919406	CaO_3S	Calcium Sulfite, 1711
119.928923	$CaHO_3P$	Calcium Phosphite, 1697
119.934512	CCl_2F_2	Dichlorodifluoromethane, 3038
119.936775	MgO_4S	Magnesium Sulfate, 5518
119.946292	$HMgO_4P$	Magnesium Phosphate, Dibasic, 5505
119.949325	$HNaO_4S$	Sodium Bisulfate, 8330
119.957461	C_3H_5Br	Allyl Bromide, 279
119.958842	H_2NaO_4P	Sodium Phosphate, Monobasic, 8422
119.981592	$C_2H_4N_2S_2$	Rubeanic Acid, 8037
120.005873	$C_3H_4O_5$	Tartronic Acid, 8848
120.024501	$C_4H_8O_2S$	Sulfolane, 8752
120.034193	C_5H_9ClO	Isovaleryl Chloride, 5096
120.035735	$C_3H_8N_2OS$	Noxythiolin, 6534
120.042259	$C_4H_8O_4$	D-Erythrose, 3612 L-Erythrose, 3613 L-Erythrulose, 3616 D-Threose, 9125 L-Threose, 9126
120.043596	$C_5H_4N_4$	Purine, 7729

Molecular Weight	Empirical Formula	Compound Name, Monograph Number
120.053492	$C_3H_8N_2O_3$	Oxymethurea, 6781
120.057515	C_8H_8O	Acetophenone, 60 Coumaran, 2544 Phenylacetaldehyde, 7066 o-Tolualdehyde, 9223
120.068748	$C_7H_8N_2$	Merimine, 5745
120.070578	$C_6H_{13}Cl$	1-Chlorohexane, 2123
120.078644	$C_5H_{12}O_3$	Methyl Carbitol®, 5911
120.093900	C_9H_{12}	Cumene, 2615 Mesitylene, 5754 n-Propylbenzene, 7632 Pseudocumene, 7700
120.128275	$[C_{15}H_{32}N_2]^{2+}$	Pentolinium (half mass), 6930
120.904745	CCuNS	Cuprous Thiocyanate, 2676
121.019750	$C_3H_7NO_2S$	Cysteine, 2779
121.052764	C_7H_7NO	Benzamide, 1060 Formanilide, 4097 Methyl Pyridyl Ketone, 5986
121.073893	$C_4H_{11}NO_3$	Tromethamine, 9435
121.089149	$C_8H_{11}N$	N,N-Dimethylaniline, 3223 Ethylaniline, 3694 3-Ethyl-4-picoline, 3784 4-Ethyl-2-picoline, 3785 5-Ethyl-2-picoline, 3786 α-Methylbenzylamine, 5902 Phenethylamine, 7016 2,4,6-Trimethylpyridine, 9387 Xylidine, 9746
121.878215	Cl_2Cr	Chromous Chloride, 2239
121.891269	$GeMg_2$	Magnesium Germanide, 5490
121.894537	O_2Zr	Zirconium Oxide, 9846
121.898390	OPd	Palladium Oxide, 6802
121.908313	F_2Kr	Krypton Difluoride, 5173
121.911104	H_2O_2Sr	Strontium Hydroxide, 8634
121.917305	$ClKO_3$	Potassium Chlorate, 7398
121.918476	NaO_3V	Sodium Vanadate(V), 8475
121.927145	CrF_2O_2	Chromyl Fluoride, 2245
121.934963	AlO_4P	Aluminum Phosphate, 358
121.936726	C_2H_3BrO	Acetyl Bromide, 74

Use in conjunction with The Merck Index, Ninth Edition

Molecular Weight	Empirical Formula	Compound Name, Monograph Number
121.938281	ClNaO$_4$	Sodium Perchlorate, 8414
121.941212	Na$_2$O$_3$Si	Sodium Metasilicate, 8400
121.973111	C$_3$H$_7$Br	Isopropyl Bromide, 5071 Propyl Bromide, 7633
121.985964	H$_{10}$Si$_4$	Tetrasilane, 8956
121.986008	C$_3$H$_6$OS$_2$	Ethylxanthic Acid, 7495, 8476
122.007396	C$_4$H$_{10}$Zn	Diethylzinc, 3119
122.013457	C$_4$H$_7$ClO$_2$	β-Chloroethyl Acetate, 2114 β-Chloropropionic Acid, Methyl ester, 2142 Ethyl Chloroacetate, 3714 Propyl Chlorocarbonate, 7636
122.036780	C$_7$H$_6$O$_2$	Benzoic Acid, 1100 p-Hydroxybenzaldehyde, 4710 Salicylaldehyde, 8087
122.040151	C$_4$H$_{10}$O$_2$S	2,2'-Thiodiethanol, 9063
122.048013	C$_6$H$_6$N$_2$O	Nicotinamide, 6340
122.053910	C$_6$H$_7$BO$_2$	Benzeneboronic Acid, 1071
122.057909	C$_4$H$_{10}$O$_4$	Erythritol, 3596
122.073165	C$_8$H$_{10}$O	Benzyl Methyl Ether, 1157 Phenethyl Alcohol, 7015 Phenetole, 7020 Phlorol, 7139 Xylenol, 9744
122.073652	[C$_5$H$_{13}$ClN]$^+$	(2-Chloroethyl)trimethylammonium, 2118
122.084398	C$_7$H$_{10}$N$_2$	Diallylcyanamide, 2931 Tolylhydrazine, 9241
122.125732	[C$_{14}$H$_{32}$N$_2$O]$^{2+}$	Plegarol, Cation (half mass), 7311
122.900010	AgO	Silver(II) Oxide, 8266
123.032028	C$_6$H$_5$NO$_2$	Isonicotinic Acid, 5044 Nicotinic Acid, 6343, 3183, 4580, 6346, 9722 Nitrobenzene, 6409 p-Nitrosophenol, 6462 Picolinic Acid, 7206
123.043262	C$_5$H$_5$N$_3$O	Pyrazinamide, 7740
123.068414	C$_7$H$_9$NO	Anisidines, 700-702 p-Methylaminophenol, 5891, 6025
123.079647	C$_6$H$_9$N$_3$	Kyanmethin, 5177
123.913889	CO$_3$Zn	Zinc Carbonate, 9788
123.927299	H$_3$Sb	Stibine, 8592

Molecular Weight	Empirical Formula	Compound Name, Monograph Number
123.941560	F_4Ti	Titanium Tetrafluoride, 9189
123.952376	C_2H_5BrO	Ethylene Bromohydrin, 3729
124.001659	$C_3H_8OS_2$	Dimercaprol, 3196
124.027277	$C_5H_4N_2O_2$	Pyrazinoic Acid, 7743
124.034672	C_7H_8S	Thiobenzyl Alcohol, 9052 Thiocresol, 9060
124.052430	$C_7H_8O_2$	Guaiacol, 4399 Orcinol, 6705 Salicyl Alcohol, 8086
124.063663	$C_6H_8N_2O$	2,4-Diaminophenol, 2940 4,6-Dimethyl-2-pyrimidinol, 6309
124.088815	$C_8H_{12}O$	1-Cyclobutyl-1-ethynylethanol, 2720 1-Ethynylcyclohexanol, 3816
124.202604	$B_{10}H_{14}$	Decaborane(14), 2828
124.875752	Cl_2Mn	Manganese Chloride, 5556
125.014665	$C_2H_7NO_3S$	Taurine, 8850
125.047679	$C_6H_7NO_2$	N-Ethylmaleimide, 3762
125.058912	$C_5H_7N_3O$	5-Methylcytosine, 5922
125.120450	$C_8H_{15}N$	β-Coniceine, 2471 γ-Coniceine, 2472 Tropane, 9440
125.872645	Cl_2Fe	Ferrous Chloride, 3968
125.896060	Na_2Se	Sodium Selenide, 8440
125.902431	F_2Sr	Strontium Fluoride, 8632
125.903498	AgF	Silver Fluoride, 8256
125.920791	F_3Ga	Gallium Trifluoride, 4194
125.927535	$C_2Cl_2O_2$	Oxalyl Chloride, 6745
125.936355	Na_2O_3S	Sodium Sulfite, 8451
125.945872	HNa_2O_3P	Sodium Phosphite, 8425
125.963920	$C_3H_4Cl_2O$	1,1-Dichloroacetone, 3022 1,3-Dichloroacetone, 3023
125.965780	F_5P	Phosphorus Pentafluoride, 7162
125.966626	BF_4K	Potassium Tetrafluoroborate, 7478
125.998681	$C_2H_6O_4S$	Dimethyl Sulfate, 3246 Ethyl Sulfate, 3800 Isethionic Acid, 4955, 6912, 6994, 7590, 8597

Use in conjunction with The Merck Index, Ninth Edition

Molecular Weight	Empirical Formula	Compound Name, Monograph Number
126.000306	$C_4H_8Cl_2$	Butylidene Chloride, 1561
126.023628	C_7H_7Cl	Benzyl Chloride, 1144 Chlorotoluenes, 2159-2161
126.029250	$C_4H_5F_3O$	Fluroxene, 4081
126.031694	$C_6H_6O_3$	1,2,4-Benzenetriol, 1076 2-Furoic Acid, Methyl ester, 4160 5-(Hydroxymethyl)-2-furaldehyde, 4738 Isomaltol, 5036 Maltol, 5540 Methyl Furoate, 5943 Pyrogallol, 7784
126.042928	$C_5H_6N_2O_2$	5-Methylpyrazole-3-carboxylic Acid, 5985 6-Methyluracil, 6006 Thymine, 9139
126.054161	$C_4H_6N_4O$	2,4-Diamino-6-hydroxypyrimidine, 2938 5-Aminoimidazole-4-carboxamide, 6703
126.065394	$C_3H_6N_6$	Melamine, 5629
126.937576	F_4V	Vanadium Tetrafluoride, 9579
126.957544	HNO_5S	Nitrosylsulfuric Acid, 6468
127.009186	C_5H_5NOS	Pyrithione, 7775
127.018877	C_6H_6ClN	Chloroanilines, 2090-2092
127.038176	$C_4H_5N_3O_2$	6-Azathymine, 916
127.063329	$C_6H_9NO_2$	Guvacine, 4435
127.099714	$C_7H_{13}NO$	Cycloheptanone, Oxime, 2727 3-Quinuclidinol, 7905
127.110948	$C_6H_{13}N_3$	Galegine, 4184
127.136100	$C_8H_{17}N$	Coniine, 2475-2477
127.873053	Cl_2Ni	Nickel Chloride, 6319
127.910905	$BrHO_3$	Bromic Acid, 1390
127.912302	HI	Hydrogen Iodide, 4690
127.934123	CrF_4	Chromium Tetrafluoride, 2232
127.943185	$C_2H_2Cl_2O_2$	Dichloroacetic Acid, 3021, 3181
127.979570	$C_3H_6Cl_2O$	1,3-Dichloro-2-propanol, 3050
127.993201	$C_5H_4O_2S$	3-Thenoic Acid, 8992 2-Thiophenecarboxylic Acid, 9091
128.002893	C_6H_5ClO	Chlorophenols, 2133-2135
128.004435	$C_4H_4N_2OS$	2-Thiouracil, 9106

128.

Molecular Weight	Empirical Formula	Compound Name, Monograph Number
128.012296	C_6N_4	Tetracyanoethylene, 8912
128.022192	$C_4H_4N_2O_3$	Barbituric Acid, 966
128.040820	$C_5H_8N_2S$	Methylmethimazole, 5841
128.047344	$C_6H_8O_3$	α-Acetylbutyrolactone, 75 Methylreductic Acid, 7925
128.062600	$C_{10}H_8$	Azulene, 931 Naphthalene, 6194
128.083730	$C_7H_{12}O_2$	Angelic Acid, Ethyl ester, 681 n-Butyl Acrylate, 1523 Cyclohexanecarboxylic Acid, 2729 Tiglic Acid, Ethyl ester, 9161
128.120115	$C_8H_{16}O$	Caprylic Aldehyde, 1765 Ethyl Amyl Ketone, 3693 Hexyl Methyl Ketone, 4586
128.870903	Cl_2Co	Cobaltous Chloride, 2389
129.017441	$C_3H_3N_3O_3$	Cyamelide, 2691 Cyanuric Acid, 2704
129.024836	C_5H_7NOS	Goitrin, 4353
129.033840	$C_4H_4FN_3O$	Flucytosine, 4021
129.039006	$C_6H_5F_2N$	2,4-Difluoroaniline, 3124
129.042593	$C_5H_7NO_3$	Dimethadione, 3200 L-Pyroglutamic Acid, 7787
129.057849	C_9H_7N	Isoquinoline, 5084 Quinoline, 7888, 7889, 7892-7895
129.061221	$C_6H_{11}NS$	Isoamyl Thiocyanate, 4976
129.078979	$C_6H_{11}NO_2$	Cycloleucine, 2739 Isonipecotic Acid, 5048 Nipecotic Acid, 6381 Pipecolic Acid, 7251
129.101445	$C_4H_{11}N_5$	Metformin, 5792
129.115364	$C_7H_{15}NO$	1-Ethyl-3-piperidinol, 3787
129.151750	$C_8H_{19}N$	n-Dibutylamine, 3006 N,1-Dimethylhexylamine, 3236 Methamexamine, 5804 Octodrine, 6567
129.151750	$[C_{16}H_{38}N_2]^{2+}$	Decamethonium (half mass), 2830
129.898275	CdO	Cadmium Oxide, 1610
129.914383	C_2HCl_3	Trichloroethylene, 9319
129.916914	H_2O_3Se	Selenious Acid, 8178

Molecular Weight	Empirical Formula	Compound Name, Monograph Number
129.952177	$CrLi_2O_4$	Lithium Chromate(VI), 5365
129.982157	$C_5H_3ClO_2$	Furoyl Chloride, 4162
130.017856	$C_4H_3FN_2O_2$	Fluorouracil, 4067
130.026609	$C_5H_6O_4$	Citraconic Acid, 2302 Fumaric Acid, Monomethyl ester, 4137 Itaconic Acid, 5101 Mesaconic Acid, 5751
130.053098	$C_8H_6N_2$	Cinnoline, 2296 Phthalazine, 7177 Quinazoline, 7840 Quinoxaline, 7902
130.062994	$C_6H_{10}O_3$	Ethyl Acetoacetate, 3686 Kethoxal, 5149 Pantolactone, 6818 Propionic Anhydride, 7616
130.099380	$C_7H_{14}O_2$	n-Butyl Propionate, 1576 Ethyl Isovalerate, 3756 Ethyl n-Valerate, 3810 Heptanoic Acid, 4522 Isoamyl Acetate, 4958 Isobutyl Propionate, 5004 Pivalic Acid, Ethyl ester, 7293 Propyl Butyrate, 7634
130.121847	$C_5H_{14}N_4$	Agmatine, 176
130.135765	$C_8H_{18}O$	n-Butyl Ether, 1557 2-Ethyl-1-hexanol, 3746 Isobutyl Ether, 4996 1-Octanol, 6563 2-Octanol, 6564
130.159575	$[C_8H_{20}N]^+$	Tetraethylammonium, 8917- 8920
131.058243	$C_5H_9NO_3$	Hydroxyproline, 4759
131.069477	$C_4H_9N_3O_2$	Creatine, 2556
131.073499	C_9H_9N	Skatole, 8296
131.094629	$C_6H_{13}NO_2$	γ-Aminobutyric Acid, Ethyl ester, 436 ε-Aminocaproic Acid, 437 Amyl Carbamate, Tertiary, 642 5-(Hydroxymethyl)tetrahydro-2-furfurylamine, 4742 Isoamyl Carbamate, 4963 Isoleucine, 5033 Leucine, 5301 Norleucine, 6513
131.888099	$AlCl_3$	Aluminum Chloride, 334
131.916805	AsF_3	Arsenic Trifluoride, 830
131.921879	H_2Te	Hydrogen Telluride, 4696

131.

Molecular Weight	Empirical Formula	Compound Name, Monograph Number
131.930033	$C_2H_3Cl_3$	1,1,1-Trichloroethane, 9316 1,1,2-Trichloroethane, 9317
131.948397	CaN_2O_4	Calcium Nitrite, 1683
132.005873	$C_4H_4O_5$	Oxalacetic Acid, 6741
132.019828	$C_3H_4F_4O$	2,2,3,3-Tetrafluoro-1-propanol, 8924
132.020479	$H_8N_2O_4S$	Ammonium Sulfate, 586
132.029995	$H_9N_2O_4P$	Ammonium Phosphate, Dibasic, 572
132.042259	$C_5H_8O_4$	α-Acetoxypropionic Acid, 68 Glutaric Acid, 4305 Methyl Malonate, 5961
132.043596	$C_6H_4N_4$	2-Amino-1,1,3-tricyanopropene, 498 Pteridine, 7716
132.053492	$C_4H_8N_2O_3$	Asparagine, 861 N-Glycylglycine, 4340
132.057515	C_9H_8O	1,2-Benzopyran, 1112 Cinnamaldehyde, 2286
132.078644	$C_6H_{12}O_3$	2,2-Dimethyl-1,3-dioxolane-4-methanol, 3232 2-Ethoxyethyl Acetate, 3679 Paraldehyde, 6832 2,5-Tetrahydrofurandimethanol, 8930
132.089878	$C_5H_{12}N_2O_2$	2,3-Diaminopropionic Acid, Ethyl ester, 2941 Ornithine, 6710
132.093900	$C_{10}H_{12}$	1-Methylindan, 4820 Tetralin®, 8936
132.113687	$[C_5H_{14}N_3O]^+$	Girard's Reagent T, Cation, 4257
132.115030	$C_7H_{16}O_2$	2,2-Diethyl-1,3-propanediol, 3111
132.867305	Cl_2Cu	Cupric Chloride, 2633
132.908169	$CAgN$	Silver Cyanide, 8254
132.965607	$C_3H_3NOS_2$	Rhodanine, 7975
132.987818	BeN_2O_6	Beryllium Nitrate, 1197
132.993056	$C_4H_4ClNO_2$	N-Chlorosuccinimide, 2150
133.019750	$C_4H_7NO_2S$	Thiazolidinecarboxylic Acid, 9040
133.037508	$C_4H_7NO_4$	Aspartic Acid, 864 Iminodiacetic Acid, 4811
133.052764	C_8H_7NO	Mandelonitrile, 5547
133.073893	$C_5H_{11}NO_3$	Isoamyl Nitrate, 4971 Pentahomoserine, 6910

Use in conjunction with The Merck Index, Ninth Edition

Molecular Weight	Empirical Formula	Compound Name, Monograph Number
133.089149	$C_9H_{11}N$	Tranylcypromine, 9263
133.866851	Cl_2Zn	Zinc Chloride, 9789
133.881849	Cl_2S_2	Sulfur Chloride, 8765
133.891312	Cl_3HSi	Trichlorosilane, 9327
133.899607	Cl_2O_2S	Sulfuryl Chloride, 8776
133.908242	F_2OSe	Selenium Oxyfluoride, 8186
133.920482	ILi	Lithium Iodide, 5373
134.019022	C_8H_6S	Thianaphthene, 9036
134.021523	$C_4H_6O_5$	Malic Acid, 5535
134.049843	$C_6H_{11}ClO$	Caproyl Chloride, 1762
134.057909	$C_5H_{10}O_4$	D-2-Deoxyribose, 2869 Monacetin, 6076
134.073165	$C_9H_{10}O$	p-Anol, 712 Chavicol, 2002 Cinnamyl Alcohol, 2293 Propiophenone, 7620
134.094294	$C_6H_{14}O_3$	Carbitol®, 1804 Diglyme, 3142
134.109551	$C_{10}H_{14}$	n-Butylbenzene, 1533 sec-Butylbenzene, 1534 tert-Butylbenzene, 1535 Cymene, 2765 Durene, 3445 Isobutylbenzene, 4987 Isodurene, 5023
134.143925	$[C_{17}H_{36}N_2]^{2+}$	Trimethidinium (half mass), 9375
134.854567	$MnSe$	Manganese Selenide, 5567
134.917269	LiO_3Se	Lithium Selenite, 5381
134.992491	$C_2H_5N_3S_2$	2,4-Dithiobiuret, 3389
134.999015	$C_3H_5NO_3S$	2-Cyanoethylsulfonic Acid, 8452
135.014271	C_7H_5NS	Benzothiazole, 1118 Phenyl Isothiocyanate, 7102
135.035400	$C_4H_9NO_2S$	Homocysteine, 4615 Mecysteine, 5611
135.054495	$C_5H_5N_5$	Adenine, 141

135.

Molecular Weight	Empirical Formula	Compound Name, Monograph Number
135.068414	C_8H_9NO	Acetanilide, 37 Aminoacetophenones, 413-415 Phenacylamine, 6990 α-Phenylacetamide, 7067 o-Toluamide, 9224
135.104799	$C_9H_{13}N$	Amphetamine, 616 Cumidine, 2619 Dextroamphetamine, 2910
135.851460	FeSe	Ferrous Selenide, 3980
135.880322	Cl_3P	Phosphorus Trichloride, 7168
135.897114	OSn	Stannous Oxide, 8567
135.904961	CCl_3F	Trichlorofluoromethane, 9320
135.913546	BrF_3	Bromine Trifluoride, 1394
135.913933	MgO_3S_2	Magnesium Thiosulfate, 5521
135.914321	CaO_4S	Calcium Sulfate, 1709
135.923263	HKO_4S	Potassium Bisulfate, 7389
135.923838	$CaHO_4P$	Calcium Phosphate, Dibasic, 1693
135.932780	H_2KO_4P	Potassium Phosphate, Monobasic, 7446
135.952376	C_3H_5BrO	Bromoacetone, 1399
135.988761	C_4H_9Br	n-Butyl Bromide, 1537 sec-Butyl Bromide, 1538 tert-Butyl Bromide, 1539 Isobutyl Bromide, 4989
136.000788	$C_3H_4O_6$	Dihydroxymalonic Acid, 5759
136.001659	$C_4H_8OS_2$	Isopropylxanthic Acid, 8390
136.029107	$C_5H_9ClO_2$	Ethyl α-Chloropropionate, 3716 Isobutyl Chlorocarbonate, 4994
136.038511	$C_5H_4N_4O$	Allopurinol, 271 Hypoxanthine, 4791
136.052430	$C_8H_8O_2$	p-Anisaldehyde, 696 Benzyl Formate, 1150 o-Methoxybenzaldehyde, 5863 Methyl Benzoate, 5899 Phenyl Acetate, 7068 Phenylacetic Acid, 7069 Toluic Acid, 9231
136.063663	$C_7H_8N_2O$	Methyl Pyridyl Ketoxime, 5986 Phenylurea, 7126
136.073559	$C_5H_{12}O_4$	Glyceraldehyde, Dimethylacetal, 4316 Pentaerythritol, 6904

Use in conjunction with The Merck Index, Ninth Edition

Molecular Weight	Empirical Formula	Compound Name, Monograph Number
136.088815	$C_9H_{12}O$	Benzyl Ethyl Ether, 1149 p,α-Dimethylbenzyl Alcohol, 3226 α-Ethylbenzyl Alcohol, 3699
136.100048	$C_8H_{12}N_2$	Betahistine, 1206 Dimethyl-p-phenylenediamine, 3242 Mebanazine, 5592 Phenelzine, 7010
136.112625	$[C_9H_{14}N]^+$	Phenyltrimethylammonium, 7125
136.125201	$C_{10}H_{16}$	Adamantane, 140 Camphene, 1732 3-Carene, 1838 Limonene, 5333 β-Myrcene, 6157 Ocimene, 6550 α-Phellandrene, 6983 β-Phellandrene, 6984 α-Pinene, 7242 Terpinene, 8885 Twistane, 9482
136.876583	Cl_2OV	Vanadyl Dichloride, 9584
136.947625	C_2H_4BrNO	N-Bromoacetamide, 1396
136.972841	H_4MgNO_4P	Ammonium Magnesium Phosphate, 556
136.985391	H_5NNaO_4P	Ammonium Sodium Phosphate, 583
137.047679	$C_7H_7NO_2$	Aminobenzoic Acids, 424-426 Carbanilic Acid, 1787 Homarine, 4604 Isonicotinic Acid, Methyl ester, 5044 Methyl Nicotinate, 5966 Nitrotoluene, 6470 3-Pyridineacetic Acid, 7753 Salicylaldoxime, 8088 Salicylamide, 8089 Trigonelline, 9360
137.058912	$C_6H_7N_3O$	Isoniazid, 5041, 6852 Nicotinic Acid Hydrazide, 6345
137.071488	$[C_7H_9N_2O]^+$	3-Carbamyl-1-methylpyridinium, 1786 Pralidoxime, 7502-7504
137.084064	$C_8H_{11}NO$	Methyridine, 6010 o-Phenetidine, 7017 p-Phenetidine, 7018 Phenylethanolamine, 7092 Tyramine, 9489
137.095297	$C_7H_{11}N_3$	N-tert-Butyl-N-cyanoglycinonitrile, 1553
137.896571	O_3Sc_2	Scandium Oxide, 8145
137.912160	CK_2O_3	Potassium Carbonate, 7397
137.912219	$ClKO_4$	Potassium Perchlorate, 7439

Molecular Weight	Empirical Formula	Compound Name, Monograph Number
137.919501	CaFO$_3$P	Calcium Fluorophosphate, 1664
137.931640	C$_2$H$_3$BrO$_2$	Bromoacetic Acid, 1398
137.966200	C$_2$H$_7$AsO$_2$	Cacodylic Acid, 1593, 8339
138.008197	C$_3$H$_7$O$_4$P	Fosfomycin, 4110
138.017775	C$_4$H$_2$N$_4$O$_2$	5-Diazouracil, 2967
138.031694	C$_7$H$_6$O$_3$	2-Furanacrylic Acid, 4147 p-Hydroxybenzoic Acid, 4712 Perbenzoic Acid, 6947 Protocatechualdehyde, 7678 β-Resorcylaldehyde, 7953 Salicylic Acid, 8093
138.035066	C$_4$H$_{10}$O$_3$S	2-(Ethylsulfonyl)ethanol, 3802
138.042928	C$_6$H$_6$N$_2$O$_2$	6-Aminonicotinic Acid, 465 N-Hydroxy-N-nitrosobenzenamine, 2622 Nitroanilines, 6403-6405 Pyrazinoic Acid, Methyl ester, 7743 Urocanic Acid, 9546
138.068080	C$_8$H$_{10}$O$_2$	Anise Alcohol, 698 Creosol, 2559 2-Phenoxyethanol, 7056 Styrene Glycol, 8658 Veratrole, 9610
138.090546	C$_6$H$_{10}$N$_4$	Pentylenetetrazole, 6935
138.104465	C$_9$H$_{14}$O	Phorone, 7144
138.128275	[C$_{18}$H$_{32}$N$_2$]$^{2+}$	N,N'-Dimethylpimetinium, 7233
138.140851	C$_{10}$H$_{18}$	Decalin®, 2829
139.026943	C$_6$H$_5$NO$_3$	Nitrophenols, 6441-6443 Oxiniacic Acid, 6760
139.030315	C$_3$H$_9$NO$_3$S	N-Methyltaurine, 5997
139.074562	C$_6$H$_9$N$_3$O	Toxopyrimidine, 9256
139.914075	H$_3$O$_3$Y	Yttrium Hydroxide, 9773
139.926677	Mg$_2$O$_4$Si	Magnesium Orthosilicate, 5514
139.935337	C$_2$MgN$_2$S$_2$	Magnesium Thiocyanate, 5520
139.945465	CH$_5$AsO$_3$	Methanearsonic Acid, 5810
139.977945	C$_2$H$_4$O$_5$S	Sulfoacetic Acid, 8750
140.002893	C$_7$H$_5$ClO	Benzoyl Chloride, 1123
140.006265	C$_4$H$_9$ClOS	Hemisulfur Mustard, 4503
140.010959	C$_6$H$_4$O$_4$	Coumalic Acid, 2542

Use in conjunction with The Merck Index, Ninth Edition

Molecular Weight	Empirical Formula	Compound Name, Monograph Number
140.015956	$C_5H_{10}Cl_2$	Amylene Dichloride, 646
140.027358	$C_7H_5FO_2$	p-Fluorobenzoic Acid, 4056
140.039278	C_8H_9Cl	Xylyl Chlorides, 9755-9757
140.040246	$C_4H_{10}FO_2P$	Sarin, 8127
140.047344	$C_7H_8O_3$	Ethyl Furoate, 3744 Gentisyl Alcohol, 4232 Sarkomycin A, 8128
140.058578	$C_6H_8N_2O_2$	5-Methylpyrazole-3-carboxylic Acid, Methyl ester, 5985
140.083730	$C_8H_{12}O_2$	5,5-Dimethyl-1,3-cyclohexanedione, 3231
140.106196	$C_6H_{12}N_4$	Methenamine, 5825, 4897, 5826-5835
140.107539	$[C_8H_{14}NO]^+$	Furtrethonium, 4165
141.053827	$C_5H_7N_3O_2$	Dimetridazole, 3256
141.078979	$C_7H_{11}NO_2$	Arecaidine, 802 Ethosuximide, 3674 Guvacoline, 4436 Hypoglycine A, 4787 Meparfynol Carbamate, 5672
141.090212	$C_6H_{11}N_3O$	Senecialdehyde, Semicarbazone, 8195
141.115364	$C_8H_{15}NO$	Hygrine, 4773 Pelletierine, 6866 Pseudotropine, 7707 Tropine, 9446
141.151750	$C_9H_{19}N$	Cyclopentamine, 2744 Isometheptene, 5040 Methylconiine, 5920
141.847935	BrCu	Cuprous Bromide, 2664
141.854113	Cu_2O	Cuprous Oxide, 2671
141.873948	AgCl	Silver Chloride, 8250
141.890125	$CaCl_2O_2$	Calcium Hypochlorite, 1671
141.907475	$MnNaO_4$	Sodium Permanganate, 8415
141.922100	O_5P_2	Phosphorus Pentoxide, 7165
141.924729	AsH_3O_4	Arsenic Acid, 821
141.927952	CH_3I	Methyl Iodide, 5955
141.931270	Na_2O_4S	Sodium Sulfate, 8449
141.940786	HNa_2O_4P	Sodium Phosphate, Dibasic, 8421
141.958835	$C_3H_4Cl_2O_2$	α,α-Dichloropropionic Acid, 3052

141.

Molecular Weight	Empirical Formula	Compound Name, Monograph Number
141.973143	B_2MgO_6	Magnesium Perborate, 5501
141.975838	$C_2H_6O_3S_2$	Mesna, Free acid, 5756
141.995220	$C_4H_8Cl_2O$	sym-Dichloroethyl Ether, 3040
142.001457	$C_4H_2N_2O_4$	Alloxan, 274
142.018543	C_7H_7ClO	4-Chloro-m-cresol, 2108
142.020085	$C_5H_6N_2OS$	Methylthiouracil, 6002
142.026609	$C_6H_6O_4$	Kojic Acid, 5165 Muconic Acid, 6126
142.049076	$C_4H_6N_4O_2$	Acetyleneurea, 86 Divicine, 3401
142.074228	$C_6H_{10}N_2O_2$	Nioxime®, 6380 Piracetam, 7282
142.099380	$C_8H_{14}O_2$	Cyclohexanecarboxylic Acid, Methyl ester, 2729
142.843642	Fe_2P	Ferrous Phosphide, 3979
142.846120	$CuSe$	Cupric Selenide, 2654
142.909632	C_2Cl_3N	Trichloroacetonitrile, 9307
143.033091	$C_4H_5N_3O_3$	Uramil, 9508
143.040486	C_6H_9NOS	4-Methyl-5-thiazoleethanol, 6000
143.058243	$C_6H_9NO_3$	Trimethadione, 9373
143.073499	$C_{10}H_9N$	Lepidine, 5295 1-Naphthylamine, 6225 2-Naphthylamine, 6226 Quinaldine, 7833
143.094629	$C_7H_{13}NO_2$	Stachydrine, 8549
143.131014	$C_8H_{17}NO$	Conhydrine, 2470 Dipropylacetamide, 3362 Pseudoconhydrine, 7699 Valnoctamide, 9573
143.131014	$[C_{16}H_{34}N_2O_2]^{2+}$	Dicolin, Cation (half mass), 3062
143.845666	$SeZn$	Zinc Selenide, 9824
143.847176	$AsGa$	Gallium Arsenide, 4192
143.858884	Cl_2Ge	Germanium Dichloride, 4241
143.885961	O_3SZn	Zinc Sulfite, 9829
143.895478	HO_3PZn	Zinc Phosphite, 9819
143.926165	C_3Al_4	Aluminum Carbide, 331

Use in conjunction with The Merck Index, Ninth Edition

Molecular Weight	Empirical Formula	Compound Name, Monograph Number
143.950311	C_2H_6Cd	Dimethylcadmium, 3229
143.974310	$C_2H_6ClO_3P$	2-Chloroethylphosphonic Acid, 2116
143.982998	F_6H_2Si	Fluosilicic Acid, 4068
144.034193	C_7H_9ClO	Ethclorvynol, 3656
144.042259	$C_6H_8O_4$	Fumaric Acid, Dimethyl ester, 4137 Lactide, 5188 Meldrum's Acid, 5636
144.057515	$C_{10}H_8O$	1-Naphthol, 6210 2-Naphthol, 6211
144.061121	$[C_{14}H_{16}N_4O_3]^{2+}$	Obidoxime (half mass), 6548
144.078644	$C_7H_{12}O_3$	Ethyl Levulinate, 3759 Isopropyl Acetoacetate, 5067
144.101111	$C_5H_{12}N_4O$	Tiformin, 9160
144.115030	$C_8H_{16}O_2$	n-Butyl n-Butyrate, 1540 Butyroin, 1587 Caprylic Acid, 1764 Ethyl Caproate, 3708 Heptanoic Acid, Methyl ester, 4522 Isoamyl Propionate, 4973 Isobutyl n-Butyrate, 4990 Isobutyl Isobutyrate, 4999 Valproic Acid, 9574
144.151415	$C_9H_{20}O$	n-Nonyl Alcohol, 6489
144.162649	$C_8H_{20}N_2$	Dibutadiamin, 3003 (1-Methylheptyl)hydrazine, 5951 Tetramethyldiaminobutane, 8941
144.901902	AgF_2	Silver Difluoride, 8255
144.938851	H_4IN	Ammonium Iodide, 553
144.994598	$C_3H_3N_3O_2S$	2-Amino-5-nitrothiazole, 466
145.016378	$C_8H_3NO_2$	Diatretyne II, 2958
145.052764	C_9H_7NO	Carbostyril, 1828 8-Hydroxyquinoline, 4761
145.073893	$C_6H_{11}NO_3$	Cyclopentanone, Semicarbazone, 2747
145.110279	$C_7H_{15}NO_2$	Emylcamate, 3513 Isobutyl Urethane, 5008
145.110279	$[C_{14}H_{30}N_2O_4]^{2+}$	Succinylcholine (half mass), 8674-8676
145.146664	$C_8H_{19}NO$	Heptaminol, 4518
145.875433	CdS	Cadmium Sulfide, 1617
145.877638	InP	Indium Phosphide, 4833

Molecular Weight	Empirical Formula	Compound Name, Monograph Number
145.890149	MoO_3	Molybdenum Trioxide, 6072
145.909298	C_2HCl_3O	2,2-Dichloroacetyl Chloride, 3024 Trichloroacetaldehyde, 9305
145.911829	H_2O_4Se	Selenic Acid, 8177
145.935979	F_5V	Vanadium Pentafluoride, 9577
145.962491	F_6S	Sulfur Hexafluoride, 8768
145.969006	$C_6H_4Cl_2$	Dichlorobenzenes, 3028-3030
146.021523	$C_5H_6O_5$	Acetonedicarboxylic Acid, 54 α-Ketoglutaric Acid, 5152
146.034335	$C_7H_5F_3$	Benzotrifluoride, 1121
146.036780	$C_9H_6O_2$	Biformin, 1233 Coumarin, 2547
146.043990	$C_3H_6N_4O_3$	Triuret, 9429
146.057909	$C_6H_{10}O_4$	Adipic Acid, 151, 7255 Diethyl Oxalate, 3109 Ethylene Glycol Diacetate, 3736 Ethylidene Diacetate, 3751 Isosorbide, 5088 Methyl Succinate, 5993 Mevaldic Acid, 6033
146.069142	$C_5H_{10}N_2O_3$	Glutamine, 4304 L-Isoglutamine, 5031
146.073165	$C_{10}H_{10}O$	Benzylideneacetone, 1153 3-Phenyl-1-butyn-3-ol, 7079
146.084398	$C_9H_{10}N_2$	5,6-Dimethylbenzimidazole, 3225
146.105528	$C_6H_{14}N_2O_2$	Lysine, 5455
146.112922	$C_8H_{18}S$	n-Butyl Sulfide, 1580 Isobutyl Sulfide, 5006
146.118104	$[C_7H_{16}NO_2]^+$	Acetylcholine, 78, 79 Oxapropanium, 6750
146.130680	$C_8H_{18}O_2$	DTBP, 3441 Ethohexadiol, 3669 Pinacol, Dimethyl ether, 7238
146.893932	CrO_4P	Chromic Phosphate, 2225
146.925977	Cl_3H_4MgN	Ammonium Magnesium Chloride, 555
147.032028	$C_8H_5NO_2$	Agrocybin, 178 Isatin, 4952 Phthalimide, 7180
147.053158	$C_5H_9NO_4$	Glutamic Acid, 4302, 807, 6091 HON, 4623

Use in conjunction with The Merck Index, Ninth Edition

Molecular Weight	Empirical Formula	Compound Name, Monograph Number
147.064391	$C_4H_9N_3O_3$	Albizziin, 199
147.068414	C_9H_9NO	Hydrocarbostyril, 4666
147.089543	$C_6H_{13}NO_3$	Daunosamine, 2815, 3428 Hexahomoserine, 4553
147.113353	$[C_6H_{15}N_2O_2]^+$	Carbachol, Cation, 1777
147.885365	AlSb	Aluminum Antimonide, 320
147.890369	CO_3Sr	Strontium Carbonate, 8628
147.892240	CrO_4S	Chromous Sulfate, 2243
147.908840	CdH_2O_2	Cadmium Hydroxide, 1607
147.924948	$C_2H_3Cl_3O$	2,2,2-Trichloroethanol, 9318
147.960681	MgN_2O_6	Magnesium Nitrate, 5497
147.997636	$H_8N_2O_3S_2$	Ammonium Thiosulfate, 595
148.000788	$C_4H_4O_6$	Dihydroxymaleic Acid, 3168
148.016044	$C_8H_4O_3$	Phthalic Anhydride, 7179
148.037173	$C_5H_8O_5$	Citramalic Acid, 2304
148.052430	$C_9H_8O_2$	Atropic Acid, 891 Cinnamic Acid, 2288
148.055802	$C_6H_{12}O_2S$	2,4-Dimethylsulfolane, 3247
148.067035	$C_5H_{12}N_2OS$	β-Aletheine, 221
148.073559	$C_6H_{12}O_4$	Digitoxose, 3140 Mevalonic Acid, 6034
148.088815	$C_{10}H_{12}O$	Anethole, 678 Cuminaldehyde, 2620 Estragole, 3631 Tetralol, 8937
148.100048	$C_9H_{12}N_2$	Nornicotine, 6521
148.109945	$C_7H_{16}O_3$	Ethyl Orthoformate, 3775, 6720
148.125201	$C_{11}H_{16}$	Amylbenzene, 636
148.159575	$[C_{19}H_{40}N_2]^{2+}$	Mebezonium (half mass), 5595
149.033293	$C_5H_{11}NS_2$	Diethyldithiocarbamic Acid, 8358
149.043656	$C_3H_7N_3O_4$	L-Alanosine, 195
149.047679	$C_8H_7NO_2$	Isonitrosoacetophenone, 5050
149.051050	$C_5H_{11}NO_2S$	Methionine, 5845 Penicillamine, 6880

Molecular Weight	Empirical Formula	Compound Name, Monograph Number
149.084064	$C_9H_{11}NO$	Acetotoluides, 62-64 p-Aminopropiophenone, 481 p-Dimethylaminobenzaldehyde, 3217 N-Methylacetanilide, 5883 Propiophenone, Oxime, 7620
149.105193	$C_6H_{15}NO_3$	Triethanolamine, 9341
149.120450	$C_{10}H_{15}N$	Carvacrylamine, 1866 Diethylaniline, 3091 p,α-Dimethylphenethylamine, 3241 Methamphetamine, 5805 Phentermine, 7062 Phenylpropylmethylamine, 7114
149.567101	$[C_{10}H_{17}N_7O_4]^{2+}$	Saxitoxin (half mass), 8141
149.862472	Cl_3Sc	Scandium Chloride, 8145
149.872669	O_3V_2	Vanadium Trioxide, 9581
149.879298	Al_2S_3	Aluminum Sulfide, 369
149.892850	$BrNaO_3$	Sodium Bromate, 8337
149.894247	INa	Sodium Iodide, 8387
149.908173	$CsHO$	Cesium Hydroxide, 1968
149.914792	F_4Ge	Germanium Tetrafluoride, 4244
149.998681	$C_4H_6O_4S$	Thiodiglycolic Acid, 9064 Thiomalic Acid, 9077
150.004412	$C_5H_{11}Br$	d-Amyl Bromide, 637 n-Amyl Bromide, 638 tert-Amyl Bromide, 639 Isoamyl Bromide, 4961
150.016438	$C_4H_6O_6$	D-Tartaric Acid, 8843 DL-Tartaric Acid, 8844 L-Tartaric Acid, 8845 meso-Tartaric Acid, 8846
150.025170	$C_7H_6N_2S$	2-Aminobenzothiazole, 428 6-Aminobenzothiazole, 429 2-Benzimidazolethiol, 1089
150.031694	$C_8H_6O_3$	Piperonal, 7269
150.035066	$C_5H_{10}O_3S$	2-Hydroxy-4-(methylthio)butyric Acid, 4743
150.044757	$C_6H_{11}ClO_2$	tert-Butyl Chloroacetate, 1549
150.052824	$C_5H_{10}O_5$	Apiose, 768 Arabinose, 789 D-Lyxose, 5463 D-Ribose, 7997 D-Ribulose, 8001 Xylose, 9749 Xylulose, 9750

Molecular Weight	Empirical Formula	Compound Name, Monograph Number
150.065394	$C_5H_6N_6$	2,6-Diaminopurine, 2942
150.068080	$C_9H_{10}O_2$	Benzyl Acetate, 1137 m-Cresyl Acetate, 2578 o-Cresyl Acetate, 2579 Ethyl Benzoate, 3697 Hydrocinnamic Acid, 4671 Paroxypropione, 6846 Phenylacetic Acid, Methyl ester, 7069
150.079313	$C_8H_{10}N_2O$	1-Acetyl-2-phenylhydrazine, 97 p-Aminoacetanilide, 411 m-Aminoacetophenone, Oxime, 413 o-Aminoacetophenone, Oxime, 414 p-Anisaldehyde, Hydrazone, 696 Benzylurea, 1174 N-Glycylaniline, 4339 p-Nitroso-N,N-dimethylaniline, 6459
150.089209	$C_6H_{14}O_4$	Triethylene Glycol, 9345
150.104465	$C_{10}H_{14}O$	p-tert-Butylphenol, 1573 Carvacrol, 1865 Carvone, 1867 Chrysanthenone, 2249 Cumic Alcohol, 2618 Perillaldehyde, 6956 Safranal, 8075 Thymol, 9140 d-Verbenone, 9615
150.115699	$C_9H_{14}N_2$	Pheniprazine, 7028
150.888361	FeO_4P	Ferric Phosphate, 3948
150.889777	MnO_4S	Manganese Sulfate, 5570
150.899293	$HMnO_4P$	Manganese Phosphate, Dibasic, 5565
150.912184	LiO_4Se	Lithium Selenate, 5380
150.986535	CHN_3O_6	Trinitromethane, 9396
151.026943	$C_7H_5NO_3$	Nitrobenzaldehyde, 6408
151.048073	$C_4H_9NO_5$	Tris(hydroxymethyl)nitromethane, 9419
151.049410	$C_5H_5N_5O$	Guanine, 4416
151.063329	$C_8H_9NO_2$	Acetaminophen, 36 p-Aminophenylacetic Acid, 473 Isonicotinic Acid, Ethyl ester, 5044 Methyl Anthranilate, 5893 4-Nitro-m-xylene, 6476 N-Phenylglycine, 7096 α-Phenylglycine, 7097 3-Pyridineacetic Acid, Methyl ester, 7753
151.074562	$C_7H_9N_3O$	Phenicarbazide, 7024 4-Phenylsemicarbazide, 7117

151.

Molecular Weight	Empirical Formula	Compound Name, Monograph Number
151.099714	$C_9H_{13}NO$	β-Amino-α-methylphenethyl Alcohol, 454 Halostachine, 4454 Hydroxyamphetamine, 4708 Norpseudoephedrine, 6524 Phenylpropanolamine, 7113
151.136100	$C_{10}H_{17}N$	Amantadine, 377
151.865763	Cr_2O_3	Chromic Oxide, 2224
151.874271	SSn	Stannous Sulfide, 8571
151.875236	Cl_3OP	Phosphorus Oxychloride, 7159
151.875411	CCl_4	Carbon Tetrachloride, 1821
151.886309	MgO_3Se	Magnesium Selenite, 5513
151.886670	FeO_4S	Ferrous Sulfate, 3982
151.891478	CaS_2O_3	Calcium Thiosulfate, 1715
151.892028	O_2Sn	Stannic Oxide, 8556
151.898859	$HNaO_3Se$	Sodium Hydroselenite, 8378
151.900167	CdF_2	Cadmium Fluoride, 1606
151.947291	$C_3H_5BrO_2$	β-Bromopropionic Acid, 1435
151.983676	C_4H_9BrO	3-Bromo-2-butanol, 1412
151.988917	$CrH_8N_2O_4$	Ammonium Chromate(VI), 529
152.015668	$C_5H_4N_4S$	6-Mercaptopurine, 5702
152.033425	$C_5H_4N_4O_2$	Xanthine, 9721
152.040820	$C_7H_8N_2S$	Phenylthiourea, 7120
152.044192	$C_4H_{12}N_2S_2$	Cystamine, 2775
152.044659	$C_4H_4N_6O$	Guanazolo, 4412
152.047344	$C_8H_8O_3$	p-Anisic Acid, 699 Cresotic Acids, 2575-2577 Mandelic Acid, 5545 Methylparaben, 5972 Methyl Salicylate, 5990 Phenoxyacetic Acid, 7054 Resacetophenone, 7932 Resorcinol Monoacetate, 7952 Vanillin, 9591
152.058578	$C_7H_8N_2O_2$	3,5-Diaminobenzoic Acid, 2934 N-(Hydroxymethyl)nicotinamide, 4741
152.068474	$C_5H_{12}O_5$	Adonitol, 155 Arabitol, 790

Use in conjunction with The Merck Index, Ninth Edition

Molecular Weight	Empirical Formula	Compound Name, Monograph Number
152.120115	$C_{10}H_{16}O$	Camphor, 1734 Citral, 2303 d-Fenchone, 3896 Homocamfin, 4611 Piperitone, 7267 Pulegone, 7725 Thujone, 9133
152.854505	Cl_3Ti	Titanium Trichloride, 9190
152.897998	$AgNO_2$	Silver Nitrite, 8263
152.907551	CIN	Cyanogen Iodide, 2702
152.908362	BeO_4Se	Beryllium Selenate, 1203
153.042593	$C_7H_7NO_3$	m-Aminosalicylic Acid, 490 p-Aminosalicylic Acid, 491, 6850, 6852 Nitroanisole, 6406 Salicylhydroxamic Acid, 8092
153.053827	$C_6H_7N_3O_2$	4-Nitro-o-phenylenediamine, 6445 p-Nitrophenylhydrazine, 6446
153.065060	$C_5H_7N_5O$	Hydracarbazine, 4642
153.078979	$C_8H_{11}NO_2$	4-Desoxypyridoxine, 2892 Dopamine, 3422 Norfenefrine, 6508 Octopamine, 6568
153.115364	$C_9H_{15}NO$	Pseudopelletierine, 7706
153.846076	Ga_2O	Gallium Suboxide, 4191
153.868044	Cl_2CrO_2	Chromyl Chloride, 2244
153.887077	NiO_4S	Nickel Sulfate, 6331
153.895755	CNa_2S_3	Sodium Thiocarbonate, 8464
153.899455	C_2CuKN_2	Cuprous Potassium Cyanide, 2672
153.900151	BaO	Barium Oxide, 991
153.958835	$C_4H_4Cl_2O_2$	Succinyl Chloride, 8673
153.961115	$C_2H_7AsO_3$	Ethanearsonic Acid, 3650
154.004624	$C_5H_3ClN_4$	6-Chloropurine, 2144
154.008851	$C_7H_6O_2S$	Thiosalicylic Acid, 9099
154.012223	$C_4H_{10}O_2S_2$	1,4-Dithiothreitol, 3393
154.018543	C_8H_7ClO	p-Chloroacetophenone, 2086 ω-Chloroacetophenone, 2087

154.

Molecular Weight	Empirical Formula	Compound Name, Monograph Number
154.026609	$C_7H_6O_4$	Coumalic Acid, Methyl ester, 2542 Gentisic Acid, 4230 Patulin, 6855 Protocatechuic Acid, 7679 β-Resorcylic Acid, 7954 Terreic Acid, 8887
154.029981	$C_4H_{10}O_4S$	Diethyl Sulfate, 3117
154.043008	$C_8H_7FO_2$	p-Fluorophenylacetic Acid, 4062
154.049076	$C_5H_6N_4O_2$	Glycarbylamide, 4314
154.062994	$C_8H_{10}O_3$	Filicinic Acid, 3997
154.067130	$C_4H_{12}FN_2OP$	Dimefox, 3191
154.074228	$C_7H_{10}N_2O_2$	5-Methylpyrazole-3-carboxylic Acid, Ethyl ester, 5985
154.078250	$C_{12}H_{10}$	Acenaphthene, 19 Diphenyl, 3322 Hemi-Dewar Biphenyl, 4500
154.110613	$C_8H_{14}N_2O$	Festucine, 3988
154.135765	$C_{10}H_{18}O$	Borneol, 1350 Cineole, 2280 Citronellal, 2310 Geraniol, 4235 Isoborneol, 4980 Linalool, 5335 l-Menthone, 5664 Nerol, 6299 Rhodinal, 7976 α-Terpineol, 8886
154.189281	$C_{10}H_{23}B$	Bis(1,2-dimethylpropyl)borane, 1269
154.884928	CoO_4S	Cobaltous Sulfate, 2401
155.021858	$C_6H_5NO_4$	Citrazinic Acid, 2306
155.026855	$C_5H_{11}Cl_2N$	Mechlorethamine, 5600
155.058243	$C_7H_9NO_3$	Aloxidone, 304
155.069477	$C_6H_9N_3O_2$	Bacimethrin, 941 Histidine, 4597
155.076871	$C_8H_{13}NS$	α-Ethyl-2-thienylethylamine, 3805
155.094629	$C_8H_{13}NO_2$	Arecolidine, 803 Arecoline, 804, 805, 3434 Bemegride, 1033 Hypoglycine A, Methyl ester, 4787 Scopoline, 8163
155.095298	$[C_{17}H_{22}N_6]^{2+}$	Quinapyramine, Cation (half mass), 7839
155.131014	$C_9H_{17}NO$	Novonal, 6532 Pelletierine, Methyl deriv, 6866

Use in conjunction with The Merck Index, Ninth Edition

Molecular Weight	Empirical Formula	Compound Name, Monograph Number
155.167400	$C_{10}H_{21}N$	Cyclexedrine, 2713 Pempidine, 6874 Propylhexedrine, 7649
155.882759	$CaCrO_4$	Calcium Chromate(VI), 1653
155.910134	F_4Se	Selenium Tetrafluoride, 8190
155.912882	$C_2CaN_2S_2$	Calcium Thiocyanate, 1713
155.914952	N_2O_4Zn	Zinc Nitrite, 9808
155.943602	C_2H_5I	Ethyl Iodide, 3753
155.957461	C_6H_5Br	Bromobenzene, 1405
155.974485	$C_4H_6Cl_2O_2$	Dichloroacetic Acid, Ethyl ester, 3021 Ethyl Dichloroacetate, 3724
155.997807	$C_7H_5ClO_2$	Chlorobenzoic Acids, 2099-2101
156.006744	$C_7H_8S_2$	Toluene-3,4-dithiol, 9227
156.010696	$C_4H_{10}ClO_2P$	Ethyl Phosphorochloridite, 3782
156.017107	$C_5H_4N_2O_4$	α,β-Imidazoledicarboxylic Acid, 4808 Nifuroxime, 6360 Orotic Acid, 6711, 6703
156.024501	$C_7H_8O_2S$	p-Toluenesulfinic Acid, 9228
156.028340	$C_4H_4N_4O_3$	Oxonic Acid Amide, 6764
156.034193	C_8H_9ClO	4-Chloro-3,5-xylenol, 2165
156.053492	$C_6H_8N_2O_3$	Pentoxyl, 6932
156.068748	$C_{10}H_8N_2$	α,α'-Dipyridyl, 3367 γ,γ'-Dipyridyl, 3368
156.089878	$C_7H_{12}N_2O_2$	Ectylurea, 3474 Heptoxime, 4527
156.126263	$C_8H_{16}N_2O$	Hygrine, Oxime, 4773
156.151415	$C_{10}H_{20}O$	β-Citronellol, 2311 Menthol, 5663 Rhodinol, 7977
156.847068	Cl_3Cr	Chromic Chloride, 2219
156.980960	$C_3H_4NNaO_3S$	Sodium β-Sulfopropionitrile, 8452
156.993056	$C_6H_4ClNO_2$	Chloronitrobenzenes, 2128-2130
157.012356	$C_4H_3N_3O_4$	Oxonic Acid, 6764 Violuric Acid, 9652
157.029267	$C_6H_8NO_2P$	(p-Aminophenyl)phosphonous Acid, 475
157.037508	$C_6H_7NO_4$	2-(Methoxymethyl)-5-nitrofuran, 5868

Molecular Weight	Empirical Formula	Compound Name, Monograph Number
157.073893	$C_7H_{11}NO_3$	Ethadione, 3644 Paramethadione, 6833
157.110279	$C_8H_{15}NO_2$	Methyl N-Methylnipecotate, 5964 Oxanamide, 6748 Tranexamic Acid, 9259
157.132746	$C_6H_{15}N_5$	Buformin, 1465
157.146664	$C_9H_{19}NO$	Isovaleryl Diethylamide, 5097
157.183050	$C_{10}H_{23}N$	Diisoamylamine, 3177
157.831270	Cu_2S	Cuprous Sulfide, 2674
157.843330	Cl_2Sr	Strontium Chloride, 8630
157.843936	K_2Se	Potassium Selenide, 7455
157.860837	Mn_2O_3	Manganese Sesquioxide, 5568
157.881413	$KMnO_4$	Potassium Permanganate, 7441
157.884232	K_2O_3S	Potassium Sulfite, 7465
157.893748	HK_2O_3P	Potassium Phosphite, 7448
157.899006	F_2Sn	Stannous Fluoride, 8563
157.908427	$Na_2O_3S_2$	Sodium Thiosulfate, 8468
157.915667	H_4O_4Zr	Zirconium Hydroxide, 9843
157.972378	$C_4H_8Cl_2S$	Mustard Gas, 6142
158.003766	$C_6H_6O_3S$	Benzenesulfonic Acid, 1073
158.032757	$C_5H_6N_2O_4$	Hydroorotic Acid, 4702 Ibotenic Acid, 4793 Maleuric Acid, 5534 Muscazone, 6136
158.036780	$C_{10}H_6O_2$	1,2-Naphthoquinone, 6221 1,4-Naphthoquinone, 6222
158.043990	$C_4H_6N_4O_3$	Allantoin, 240
158.057909	$C_7H_{10}O_4$	Mesaconic Acid, Dimethyl ester, 5751 Terebic Acid, 8881
158.069142	$C_6H_{10}N_2O_3$	1-(Hydroxymethyl)-5,5-dimethylhydantoin, 4736
158.073165	$C_{11}H_{10}O$	2-Methoxynaphthalene, 5869
158.084398	$C_{10}H_{10}N_2$	1,8-Naphthalenediamine, 6196 α-Nicotyrine, 6348 β-Nicotyrine, 6349
158.094294	$C_8H_{14}O_3$	Butyric Anhydride, 1586

Use in conjunction with The Merck Index, Ninth Edition

Molecular Weight	Empirical Formula	Compound Name, Monograph Number
158.130680	$C_9H_{18}O_2$	n-Amyl Butyrate, 640 Ethyl Oenanthate, 3773 Isoamyl Butyrate, 4962 Isobutyl Isovalerate, 5000 Pelargonic Acid, 6864
158.167065	$C_{10}H_{22}O$	n-Amyl Ether, 647 n-Decyl Alcohol, 2835 Isoamyl Ether, 4966
158.178299	$C_9H_{22}N_2$	Novoldiamine, 6531
158.881330	CuO_4S	Cupric Sulfate, 2657
159.035400	$C_6H_9NO_2S$	Citiolone, 2301
159.068414	$C_{10}H_9NO$	4-Amino-1-naphthol, 462 Echinopsine, 3467
159.089543	$C_7H_{13}NO_3$	Betonicine, 1218 Turicine, 9473
159.104799	$C_{11}H_{13}N$	Pargyline, 6843
159.125929	$C_8H_{17}NO_2$	Butyroin, Oxime, 1587 Octanohydroxamic Acid, 6562
159.125929	$[C_{16}H_{34}N_2O_4]^{2+}$	Suxethonium (half mass), 8798
159.854623	Fe_2O_3	Ferric Oxide, 3946
159.880876	O_4SZn	Zinc Sulfate, 9827
159.894592	O_5STi	Titanium Sulfate, 9186
159.948145	$C_2H_6AsNaO_2$	Sodium Cacodylate, 8339
159.979697	$C_2H_4CaN_2O_4$	Calcium Carbamate, 1647
159.984656	$C_7H_6Cl_2$	Benzal Chloride, 1056
160.034672	$C_{10}H_8S$	1-Naphthalenethiol, 6202 2-Naphthalenethiol, 6203
160.037173	$C_6H_8O_5$	6-Desoxy-L-ascorbic Acid, 2889
160.052430	$C_{10}H_8O_2$	Naphthoresorcinol, 6223 Tricromyl, 9337
160.073559	$C_7H_{12}O_4$	n-Butylmalonic Acid, 1564 Diethylmalonic Acid, 3104 Ethyl Malonate, 3763 Glutaric Acid, Dimethyl ester, 4305 Pimelic Acid, 7231
160.074896	$C_8H_8N_4$	Hydralazine, 4645
160.084792	$C_6H_{12}N_2O_3$	Succinic Acid 2,2-Dimethylhydrazide, 8669

Molecular Weight	Empirical Formula	Compound Name, Monograph Number
160.100048	$C_{10}H_{12}N_2$	Anatabine, 665 N-Methylmyosmine, 5965 Tolazoline, 9207 Tryptamine, 9456
160.133754	$[C_8H_{18}NO_2]^+$	Carpronium, 1860 Methacholine, 5793-5795
160.841497	Cl_3Fe	Ferric Chloride, 3933
160.979905	$C_6H_5Cl_2N$	3,4-Dichloroaniline, 3026
161.006599	C_6H_8ClNS	Clomethiazole, 2348
161.032422	$C_5H_7NO_5$	α-Ketoglutaric Acid, Oxime, 5152
161.068808	$C_6H_{11}NO_4$	α-Aminoadipic Acid, 418
161.070145	$C_7H_7N_5$	Fenamole, 3890
161.084064	$C_{10}H_{11}NO$	Abikoviromycin, 2 Tryptophol, 9459
161.105193	$C_7H_{15}NO_3$	Carnitine, 1849
161.129003	$[C_7H_{17}N_2O_2]^+$	Bethanechol, 1215
161.849549	MoS_2	Molybdenum Disulfide, 6069
161.873330	ClI	Iodine Monochloride, 4878
161.894022	B_2Cl_4	Diboron Tetrachloride, 2982
161.896058	O_2Te	Tellurium Dioxide, 8867
161.899708	$CrNa_2O_4$	Sodium Chromate(VI), 8346
161.904213	$C_2HCl_3O_2$	Trichloroacetic Acid, 9306
161.909541	H_4IP	Phosphonium Iodide, 7152
161.914584	Al_2O_5Si	Aluminum Silicate, 364
161.930907	$C_2H_4Cl_2O_2S$	Bis(chloromethyl) Sulfone, 1265
161.940598	$C_3H_5Cl_3O$	1,1,1-Trichloro-2-propanol, 9325
161.948315	$H_4O_6P_2$	Hypophosphoric Acid, 4789
162.004412	$C_6H_{11}Br$	Cyclohexyl Bromide, 2735
162.017309	$C_6H_{10}OS_2$	Allicin, 243
162.023628	$C_{10}H_7Cl$	1-Chloronaphthalene, 2126 2-Chloronaphthalene, 2127
162.028542	$C_5H_{10}N_2S_2$	Dazomet, 2818 Picadex, 7200

Use in conjunction with The Merck Index, Ninth Edition

Molecular Weight	Empirical Formula	Compound Name, Monograph Number
162.031694	$C_9H_6O_3$	Coumarilic Acid, 2546 Phloroglucinol, 7138 Umbelliferone, 9503
162.042928	$C_8H_6N_2O_2$	Fenadiazole, 3887 p-Nitrobenzyl Cyanide, 6415
162.046299	$C_5H_{10}N_2O_2S$	Methomyl, 5854
162.052824	$C_6H_{10}O_5$	Lactic Acid Lactate, 5187 Lichenin, 5322 Pyrocarbonic Acid Diethyl Ester, 7779 L-Streptose, 8616
162.054161	$C_7H_6N_4O$	N,N'-Carbonyldiimidazole, 1824
162.055816	$C_5H_{11}N_2O_2P$	Tabun, 8809
162.068080	$C_{10}H_{10}O_2$	Cinnamic Acid, Methyl ester, 2288 Isosafrole, 5087 Safrole, 8076
162.079313	$C_9H_{10}N_2O$	Aminorex, 489 1-Phenyl-3-pyrazolidinone, 7115
162.089209	$C_7H_{14}O_4$	D-Chalcose, 1993 Cymarose, 2764 Mycarose, 6145 Sarmentose, 8130
162.104465	$C_{10}H_{11}OCH_3$	Tetralol, Methyl ether, 8937
162.115699	$C_{10}H_{14}N_2$	Anabasine, 658 Isonicotine, 5043 Nicotine, 6342
162.125595	$C_8H_{18}O_3$	Butyl Carbitol®, 1541 Diethyl Carbitol®, 3097 4,4'-Oxydi-2-butanol, 6775
162.890608	O_5SV	Vanadyl Sulfate, 9585
162.899461	CCl_3NO_2	Chloropicrin, 2138
162.933211	BeF_4K_2	Beryllium Potassium Fluoride, 1201
162.935847	$C_2H_4Cl_3NO$	Chloral Ammonia, 2028
162.938378	H_5NO_4Se	Ammonium Hydroselenate, 551
162.998557	F_6H_4NP	Ammonium Hexafluorophosphate, 549
163.009186	C_8H_5NOS	Benzoyl Isothiocyanate, 1125
163.026943	$C_8H_5NO_3$	Carsalam, 1862 Diatretyne I, 2957
163.030315	$C_5H_9NO_3S$	Acetylcysteine, 80 N-(2-Mercaptopropionyl)glycine, 5701
163.048073	$C_5H_9NO_5$	Hydroxyglutamic Acid, 4727

Molecular Weight	Empirical Formula	Compound Name, Monograph Number
163.063329	$C_9H_9NO_2$	Epinochrome, 3544
163.066701	$C_6H_{13}NO_2S$	Ethionine, 3665 α-Methylmethionine, 5962
163.084458	$C_6H_{13}NO_4$	6-Desoxy-D-glucosamine, 2890 Diethylolglycine, 3108 Fucosamine, 4128 Mycosamine, 6153 2-Nitro-2-propyl-1,3-propanediol, 6453
163.099714	$C_{10}H_{13}NO$	Ethylacetanilide, 3684 α-Phenylbutyramide, 7080 Thalline, 8969
163.136100	$C_{11}H_{17}N$	N-Ethylamphetamine, 3692 Mephentermine, 5681 Phenpentermine, 7058 Xylopropamine, 9748
163.830136	BrRb	Rubidium Bromide, 8041
163.855438	Ni_2O_3	Nickel Sesquioxide, 6330
163.875411	C_2Cl_4	Tetrachloroethylene, 8907
163.877942	$ClHO_3Se$	Chloroselenic Acid, 2149
163.900714	H_2MoO_4	Molybdic(VI) Acid, 6073
163.919863	$C_2H_3Cl_3O_2$	Chloral Hydrate, 2033, 2029
163.922731	Na_3O_4P	Sodium Phosphate, Tribasic, 8424
163.938227	CaN_2O_6	Calcium Nitrate, 1682
163.960727	$C_3H_4Cl_2F_2O$	Methoxyflurane, 5866
163.970013	$H_8N_2O_3Se$	Ammonium Selenite, 582
164.010308	$H_8N_2O_6S$	Hydroxylamine Sulfate, 4732
164.033425	$C_6H_4N_4O_2$	Lumazine, 5414
164.047344	$C_9H_8O_3$	Carbic Anhydride, 1799 p-Coumaric Acid, 2545
164.058578	$C_8H_8N_2O_2$	Phthalamide, 7176 Ricinine, 8004
164.068474	$C_6H_{12}O_5$	D-Fucose, 4129 L-Fucose, 4130 d-Quercitol, 7823 Quinovose, 7901 Rhamnose, 7963
164.074526	$[C_6H_{14}NO_2S]^+$	Vitamin U, 9693

Use in conjunction with The Merck Index, Ninth Edition

Molecular Weight	Empirical Formula	Compound Name, Monograph Number
164.083730	$C_{10}H_{12}O_2$	Cumic Acid, 2617 Duroquinone, 3447 Ethyl Phenylacetate, 3780 Eugenol, 3843 Isoeugenol, 5027 Thujic Acid, 9132 Thymoquinone, 9147
164.094963	$C_9H_{12}N_2O$	Fenuron, 3923
164.106196	$C_8H_{12}N_4$	2,2'-Azobisisobutyronitrile, 925
164.120115	$C_{11}H_{16}O$	Fenipentol, 3904 Jasmone, 5110 p-tert-Pentylphenol, 6937
164.131349	$C_{10}H_{16}N_2$	Tetramethyl-p-phenylenediamine, 8945
165.042593	$C_8H_7NO_3$	4-Pyridoxic Acid, Lactone, 7765
165.078979	$C_9H_{11}NO_2$	p-Acetanisidine, 38 Atrolactamide, 889 4-(Dimethylamino)benzoic Acid, 3219 Ethenzamide, 3658 Ethyl p-Aminobenzoate, 3691 Ethyl m-Aminobenzoate, 9300 Isopropyl Nicotinate, 5076 Phenylalanine, 7071 α-Phenylglycine, Methyl ester, 7097 Phenylurethan(e), 7127
165.090212	$C_8H_{11}N_3O$	1-(m-Tolyl)semicarbazide, 9243
165.115364	$C_{10}H_{15}NO$	α-(α-Aminopropyl)benzyl Alcohol, 482 Carvone, Oxime, 1867 Ephedrine, 3534, 2616 Hordenine, 4625 Perillaldehyde, Oxime, 6956 Pholedrine, 7141 Safranal, Oxime, 8075
165.849141	Cl_2OSe	Selenium Oxychloride, 8185
165.866788	$BrKO_3$	Potassium Bromate, 7394
165.868185	IK	Potassium Iodide, 7426
165.884006	O_4Ru	Ruthenium Tetroxide, 8060
165.891061	$C_2H_2Cl_4$	Tetrachloroethane, 8906
165.898321	F_4Zr	Zirconium Fluoride, 9841
165.909509	C_2CdN_2	Cadmium Cyanide, 1605
165.952393	F_6MgSi	Magnesium Hexafluorosilicate, 5491
165.962941	$C_4H_7BrO_2$	α-Bromobutyric Acid, 1413 α-Bromoisobutyric Acid, 1419 β-Bromopropionic Acid, Methyl ester, 1435

166.

Molecular Weight	Empirical Formula	Compound Name, Monograph Number
166.018543	C_9H_7ClO	Cinnamoyl Chloride, 2291
166.023136	$[C_5H_{13}BrN]^+$	Bromcholine, 8997
166.026609	$C_8H_6O_4$	Isophthalic Acid, 5058 Phthalic Acid, 7178 Piperonylic Acid, 7271 Terephthalic Acid, 8882
166.049076	$C_6H_6N_4O_2$	2,3-Pyrazinedicarboxylic Acid, Diamide, 7742
166.056470	$C_8H_{10}N_2S$	S-Benzylpseudothiourea, 1173 Ethionamide, 3664
166.062994	$C_9H_{10}O_3$	Apocynin, 771 Atrolactic Acid, 890 Ethylparaben, 3777 Ethyl Salicylate, 3793 Ethyl Vanillin, 3811 Tropic Acid, 9444 Veratraldehyde, 9605
166.078250	$C_{13}H_{10}$	Fluorene, 4037
166.099380	$C_{10}H_{14}O_2$	p-Butoxyphenol, 1516 Durohydroquinone, 3446 Nepetalactone, 6293
166.110613	$C_9H_{14}N_2O$	5-Amino-2-butoxypyridine, 433 p-Aminonorephedrine, 467 Phenoxypropazine, 7057
166.123189	$[C_{10}H_{16}NO]^+$	Edrophonium, 3477
166.848855	$BeBr_2$	Beryllium Bromide, 1189
166.986343	$C_7H_5NS_2$	2-Mercaptobenzothiazole, 5698
167.021858	$C_7H_5NO_4$	Cinchomeronic Acid, 2272 Isocinchomeronic Acid, 5015 Nitrobenzoic Acids, 6411-6413 Quinolinic Acid, 7896
167.026567	$C_5H_5N_5S$	Thioguanine, 9071
167.058243	$C_8H_9NO_3$	N-(p-Hydroxyphenyl)glycine, 4752 Orthocaine, 6719 Pyridoxal, 7762
167.069477	$C_7H_9N_3O_2$	p-Aminosalicylic Acid Hydrazide, 492
167.073499	$C_{12}H_9N$	Carbazole, 1793
167.080710	$C_6H_9N_5O$	4-Amino-2-methyl-5-pyrimidinecarboxylic Acid Hydrazide, 457

Use in conjunction with The Merck Index, Ninth Edition

Molecular Weight	Empirical Formula	Compound Name, Monograph Number
167.094629	$C_9H_{13}NO_2$	Phenylephrine, 7091 Deoxyepinephrine, 2866 Ecgonidine, 3460 Ethinamate, 3661 Metaraminol, 5783 Pyrithyldione, 7777 Synephrine, 8803
167.131014	$C_{10}H_{17}NO$	Camphor, Oxime, 1734
167.167400	$C_{11}H_{21}N$	Mecamylamine, 5599
167.852339	Cl_4Si	Silicon Tetrachloride, 8241
167.852393	Cl_3PS	Phosphorus Sulfochloride, 7166
167.872289	Mg_2Sn	Magnesium Stannide, 5516
167.872797	HKO_3Se	Potassium Biselenite, 7388
167.874286	$ClCs$	Cesium Chloride, 1967
167.881224	MgO_4Se	Magnesium Selenate, 5511
167.938099	$C_4H_2Cl_2O_3$	Mucochloric Acid, 6124
167.943602	C_3H_5I	Allyl Iodide, 288
167.955635	C_6H_5AsO	Oxophenylarsine, 6766
167.976765	$C_3H_9AsO_3$	1-Propanearsonic Acid, 7592
167.988116	$C_7H_4O_3S$	Tioxolone, 9176
168.009041	$C_7H_5ClN_2O$	Zoxazolamine, 9853
168.017107	$C_6H_4N_2O_4$	Dinitrobenzene, 3269 2,3-Pyrazinedicarboxylic Acid, 7742
168.028340	$C_5H_4N_4O_3$	Uric Acid, 9538
168.034193	C_9H_9ClO	Chlorindanol, 2064
168.042259	$C_8H_8O_4$	Coumalic Acid, Ethyl ester, 2542 Dehydroacetic Acid, 2840 2,6-Dimethoxyquinone, 3212 Fumigatin, 4139 Gallacetophenone, 4186 Homogentisic Acid, 4619 Monoacetyl Pyrogallol, 6083 o-Orsellinic Acid, 6717 Vanillic Acid, 9590
168.058658	$C_9H_9FO_2$	p-Fluorobenzoic Acid, Ethyl ester, 4056
168.068748	$C_{11}H_8N_2$	γ-Carboline, 1810
168.070578	$C_{10}H_{13}Cl$	Neophyl Chloride, 6284
168.078644	$C_9H_{12}O_3$	1,2,4-Benzenetriol, Trimethyl ether, 1076 Phenylglyceryl Ether, 7095

168.

Molecular Weight	Empirical Formula	Compound Name, Monograph Number
168.089878	$C_8H_{12}N_2O_2$	Pyridoxamine, 7764
168.093900	$C_{13}H_{12}$	Diphenylmethane, 3339
168.115030	$C_{10}H_{16}O_2$	Ascaridole, 851 Chrysanthemic Acid, 2248 Cicrotoic Acid, 2265 Diosphenol, 3297 3-Hydroxycamphor, 4717 Iridomyrmecin, 4938 Oxypinocamphone, 6789 Sebacil®, 8168
168.892913	$AgNO_3$	Silver Nitrate, 8261
168.993056	$C_7H_4ClNO_2$	Chlorzoxazone, 2185
169.004494	$C_3H_7NO_5S$	Cysteic Acid, 2778
169.029442	C_8H_8ClNO	Chloroacetanilides, 2080-2082
169.037508	$C_7H_7NO_4$	Gallamide, 4187
169.042505	$C_6H_{13}Cl_2N$	HN1, 4599
169.050396	$C_4H_{12}NO_4P$	Demanyl Phosphate, 2854
169.052764	$C_{11}H_7NO$	1-Naphthylisocyanate, 6238
169.073893	$C_8H_{11}NO_3$	Norepinephrine, 6504 Pyridoxol, 7759 Pyridoxine, 7766
169.085127	$C_7H_{11}N_3O_2$	Ipronidazole, 4933
169.089149	$C_{12}H_{11}N$	p-Biphenylamine, 1248 Diphenylamine, 3325
169.110279	$C_9H_{15}NO_2$	Diacetone Acrylamide, 2920 Homoarecoline, 4610 Piperidione, 7263 3-Quinuclidinol, dl-Form acetate (ester), 7905
169.146664	$C_{10}H_{19}NO$	Lupinine, 5425
169.146664	$[C_{20}H_{38}N_2O_2]^{2+}$	Dipropamine, Cation (half mass), 3360
169.877308	BaS	Barium Sulfide, 1002
169.895065	BaO_2	Barium Peroxide, 994
169.900955	F_2Xe	Xenon difluoride, 9738
169.913612	AsF_5	Arsenic Pentafluoride, 824
169.915006	C_4NiO_4	Nickel Carbonyl, 6318
169.921076	$CaH_4O_4P_2$	Calcium Hypophosphite, 1672
169.922867	C_2H_3IO	Acetyl Iodide, 89

Use in conjunction with The Merck Index, Ninth Edition

Molecular Weight	Empirical Formula	Compound Name, Monograph Number
169.930248	$C_4H_2FeO_4$	Hydrogen Tetracarbonylferrate(II), 4697
169.931318	$C_2Cl_2F_4$	Cryofluorane, 2599
169.953750	$C_4H_4Cl_2O_3$	Chloroacetic Anhydride, 2084
169.959252	C_3H_7I	Isopropyl Iodide, 5074 Propyl Iodide, 7651
169.973111	C_7H_7Br	Benzyl Bromide, 1142 Bromotoluenes, 1443-1445
169.986008	$C_7H_6OS_2$	Dithiosalicylic Acid, 3392
169.990135	$C_5H_8Cl_2O_2$	Caldariomycin, 1718
169.998026	$C_3H_7O_6P$	Glyceraldehyde 3-Phosphate, 4317
170.013457	$C_8H_7ClO_2$	p-Anisoyl Chloride, 708 Carbobenzoxy Chloride, 1805 m-Chlorobenzoic Acid, Methyl ester, 2099 p-Chlorobenzoic Acid, Methyl ester, 2101
170.021523	$C_7H_6O_5$	Gallic Acid, 4190
170.024691	$C_7H_7ClN_2O$	p-Chlorobenzhydrazide, 2097
170.032757	$C_6H_6N_2O_4$	Orotic Acid, Methyl ester, 6711
170.033633	$B_4Li_2O_7$	Lithium Borate, 5360
170.037922	$C_8H_7FO_3$	3-Fluoro-4-hydroxyphenylacetic Acid, 4059
170.051385	$C_7H_{10}N_2OS$	Propylthiouracil, 7658
170.057909	$C_8H_{10}O_4$	Muconic Acid, Dimethyl ester, 6126 Penicillic Acid, 6884
170.073165	$C_{12}H_{10}O$	Phenyl Ether, 7093 Phenylphenols, 7110, 7111
170.080376	$C_6H_{10}N_4O_2$	α-Hydrazinoimidazole-4(or 5)-propionic Acid, 4660
170.094294	$C_9H_{14}O_3$	Castelamarin, 1891
170.130680	$C_{10}H_{18}O_2$	Sebacoin, 8169 Sobrerol, 8305 Hexahydrothujic Acid, 9132
170.154489	$[C_{30}H_{60}N_3O_3]^{3+}$	Triethylgallaminium (one-third mass), 4188
170.964255	$C_7H_3Cl_2N$	2,6-Dichlorobenzonitrile, 3033
170.968360	C_6H_6BrN	p-Bromoaniline, 1403
171.021770	$C_5H_{11}Cl_2NO$	Mechlorethamine Oxide, 5601
171.046634	$C_6H_9N_3OS$	Methioprim, 5847
171.056325	$C_7H_{10}ClN_3$	Crimidine, 2581

171.

Molecular Weight	Empirical Formula	Compound Name, Monograph Number
171.064391	$C_6H_9N_3O_3$	6-Diazo-5-oxo-L-norleucine, 2966 Metronidazole, 6029
171.089543	$C_8H_{13}NO_3$	Diethadione, 3077
171.104799	$C_{12}H_{13}N$	N,N-Dimethyl-1-naphthylamine, 3240
171.112010	$C_6H_{13}N_5O$	Moroxydine, 6103
171.137162	$C_8H_{17}N_3O$	2-Heptanone, Semicarbazone, 4525
171.198700	$C_{11}H_{25}N$	Metron S, 6030
171.833041	CSe_2	Carbon Diselenide, 1816
171.845435	Cl_3OV	Vanadyl Trichloride, 9586
171.852322	CH_2Br_2	Methylene Bromide, 5930
171.867591	$ClOSb$	Antimony Chloride Oxide, 731
171.885231	$C_2FeN_2S_2$	Ferrous Thiocyanate, 3984
171.895271	CeO_2	Ceric Oxide, 1948
171.899085	F_3In	Indium Trifluoride, 4838
171.910715	BaH_2O_2	Barium Hydroxide, 982
171.920681	C_4HCoO_4	Cobalt Carbonyl Hydride, 3061
171.920994	CoH_4NO_4P	Ammonium Cobaltous Phosphate, 532
171.952376	C_6H_5BrO	Bromophenols, 1430-1432
172.000788	$C_6H_4O_6$	Tetroquinone, 8963
172.005785	$C_5H_{10}Cl_2O_2$	Pentaerythritol Dichlorohydrin, 6906
172.013677	$C_3H_9O_6P$	Glycerophosphoric Acid, 4320, 3940
172.019416	$C_7H_8O_3S$	p-Toluenesulfonic Acid, 9229
172.028933	$C_7H_9O_3P$	α-Hydroxybenzylphosphinic Acid, 4714
172.030649	$C_6H_8N_2O_2S$	Porofor® BSH, 7371 Sulfanilamide, 8717, 409
172.052430	$C_{11}H_8O_2$	Menadione, 5653 1-Naphthoic Acid, 6208 2-Naphthoic Acid, 6209 Nemotin, 6268
172.073559	$C_8H_{12}O_4$	Ethyl Maleate, 3761 Terpenylic Acid, 8883
172.081856	$[C_{21}H_{20}N_4O]^{2+}$	Quinuronium, 3371
172.088815	$C_{12}H_{12}O$	2-Ethoxynaphthalene, 3680
172.100048	$C_{11}H_{12}N_2$	Vitamin K_6, 9688

Use in conjunction with The Merck Index, Ninth Edition

Molecular Weight	Empirical Formula	Compound Name, Monograph Number
172.101878	$C_{10}H_{17}Cl$	Bornyl Chloride, 1353 α-Pinene, Hydrochloride, 7242
172.109945	$C_9H_{16}O_3$	Hexacyclonic Acid, 4547
172.133754	$[C_9H_{18}NO_2]^+$	Methyl N,N-Dimethylnipecotate, 5964
172.146330	$C_{10}H_{20}O_2$	tert-Amyl Isovalerate, 648 n-Capric Acid, 1757 Ethyl Caprylate, 3709 Isoamyl Isovalerate, 4969 Octyl Acetate, 6570 Terpin, 8884
172.193949	$C_{10}H_{24}N_2$	Hexamethonium Bromide, Free base, 4557
172.920197	$C_3H_2Cl_3NO$	Chlorocyanohydrin, 2109
172.989513	$C_4H_3N_3O_3S$	Forminitrazole, 4100
173.007270	$C_4H_3N_3O_5$	5-Nitrobarbituric Acid, 6407
173.014665	$C_6H_7NO_3S$	Metanilic Acid, 5779 Orthanilic Acid, 6718 Sulfanilic Acid, 8720
173.043656	$C_5H_7N_3O_4$	Azaserine, 914
173.047679	$C_{10}H_7NO_2$	1-Nitronaphthalene, 6436 1-Nitroso-2-naphthol, 6461 N-Phenylmaleimide, 7104 Quinaldic Acid, 7832 8-Quinolinecarboxylic Acid, 7891
173.060742	$C_8H_{12}ClNO$	CDAA, 1906
173.064809	$C_9H_8BNO_2$	8-Quinolineboronic Acid, 7890
173.068808	$C_7H_{11}NO_4$	N-Acetylhydroxyproline, 88
173.071801	$C_6H_{12}N_3OP$	Triethylenephosphoramide, 9347
173.084064	$C_{11}H_{11}NO$	3-Indolylacetone, 4844 Vitamin K_5, 9687 Vitamin K_7, 9689
173.105193	$C_8H_{15}NO_3$	N-Acetylleucine, 8825 ϵ-Acetamidocaproic Acid, 34
173.116427	$C_7H_{15}N_3O_2$	Indospicine, 4846
173.120450	$C_{12}H_{15}N$	6,7-Benzomorphan, 1106
173.832139	$GaCl_3$	Gallium Chloride, 4191
173.863777	$AgClO_2$	Silver Chlorite, 8251
173.879146	K_2O_4S	Potassium Sulfate, 7463
173.880804	Na_2O_3Se	Sodium Selenite, 8441

173.

Molecular Weight	Empirical Formula	Compound Name, Monograph Number
173.888105	$CCdO_3$	Cadmium Carbonate, 1603
173.888663	HK_2O_4P	Potassium Phosphate, Dibasic, 7445
173.903342	$Na_2O_4S_2$	Sodium Hydrosulfite, 8379
173.910352	BrF_5	Bromine Pentafluoride, 1393
173.940598	$C_4H_5Cl_3O$	α,α,β-Trichloro-n-butyraldehyde, 9312
173.998681	$C_6H_6O_4S$	p-Phenolsulfonic Acid, 7044
174.008197	$C_6H_7O_4P$	Phenyl Phosphate, 3376
174.016438	$C_6H_6O_6$	Aconitic Acid, 112 Dehydroascorbic Acid, 2841 Ketipic Acid, 5150
174.031694	$C_{10}H_6O_3$	Juglone, 5119 Lawsone, 5239
174.042928	$C_9H_6N_2O_2$	Toluene 2,4-Diisocyanate, 9226
174.052824	$C_7H_{10}O_5$	Mesoxalic Acid, Diethyl ester, 5759 Shikimic Acid, 8222
174.089209	$C_8H_{14}O_4$	Adipic Acid, Dimethyl ester, 151 Dimethoxane, 3210 Ethyl Succinate, 3799 Suberic Acid, 8661
174.111676	$C_6H_{14}N_4O_2$	Arginine, 806, 807
174.115699	$C_{11}H_{14}N_2$	Gramine, 4381
174.125595	$C_9H_{18}O_3$	n-Butyl Carbonate, 1542 Isobutyl Carbonate, 4992
174.144223	$C_{10}H_{22}S$	Isoamyl Sulfide, 4975
174.149404	$[C_9H_{20}NO_2]^+$	Muscarine, 6135
174.161980	$C_{10}H_{22}O_2$	Decamethylene Glycol, 2832
174.883489	$C_2CoN_2S_2$	Cobaltous Thiocyanate, 2403
175.010329	$C_6H_6FNO_2S$	Sulfanilyl Fluoride, 8723
175.012558	$C_6H_9NOS_2$	Sulforaphen, 8759
175.063329	$C_{10}H_9NO_2$	α-Ethynylbenzyl Carbamate, 3815 Gentianine, 4226 Indoleacetic Acid, 4841 Succinanil, 8666
175.084458	$C_7H_{13}NO_4$	L-Glutamic Acid 5-Ethyl Ester, 4303
175.095691	$C_6H_{13}N_3O_3$	Citrulline, 2312
175.099714	$C_{11}H_{13}NO$	β-Benzalbutyramide, 1055

Use in conjunction with The Merck Index, Ninth Edition

Molecular Weight	Empirical Formula	Compound Name, Monograph Number
175.110948	$C_{10}H_{13}N_3$	Debrisoquine, 2827
175.120844	$C_8H_{17}NO_3$	Desosamine, 2887
175.837328	Cl_2OZr	Zirconyl Chloride, 9850
175.841180	Cl_2Pd	Palladium Chloride, 6799
175.885283	C_2O_4Sr	Strontium Oxalate, 8638
175.897046	HIO_3	Iodic Acid, 4872
175.902043	BaF_2	Barium Fluoride, 979
175.944932	$CH_4O_6S_2$	Methionic Acid, 5844
175.956248	$C_4H_7Cl_3O$	Chlorobutanol, 2103
175.969879	$C_6H_5ClO_2S$	Benzenesulfonyl Chloride, 1075
175.979570	$C_7H_6Cl_2O$	Dybenal®, 3449
175.983676	C_6H_9BrO	1-Bromo-3-methyl-1-pentyn-3-ol, 1425
176.032088	$C_6H_8O_6$	Ascorbic Acid, 855 D-Glucuronolactone, 4300 Isoascorbic Acid, 4978 Tricarballylic Acid, 9301
176.043322	$C_5H_8N_2O_5$	N-Carbamoylaspartic Acid, 1648
176.047344	$C_{10}H_8O_3$	Erythrocentaurin, 3599 Hymecromone, 4777
176.058578	$C_9H_8N_2O_2$	Pemoline, 6872, 6873
176.068474	$C_7H_{12}O_5$	Diacetin, 2918 Lactic Acid Lactate, Methyl ester, 5187
176.083730	$C_{11}H_{12}O_2$	Centalun, 1923 Ethyl Cinnamate, 3718, 2288 Isocrotonic Acid, Benzyl ester, 5019 6-Methoxy-α-tetralone, 5877
176.090940	$C_5H_{12}N_4O_3$	Canavanine, 1741
176.094963	$C_{10}H_{12}N_2O$	Cotinine, 2536 Serotonin, 8209
176.104859	$C_8H_{16}O_4$	Carbitol®, Acetate, 1804 Methyl chalcoside, 1993 3-O-Methylmycarose, 6145
176.188864	$[C_9H_{24}N_2O]^{2+}$	Prolonium, 7576
176.942540	$C_4H_4BrNO_2$	N-Bromosuccinimide, 1441
176.974819	$C_6H_5Cl_2NO$	2-Amino-4,6-dichlorophenol, 442
177.045965	$C_6H_{11}NO_3S$	Alliin, 244

177.

Molecular Weight	Empirical Formula	Compound Name, Monograph Number
177.053827	$C_8H_7N_3O_2$	Luminol, 5417
177.078979	$C_{10}H_{11}NO_2$	Acetoacetanilide, 45 Dihydrogentianine, 4226 Tuberin, 9465
177.101445	$C_8H_{11}N_5$	Phenyl Biguanide, 7077 6-Trimethylammoniopurinide, 9380
177.115364	$C_{11}H_{15}NO$	m-Aminovalerophenone, 499 Metamfepyramone, 5775 Phenmetrazine, 7031
177.126598	$C_{10}H_{15}N_3$	Bethanidine, 1216 Modaline, 6064
177.899033	F_3Sb	Antimony Trifluoride, 747
177.924196	$H_2O_7S_2$	Pyrosulfuric Acid, 7795
177.943229	$H_4O_7P_2$	Pyrophosphoric Acid, 7794
178.019030	$[C_{14}H_{20}Cl_4N_2]^{2+}$	Chlorisondamine (half mass), 2072
178.026609	$C_9H_6O_4$	Daphnetin, 2805 Esculetin, 3620 Ninhydrin, 6373
178.029981	$C_6H_{10}O_4S$	3,3'-Thiodipropionic Acid, 9065
178.036096	$F_6H_8N_2Si$	Ammonium Hexafluorosilicate, 550
178.047738	$C_6H_{10}O_6$	Chitaric Acid, 2020 Ethyl Tartrate, Acid, 3804 Gluconolactone, 4288
178.056470	$C_9H_{10}N_2S$	Etisazol, 3826
178.062994	$C_{10}H_{10}O_3$	5-Acetyl-2-methoxybenzaldehyde, 93
178.063295	$C_{12}H_{10}Mg$	Diphenylmagnesium, 3338
178.074228	$C_9H_{10}N_2O_2$	Phenacemide, 6986
178.078250	$C_{14}H_{10}$	Anthracene, 718 Phenanthrene, 6996 Tolan, 9205
178.084124	$C_7H_{14}O_5$	Digitalose, 3135 Thevetose, 9016
178.099380	$C_{11}H_{14}O_2$	n-Butyl Benzoate, 1536 Isobutyl Benzoate, 4988 Thujic Acid, Methyl ester, 9132
178.110613	$C_{10}H_{14}N_2O$	Isonicotinic Acid Diethylamide, 5046 Nikethamide, 6366
178.120509	$C_8H_{18}O_4$	Triglyme, 9359
178.135765	$C_{12}H_{18}O$	6-n-Amyl-m-cresol, 644

Use in conjunction with The Merck Index, Ninth Edition

Molecular Weight	Empirical Formula	Compound Name, Monograph Number
178.849767	CrCuO$_4$	Cupric Chromate(VI), 2634
178.913682	MnN$_2$O$_6$	Manganese Nitrate, 5561
179.007472	C$_5$H$_9$NO$_2$S$_2$	Cheirolin, 2006
179.013792	C$_9$H$_6$ClNO	5-Chloro-8-quinolinol, 2147, 4457 7-Chloro-8-quinolinol, 4457
179.025230	C$_5$H$_9$NO$_4$S	Carbocysteine, 1807
179.044325	C$_6$H$_5$N$_5$O$_2$	Xanthopterin, 9727
179.058243	C$_9$H$_9$NO$_3$	4-(Acetylamino)benzoic Acid, 2826, 4853 Adrenochrome, 159, 441 Adrenolutin, 161, 441 Hippuric Acid, 4591 Salacetamide, 8079
179.061615	C$_6$H$_{13}$NO$_3$S	Cyclamic Acid, 2707, 1658
179.073499	C$_{13}$H$_9$N	Acridine, 118 Benzo[f]quinoline, 1115
179.079373	C$_6$H$_{13}$NO$_5$	Chondrosamine, 2213 D-Fucose, Oxime, 4129 L-Fucose, Oxime, 4130 Glucosamine, 4289
179.094629	C$_{10}$H$_{13}$NO$_2$	Acetophenetidin, 59 p-Aminophenylacetic Acid, Ethyl ester, 473 4-Amino-3-phenylbutyric Acid, 474 Fusaric Acid, 4168 Homarylamine, 4605 4'-Hydroxybutyranilide, 4715 IPC, 4925 N-Isopropylsalicylamide, 5078 Phenprobamate, 7059 N-Phenylglycine, Ethyl ester, 7096 α-Phenylglycine, Ethyl ester, 7097 Risocaine, 8013 Dihydrotuberin, 9465
179.105862	C$_9$H$_{13}$N$_3$O	Iproniazid, 4932
179.118751	C$_6$H$_{18}$N$_3$OP	Hexametapol, 4556
179.131014	C$_{11}$H$_{17}$NO	Methoxyphenamine, 5871 N-Methylephedrine, 5936 Tecomine, 8859
179.142248	C$_{10}$H$_{17}$N$_3$	Hexazole, 4568
179.828154	AsCl$_3$	Arsenic Trichloride, 829
179.879437	C$_2$N$_2$S$_2$Zn	Zinc Thiocyanate, 9833
179.880612	AsH$_2$KO$_4$	Potassium Arsenate, 7381
179.888892	O$_3$Xe	Xenon trioxide, 9738

Molecular Weight	Empirical Formula	Compound Name, Monograph Number
179.899888	Na_3O_3PS	Sodium Thiophosphate, 8467
179.906623	H_2O_3Te	Tellurous Acid, 8872
179.919774	C_2HBrF_4	Teflurane, 8861
179.930033	$C_6H_3Cl_3$	1,2,3-Trichlorobenzene, 9309 1,2,4-Trichlorobenzene, 9310 1,3,5-Trichlorobenzene, 9311
179.964927	$H_8N_2O_4Se$	Ammonium Selenate, 581
179.978591	$C_5H_9BrO_2$	α-Bromoisovaleric Acid, 1420 β-Bromoisovaleric Acid, 1421 Ethyl α-Bromopropionate, 3704 Ethyl β-Bromopropionate, 3705
180.010116	$C_6H_{12}S_3$	Thioacetaldehyde, 9047
180.035735	$C_8H_8N_2OS$	1-Benzoyl-2-thiourea, 1128
180.042259	$C_9H_8O_4$	Acetozone, 69 Aspirin, 874 Caffeic Acid, 1622 Escorin, 3619 Piperonylic Acid, Methyl ester, 7271
180.046968	$C_7H_8N_4S$	Nicothiazone, 6339
180.053492	$C_8H_8N_2O_3$	Nitroacetanilides, 6399-6401
180.063388	$C_6H_{12}O_6$	D-Allose, 272 D-Altrose, 313 Fructose, 4121 DL-Fructose, 4122 Galactose, 4179 Glucose, 4290 D-Gulose, 4429 L-Gulose, 4430 Hamamelose, 4460 Idose, 4801 Inositol, 4854 D-Mannose, 5578 D-Psicose, 7710 Sorbose, 8498 D-Tagatose, 8814
180.064726	$C_7H_8N_4O_2$	Theobromine, 8995 Theophylline, 9004, 391, 476, 4408, 5705, 9006-9012
180.068748	$C_{12}H_8N_2$	o-Phenanthroline, 6998 Phenazine, 7002
180.072120	$C_9H_{12}N_2S$	Prothionamide, 7673
180.078644	$C_{10}H_{12}O_3$	Coniferyl Alcohol, 2474 2-Phenoxyethanol, Acetate, 7056 Propylparaben, 7655
180.089878	$C_9H_{12}N_2O_2$	3,5-Diaminobenzoic Acid, Ethyl ester, 2934 Dulcin, 3444

Molecular Weight	Empirical Formula	Compound Name, Monograph Number
180.093900	$C_{14}H_{12}$	1,1-Diphenylethene, 3331 Stilbene, 8600
180.115030	$C_{11}H_{16}O_2$	4-Amylresorcinol, 655 Butylated Hydroxyanisole, 1531 p-Pentyloxyphenol, 6936
180.138839	$[C_{11}H_{18}NO]^+$	Leptodactyline, 5298 Methylhordeninium, 4625
181.037508	$C_8H_7NO_4$	m-Nitrobenzoic Acid, Methyl ester, 6411 p-Nitrophenylacetic Acid, 6444
181.048741	$C_7H_7N_3O_3$	p-Nitrophenylurea, 6448
181.073893	$C_9H_{11}NO_3$	Adrenalone, 158 p-Aminosalicylic Acid, Ethyl ester, 491 Styramate, 8656 Tyrosine, 9493 m-Tyrosine, 9494
181.089149	$C_{13}H_{11}N$	Benzylideneaniline, 1154 2-Fluorenamine, 4035 9-Fluorenamine, 4036 4-Stilbazole, 8599
181.095023	$C_6H_{15}NO_5$	Glucamine, 4278
181.097703	$[C_9H_{13}N_2O_2]^+$	Pyridostigmine, 7761
181.110279	$C_{10}H_{15}NO_2$	Etilefrin, 3821 Hexapropymate, 4567 p-Hydroxyephedrine, 4725
181.183050	$C_{12}H_{23}N$	Dicyclohexylamine, 3071 Dimecamine, 3188
181.821717	Br_2Mg	Magnesium Bromide, 5482
181.835299	Ca_3P_2	Calcium Phosphide, 1696
181.861295	O_4SiZr	Zirconium Silicate, 9847
181.862498	O_5V_2	Vanadium Pentoxide, 9578
181.902108	Cl_2O_7	Chlorine Heptoxide, 2067
181.905225	$ILiO_3$	Lithium Iodate, 5372
181.910983	N_2NiO_6	Nickel Nitrate, 6327
181.929939	CaF_6Si	Calcium Hexafluorosilicate, 1668
181.987450	$C_2H_4Cl_2N_6$	Chloroazodin, 2094
181.988510	$C_4H_6O_6S$	Sulfonyldiacetic Acid, 8757
182.006267	$C_4H_6O_8$	Dihydroxytartaric Acid, 3170
182.016634	$C_4H_4F_6O$	Flurothyl, 4080

Molecular Weight	Empirical Formula	Compound Name, Monograph Number
182.021523	$C_8H_6O_5$	4-Hydroxyisophthalic Acid, 4729
182.057909	$C_9H_{10}O_4$	Flopropione, 4014 Fumigatin, Monomethyl ether, 4139 Glycol Salicylate, 4336 Mesotan, 5758 o-Orsellinic Acid, Methyl ester, 6717 Syringaldehyde, 8805 Veratric Acid, 9607
182.069142	$C_8H_{10}N_2O_3$	5-Nitro-o-phenetidine, 6439 Oxiniacic Acid, Ethanolamine ester, 6760 Pyridoxal, Oxime, 7762
182.070798	$C_6H_{15}O_4P$	Triethyl Phosphate, 9349
182.073165	$C_{13}H_{10}O$	Benzophenone, 1109
182.079038	$C_6H_{14}O_6$	Galactitol, 4177 Mannitol, 5575 Sorbitol, 8497
182.084398	$C_{12}H_{10}N_2$	Azobenzene, 924 Harman, 4470
182.087197	$C_7H_{16}FO_2P$	Soman, 8489
182.091609	$C_6H_{10}N_6O$	Dacarbazine, 2793
182.094294	$C_{10}H_{14}O_3$	Mephenesin, 5676
182.098430	$C_6H_{16}FN_2OP$	Mipafox, 6054
182.109551	$C_{14}H_{14}$	Bibenzyl, 1225
182.130680	$C_{11}H_{18}O_2$	Geraniol, Formate, 4235
182.153147	$C_9H_{18}N_4$	Guanacline, 4410
182.908834	CoN_2O_6	Cobaltous Nitrate, 2396
182.999015	$C_7H_5NO_3S$	Saccharin, 8070, 8071
183.011903	$C_4H_{10}NO_3PS$	Acephate, 22
183.016772	$C_7H_5NO_5$	3-Nitrosalicylic Acid, 6455 5-Nitrosalicylic Acid, 6456
183.028006	$C_6H_5N_3O_4$	2,4-Dinitroaniline, 3266 2,6-Dinitroaniline, 3267
183.053158	$C_8H_9NO_4$	4-Pyridoxic Acid, 7765
183.068414	$C_{12}H_9NO$	Phenoxazine, 7053
183.081477	$C_{10}H_{14}ClN$	Chlorphentermine, 2172 Clortermine, 2366
183.089543	$C_9H_{13}NO_3$	Epinephrine, 3543 Nordefrin, 6501 Normetanephrine, 6516

Use in conjunction with The Merck Index, Ninth Edition

Molecular Weight	Empirical Formula	Compound Name, Monograph Number
183.104799	$C_{13}H_{13}N$	Benzhydrylamine, 1080, 1159 Benzylaniline, 1140 Methyldiphenylamine, 5924
183.125929	$C_{10}H_{17}NO_2$	Methyprylon, 6009
183.841066	$CdCl_2$	Cadmium Chloride, 1604
183.846343	S_2Sn	Stannic Sulfide, 8559
183.852322	$C_2H_2Br_2$	Acetylene Dibromide, 83
183.857355	O_4SSr	Strontium Sulfate, 8643
183.858770	CaO_4Se	Calcium Selenate, 1703
183.875790	NaO_4Tc	Sodium Pertechnetate, 8418
183.900583	N_4S_4	Tetrasulfur Tetranitride, 8958
183.935294	$C_2H_5AsO_5$	Arsonoacetic Acid, 839
183.971434	$C_3H_2ClF_5O$	Enflurane, 3524 Isoflurane, 5030
183.974902	C_4H_9I	n-Butyl Iodide, 1562 sec-Butyl Iodide, 1563 Isobutyl Iodide, 4998
183.988699	$C_3H_5ClN_2O_5$	Clonitrate, 2355
183.988761	C_8H_9Br	Xylyl Bromides, 9752-9754
183.994264	$C_6H_4N_2O_3S$	p-Diazobenzenesulfonic Acid, 2964
184.000788	$C_7H_4O_6$	Chelidonic Acid, 2008
184.012021	$C_6H_4N_2O_5$	Dinitrophenols, 3277-3279
184.037173	$C_8H_8O_5$	Fomecin A, 4088 Methyl Gallate, 5944 Spinulosin, 8523
184.048407	$C_7H_8N_2O_4$	Orotic Acid, Ethyl ester, 6711
184.065493	$C_{10}H_{13}ClO$	Chlorothymol, 2157
184.066461	$C_6H_{14}FO_3P$	Isofluorphate, 5029
184.073872	$[C_5H_{15}NO_4P]^+$	Phosphorylcholine, Cation, 7172
184.084792	$C_8H_{12}N_2O_3$	Barbital, 965 Primocarcin, 7545
184.088815	$C_{13}H_{12}O$	Benzohydrol, 1099 Benzylphenols, 1164, 1165 Phentydrone, 7065
184.100048	$C_{12}H_{12}N_2$	Benzidine, 1083 2,4'-Biphenyldiamine, 1249 1,1-Diphenylhydrazine, 3335

Use in conjunction with The Merck Index, Ninth Edition

184.

Molecular Weight	Empirical Formula	Compound Name, Monograph Number
184.100048	$[C_{12}H_{12}N_2]^{2+}$	Diquat, Cation, 3370
184.109945	$C_{10}H_{16}O_3$	Queen Substance, 7819
184.118493	$C_5H_{12}N_8$	Methylglyoxal Bis(guanylhydrazone), 5948
184.121178	$C_9H_{16}N_2O_2$	Apronalide, 784
184.125201	$C_{14}H_{16}$	Chamazulene, 1994
184.146330	$C_{11}H_{20}O_2$	Undecylenic Acid, 9505
184.891326	$AlCl_4H_4N$	Ammonium Tetrachloroaluminate, 592
184.896532	$H_4MnO_4P_2$	Manganese Hypophosphite, 5559
184.987971	$C_7H_4ClNO_3$	p-Nitrobenzoyl Chloride, 6414
185.008926	$C_3H_8NO_6P$	Serine Phosphate, 8208
185.042297	$C_7H_8FN_3S$	4-(p-Fluorophenyl)-3-thiosemicarbazide, 4063
185.105193	$C_9H_{15}NO_3$	5,5-Dipropyl-2,4-oxazolidinedione, 3365 Ecgonine, 3461
185.141579	$C_{10}H_{19}NO_2$	Equipax, 3554
185.152812	$C_9H_{19}N_3O$	N,N-Diethyl-1-piperazinecarboxamide, 3110
185.214350	$C_{12}H_{27}N$	Tributylamine, 9297
185.823431	AgBr	Silver Bromide, 8247
185.835906	Ga_2O_3	Gallium, Sesquioxide, 4191
185.843631	CK_2S_3	Potassium Thiocarbonate, 7484
185.867972	$C_2H_4Br_2$	Ethylene Dibromide, 3732
185.888618	$AsHNa_2O_4$	Sodium Arsenate, Dibasic, 8320
185.917781	$C_2H_3IO_2$	Iodoacetic Acid, 4886
185.968026	C_7H_7BrO	p-Bromophenol, Methyl ether, 1432
185.979259	$C_6H_7BrN_2$	p-Bromophenylhydrazine, 1433
186.007685	$C_6H_3FN_2O_4$	1-Fluoro-2,4-dinitrobenzene, 4057
186.008372	$C_8H_7ClO_3$	(4-Chlorophenoxy)acetic Acid, 617
186.013190	$C_{10}H_{10}Fe$	Ferrocene, 3959
186.031694	$C_{11}H_6O_3$	Psoralen, 7713
186.035066	$C_8H_{10}O_3S$	Ethyl Benzenesulfonate, 3696
186.046299	$C_7H_{10}N_2O_2S$	Carbimazole, 1801 Mafenide, 5472, 8742
186.050322	$C_{12}H_{10}S$	Phenyl Sulfide, 7119

Use in conjunction with The Merck Index, Ninth Edition

Molecular Weight	Empirical Formula	Compound Name, Monograph Number
186.052824	$C_8H_{10}O_5$	Endothall, Free acid, 3520
186.068080	$C_{12}H_{10}O_2$	Chimaphilin, 2014 1-Naphthaleneacetic Acid, 6195
186.079313	$C_{11}H_{10}N_2O$	Amphenidone, 614 Furfuralphenylhydrazone, 4155
186.081143	$C_{10}H_{15}ClO$	3-Chloro-d-camphor, 2107
186.089209	$C_9H_{14}O_4$	Mesaconic Acid, Diethyl ester, 5751
186.100442	$C_8H_{14}N_2O_3$	N-2-Ethylcrotonyl-N'-methylolurea, 3721 Dihydroprimocarcin, 7545
186.115699	$C_{12}H_{14}N_2$	N-(1-Naphthyl)ethylenediamine, 6237
186.125595	$C_{10}H_{18}O_3$	Cyclobutyrol, 2721 Royal Jelly Acid, 8035
186.136828	$C_9H_{18}N_2O_2$	Capuride, 1772
186.161980	$C_{11}H_{22}O_2$	n-Amyl Caproate, 641 Ethyl Pelargonate, 3778 n-Nonyl Acetate, 6488
186.198366	$C_{12}H_{26}O$	1-Dodecanol, 3410
186.905235	CuN_2O_6	Cupric Nitrate, 2645
187.005163	$C_5H_5N_3O_3S$	Aminitrozole, 410
187.041548	$C_6H_9N_3O_2S$	2-Thiolhistidine, 9075
187.048073	$C_7H_9NO_5$	Fumaroalanide, 4138
187.059306	$C_6H_9N_3O_4$	Enteromycin Carboxamide, 3529
187.063329	$C_{11}H_9NO_2$	3-Amino-2-naphthoic Acid, 461 Menadione Monoxime, 5654
187.085795	$C_9H_9N_5$	Amanozine, 376 Benzoguanamine, 1098
187.823358	Cl_4Ti	Titanium Tetrachloride, 9188
187.891659	$C_2H_4S_5$	Lenthionine, 5293
187.898394	F_5Nb	Niobium Pentafluoride, 6377
187.904781	N_2O_6Zn	Zinc Nitrate, 9806
187.946887	F_6Na_2Si	Sodium Hexafluorosilicate, 8375
188.030730	$C_8H_9FO_2S$	Fluoresone, 4044
188.043322	$C_6H_8N_2O_5$	Enteromycin, 3528

188.

Molecular Weight	Empirical Formula	Compound Name, Monograph Number
188.047344	$C_{11}H_8O_3$	6-Hydroxy-2-methyl-1,4-naphthoquinone, 4740 1-Hydroxy-2-naphthoic Acid, 4744 3-Hydroxy-2-naphthoic Acid, 4745 Phthiocol, 7185 Plumbagin, 7316
188.068474	$C_8H_{12}O_5$	Ethyl Oxalacetate, 3776, 6741
188.083730	$C_{12}H_{12}O_2$	2-(2-Naphthyloxy)ethanol, 6241
188.094963	$C_{11}H_{12}N_2O$	Antipyrine, 754, 755, 2029, 3017 Vasicine, 9595
188.104859	$C_9H_{16}O_4$	Azelaic Acid, 918
188.116093	$C_8H_{16}N_2O_3$	Tetrahydroprimocarcin, 7545
188.120115	$C_{13}H_{16}O$	Benzylidenepinacolone, 1155
188.131349	$C_{12}H_{16}N_2$	N,N-Dimethyltryptamine, 3253 Etryptamine, 3835
188.141245	$C_{10}H_{20}O_3$	2-Methyl-2-pentyl-1,3-dioxolane-4-methanol, 5974 Promoxolane, 7586
188.152478	$[C_{18}H_{40}N_4O_4]^{2+}$	Hexacarbacholine (half mass), 4543
188.156501	$C_{14}H_{20}$	Congressane, 2469
188.225249	$[C_{11}H_{28}N_2]^{2+}$	Pentamethonium, 6911
189.042593	$C_{10}H_7NO_3$	Kynurenic Acid, 5178 1-Nitro-2-naphthol, 6437
189.048958	$C_6H_{12}N_3PS$	Triethylenethiophosphoramide, 9348
189.063723	$C_7H_{11}NO_5$	N-Acetyl-L-glutamic acid, 2825
189.078979	$C_{11}H_{11}NO_2$	Phensuximide, 7061
189.088869	$[C_8H_9N_6]^+$	2,4-Diamino-6-pyridiniumTriazine, 2943
189.115364	$C_{12}H_{15}NO$	N-Benzoylpiperidine, 7261 N-Ethylcrotonanilide, 3720
189.139174	$[C_{12}H_{17}N_2]^+$	Methylgraminium, 4381
189.825471	AsIn	Indium Arsenide, 4831
189.834076	CuI	Cuprous Iodide, 2668
189.839904	Cl_2Sn	Stannous Chloride, 8562
189.846722	$AlCl_4Na$	Sodium Tetrachloroaluminate, 8458
189.856303	$K_2O_3S_2$	Potassium Thiosulfate, 7486
189.858692	$AgClO_3$	Silver Chlorate, 8249
189.875718	Na_2O_4Se	Sodium Selenate, 8439

Use in conjunction with The Merck Index, Ninth Edition

Molecular Weight	Empirical Formula	Compound Name, Monograph Number
189.892238	Cl_2MgO_6	Magnesium Chlorate, 5484
189.898256	$Na_2O_5S_2$	Sodium Metabisulfite, 8396
189.911384	C_2BaN_2	Barium Cyanide, 976
189.914574	H_3LaO_3	Lanthanum Hydroxide, 5209
189.935513	$C_4H_5Cl_3O_2$	Ethyl Trichloroacetate, 3809
189.946622	$C_2HNa_3O_6$	Sodium Sesquicarbonate, 8442
189.960582	$C_2H_6O_6S_2$	1,2-Ethanedisulfonic Acid, 3651, 1774
189.985529	$C_7H_7ClO_2S$	p-Toluenesulfonyl Chloride, 9230
189.993595	$C_6H_6SO_5$	2,5-Dihydroxybenzenesulfonic Acid, 3406, 3648
189.995220	$C_8H_8Cl_2O$	Dichloroxylenol, 3055
190.026609	$C_{10}H_6O_4$	Coumarin-3-carboxylic Acid, 2548, 2616
190.037842	$C_9H_6N_2O_3$	Nitroxoline, 6475
190.059407	$C_{12}H_8F_2$	4,4'-Difluorodiphenyl, 3126
190.074228	$C_{10}H_{10}N_2O_2$	Ethylenimine Quinone, 3740 3-Methyl-5-phenylhydantoin, 5978
190.077600	$C_7H_{14}N_2O_2S$	Aldicarb, 217
190.096694	$C_8H_{10}N_6$	Dihydralazine, 3146
190.099380	$C_{12}H_{14}O_2$	p-(2-Methylpropenyl)phenol Acetate, 5981
190.110613	$C_{11}H_{14}N_2O$	Cytisine, 2786
190.135765	$C_{13}H_{18}O$	2-(o-Tolyl)cyclohexanol, 9240
190.830864	CuO_3Se	Cupric Selenite, 2655
190.930762	$C_3H_4Cl_3NO_2$	Chloral Formamide, 2032 Trichlorourethan, 9329
190.980778	$C_6H_6ClNO_2S$	N-Chlorobenzenesulfonamide, 2037
190.990469	$C_7H_7Cl_2NO$	Clopidol, 2359
191.021858	$C_9H_5NO_4$	o-Nitrophenylpropiolic Acid, 6447
191.042987	$C_6H_9NO_6$	Nitrilotriacetic Acid, 6397
191.051719	$C_9H_9N_3S$	Amiphenazole, 502
191.058243	$C_{10}H_9NO_3$	Maleanilic Acid, 5530
191.061615	$C_7H_{13}NO_3S$	N-Acetylmethionine, 92 N-Acetylpenicillamine, 95
191.094629	$C_{11}H_{13}NO_2$	Hydrohydrastinine, 4699 Idrocilamide, 4803

Molecular Weight	Empirical Formula	Compound Name, Monograph Number
191.105862	$C_{10}H_{13}N_3O$	Propiophenone, Semicarbazone, 7620
191.115758	$C_8H_{17}NO_4$	Mycaminose, 6144
191.131014	$C_{12}H_{17}NO$	N,N-Diethyl-m-toluamide, 3118 Phendimetrazine, 7009
191.167400	$C_{13}H_{21}N$	2,6-Di-tert-butylpyridine, 3009
191.915876	H_3O_3Pr	Praseodymium Hydroxide, 7506
191.951163	$C_4H_7Cl_3O_2$	Butylchloral Hydrate, 1545 Chloral Alcoholate, 2027
191.964794	$C_6H_5ClO_3S$	p-Chlorobenzenesulfonic Acid, 2096, 8991
191.981592	$C_8H_4N_2S_2$	Phenylene-1,4-diisothiocyanate, 7090
192.005873	$C_9H_4O_5$	Trimellitic Anhydride, 9370
192.027003	$C_6H_8O_7$	Citric Acid, 2307 D-Glucaric Acid, 1,4-Lactone, 4279
192.042259	$C_{10}H_8O_4$	Anemonin, 677 Scopoletin, 8161
192.051362	$C_8H_{17}Br$	n-Octyl Bromide, 6571 sec-Octyl Bromide, 6572
192.063388	$C_7H_{12}O_6$	Chitaric Acid, Methyl ester, 2020 Quinic Acid, 7849
192.070372	$B_4H_8N_2O_7$	Ammonium Borate, 521
192.078644	$C_{11}H_{12}O_3$	Ethyl Benzoylacetate, 3698
192.089878	$C_{10}H_{12}N_2O_2$	Acetoacetanilide, Oxime, 45 Nicopholine, 6337
192.115030	$C_{12}H_{16}O_2$	Ibufenac, 4795 Isoamyl Benzoate, 4960 Thymol Acetate, 9141
192.126263	$C_{11}H_{16}N_2O$	1-(p-Amino-m-tolyl)morpholine, 495
192.151415	$C_{13}H_{20}O$	Ionone, 4915 4-(5-Isopropenyl-2-methyl-1-cyclopenten-1-yl)-2-buta-none, 5065 Pseudoionone, 7702
192.162649	$C_{12}H_{20}N_2$	Tremorine, 9270
192.915950	H_3NdO_3	Neodymium Hydroxide, 6273
193.010225	$C_6H_{12}BrNO$	Diethylbromoacetamide, 3094
193.037508	$C_9H_7NO_4$	m-Nitrocinnamic Acid, 6418
193.059975	$C_7H_7N_5O_2$	Fervenulin, 3987 Toxoflavin, 9254

Use in conjunction with The Merck Index, Ninth Edition

Molecular Weight	Empirical Formula	Compound Name, Monograph Number
193.073893	$C_{10}H_{11}NO_3$	Succinanilic Acid, 8667
193.089149	$C_{14}H_{11}N$	Diphenylacetic Acid, Nitrile, 3324
193.095023	$C_7H_{15}NO_5$	N-Methyl-α-L-glucosamine, 5946
193.110279	$C_{11}H_{15}NO_2$	Butamben, 1495 Isobutyl p-Aminobenzoate, 4986 Salsoline, 8099 o-Thymotic Acid, Amide, 9150
193.146664	$C_{12}H_{19}NO$	Etafedrine, 3637
193.812414	Cl_3Y	Yttrium Chloride, 9773
193.819881	$CdSe$	Cadmium Selenide, 1614
193.835374	$TeZn$	Zinc Telluride, 9832
193.847584	CrK_2O_4	Potassium Chromate(VI), 7400
193.859093	$BaSi_2$	Barium Silicide, 1000
193.887631	$H_4O_4P_2Zn$	Zinc Hypophosphite, 9798
193.906940	F_6Se	Selenium Hexafluoride, 8182
193.935424	$C_3H_3BrF_4$	Halopropane, 4453
193.945683	$C_7H_5Cl_3$	Benzotrichloride, 1120
193.994241	$C_6H_{11}BrO_2$	α-Bromo-n-caproic Acid, 1415 Ethyl α-Bromobutyrate, 3703
194.042653	$C_6H_{10}O_7$	D-Galacturonic Acid, 4180 D-Glucuronic Acid, 4298, 2213 2-Keto-L-gulonic acid, 5153
194.057909	$C_{10}H_{10}O_4$	Caffeic Acid, Methyl ester, 1622 Dimethyl Phthalate, 3244 Ferulic Acid, 3986 Meconin, 5609 Methyl Acetylsalicylate, 5886, 874
194.062618	$C_8H_{10}N_4S$	Methyl 4-Pyridyl Ketone Thiosemicarbazone, 5987
194.069142	$C_9H_{10}N_2O_3$	Adrenochrome, Oxime, 159 p-Aminohippuric Acid, 449
194.073165	$C_{14}H_{10}O$	Anthranol, 721 Anthrone, 726 Diphenylketene, 3337
194.079038	$C_7H_{14}O_6$	α-Methylglucoside, 5947
194.080376	$C_8H_{10}N_4O_2$	Caffeine, 1623-1629 3-Semicarbazidobenzamide, 8192
194.084398	$C_{13}H_{10}N_2$	Aminacrine, 408, 122 2-Phenylbenzimidazole, 7074

Molecular Weight	Empirical Formula	Compound Name, Monograph Number
194.094294	$C_{11}H_{14}O_3$	Butylparaben, 1572 o-Thymotic Acid, 9150 Zingerone, 9836
194.105528	$C_{10}H_{14}N_2O_2$	Phenocoll, 7035 Pilocarpidine, 7223
194.130680	$C_{12}H_{18}O_2$	4-Hexylresorcinol, 4588, 122
194.154489	$[C_{12}H_{20}NO]^+$	Methyltecominium, 8859
194.167065	$C_{13}H_{22}O$	Solanone, 8481
194.820396	InSe	Indium Selenide, 4834
194.893251	$CsNO_3$	Cesium Nitrate, 1970
194.968360	C_8H_6BrN	α-Bromobenzyl Cyanide, 1411
194.996610	$C_6H_6NNaO_3S$	Sodium Sulfanilate, 8448
195.039239	$C_6H_5N_5O_3$	Leucopterin, 5308
195.053158	$C_9H_9NO_4$	o-(Carbamoylphenoxy)acetic Acid, 1783 Cinchomeronic Acid, Dimethyl ester, 2272 Ethyl Nitrobenzoates, 3771, 6411 p-Nitrophenylacetic Acid, Methyl ester, 6444
195.064391	$C_8H_9N_3O_3$	Anot, 713
195.068414	$C_{13}H_9NO$	p-Xenylcarbimide, 9739
195.075083	$AlF_6H_{12}N_3$	Ammonium Hexafluoroaluminate, 547
195.089543	$C_{10}H_{13}NO_3$	Buramate, 1483 Damascenine, 2798 α-Methyl-m-tyrosine, 6004 α-Methyl-p-tyrosine, 6005 Pyridoxal, Monoethylacetal, 7764 Surinamine, 8797
195.092223	$[C_{13}H_{11}N_2]^+$	N-Methylphenazonium, 5976
195.100777	$C_9H_{13}N_3O_2$	Aminometradine, 460 Amisometradine, 503
195.108171	$C_{11}H_{17}NS$	1-Naphthylisothiocyanate, 6239
195.108171	$[C_{22}H_{34}N_2S_2]^{2+}$	Brintobal, Cation (half mass), 1379
195.110673	$C_7H_{17}NO_5$	N-Methylglucamine, 5945, 4869, 4883, 4913, 4922, 5627, 5628, 6477
195.162314	$C_{12}H_{21}NO$	2-(Piperidinomethyl)cyclohexanone, 7262
195.870157	$Al_2K_2O_4$	Potassium Aluminate, 7379
195.890224	$C_2HBrClF_3$	Halothane, 4455
195.895812	F_4Sn	Stannic Fluoride, 8554

Use in conjunction with The Merck Index, Ninth Edition

Molecular Weight	Empirical Formula	Compound Name, Monograph Number
195.909513	C_5FeO_5	Iron Pentacarbonyl, 4947
195.924948	$C_6H_3Cl_3O$	2,4,5-Trichlorophenol, 9322 2,4,6-Trichlorophenol, 9323
195.937120	$C_4H_5BrO_4$	Bromosuccinic Acid, 1440
195.964214	B_4CaO_7	Calcium Borate, 1643
195.971613	CN_4O_8	Tetranitromethane, 8951
195.980633	$C_5H_6Cl_2N_2O_2$	1,3-Dichloro-5,5-dimethylhydantoin, 3039
196.012021	$C_7H_4N_2O_5$	2,4-Dinitrobenzaldehyde, 3268
196.037173	$C_9H_8O_5$	Acetylpatulin, 6855 Croweacic Acid, 2596
196.040545	$C_6H_{12}O_5S$	5-Thio-D-glucose, 9068
196.048407	$C_8H_8N_2O_4$	2,3-Pyrazinedicarboxylic Acid, Dimethyl ester, 7742
196.052430	$C_{13}H_8O_2$	Xanthone, 9725
196.058303	$C_6H_{12}O_7$	Gluconic Acid, 4287 D-Gulonic Acid, 4427 L-Gulonic Acid, 4428
196.063663	$C_{12}H_8N_2O$	Hemipyocyanine, 4502
196.065493	$C_{11}H_{13}ClO$	Dowicide 9, 3425
196.073559	$C_{10}H_{12}O_4$	Aurantiogliocladin, 897 Cantharidin, 1752 Homogentisic Acid, Dimethyl ether, 4619 o-Orsellinic Acid, Ethyl ester, 6717 Sparassol, 8508 Xanthoxylin, 9731
196.076726	$C_{10}H_{13}ClN_2$	Chlordimeform, 2055
196.084792	$C_9H_{12}N_2O_3$	5-Nitro-2-propoxyaniline, 6452
196.088815	$C_{14}H_{12}O$	Phenyl Tolyl Ketones, 7123, 7124
196.100048	$C_{13}H_{12}N_2$	2,7-Fluorenediamine, 4038
196.109945	$C_{11}H_{16}O_3$	d-Camphocarboxylic Acid, 1733
196.121178	$C_{10}H_{16}N_2O_2$	N,N-Diethyl-3,5-dimethyl-4-isoxazolecarboxamide, 3098
196.146330	$C_{12}H_{20}O_2$	Bornyl Acetate, 1351 Geraniol, Acetate, 4235 Linalyl Acetate, 5336
196.845121	K_2MnO_4	Potassium Manganate(VI), 7427
196.900652	CeF_3	Cerous Fluoride, 1954
196.939496	$C_3HCl_2N_3O_3$	Troclosene, 9430

196.

Molecular Weight	Empirical Formula	Compound Name, Monograph Number
196.947625	C_7H_4BrNO	p-Bromophenyl Isocyanate, 1434
197.025898	$C_7H_7N_3O_2S$	Sulcimide, 8687
197.043656	$C_7H_7N_3O_4$	Nihydrazone, 6365
197.060742	$C_{10}H_{12}ClNO$	Beclamide, 1024
197.068808	$C_9H_{11}NO_4$	Dopa, 3421 Levodopa, 5310
197.084064	$C_{13}H_{11}NO$	Benzanilide, 1062 Benzophenone, Oxime, 1109
197.095297	$C_{12}H_{11}N_3$	p-Aminoazobenzene, 421 Diazoaminobenzene, 2963
197.105193	$C_{10}H_{15}NO_3$	Ethylnorepinephrine, 3772 Metanephrine, 5778 N-Methylepinephrine, 5937 Scopoline, Acetate, 8163 Tenuazonic Acid, 8874
197.116427	$C_9H_{15}N_3O_2$	Hercynine, 4531
197.120450	$C_{14}H_{15}N$	Dibenzylamine, 2976
197.127660	$C_8H_{15}N_5O$	Noformicin, 6485
197.799263	Br_2Ca	Calcium Bromide, 1645
197.804140	S_3V_2	Vanadium Trisulfide, 9583
197.827935	As_2O_3	Arsenic Trioxide, 832
197.867972	$C_3H_4Br_2$	2,3-Dibromopropene, 2993
197.874951	C_2AgKN_2	Potassium Silver Cyanide, 7457
197.878991	$INaO_3$	Sodium Iodate, 8386
197.889980	$CBaO_3$	Barium Carbonate, 972
197.896904	$C_2K_2O_6$	Potassium Percarbonate, 7438
197.968026	C_8H_7BrO	p-Bromoacetophenone, 1400 ω-Bromoacetophenone, 1401
197.990552	$C_5H_{11}I$	Isoamyl Iodide, 4968
198.014033	$C_6H_7NaO_6$	Sodium Ascorbate, 8323
198.016438	$C_8H_6O_6$	Puberulic Acid, 7721
198.027671	$C_7H_6N_2O_5$	4,6-Dinitro-o-cresol, 3275
198.032669	$C_6H_{12}Cl_2N_2O$	Glycine Bis(chloroethyl)amide, 4326
198.038905	$C_6H_6N_4O_4$	2,4-Dinitrophenylhydrazine, 3280 Nitrofurazone, 6422

Use in conjunction with The Merck Index, Ninth Edition

Molecular Weight	Empirical Formula	Compound Name, Monograph Number
198.050322	$C_{13}H_{10}S$	Thioxanthene, 9108
198.052824	$C_9H_{10}O_5$	4-Hydroxy-3-methoxymandelic Acid, 4735
198.055991	$C_9H_{11}ClN_2O$	Monuron, 6096
198.056196	$C_6H_{14}O_5S$	1-Thiosorbitol, 9102
198.064057	$C_8H_{10}N_2O_4$	Mimosine, 6051
198.068080	$C_{13}H_{10}O_2$	Mycomycin, 6151 Phenyl Benzoate, 7075 Xanthydrol, 9734
198.079313	$C_{12}H_{10}N_2O$	Azoxybenzene, 930 p-Nitrosodiphenylamine, 6460
198.089209	$C_{10}H_{14}O_4$	Aurantiogliocladin, Dihydro deriv, 897 Guaiacol Glyceryl Ether, 4402, 4408 Muconic Acid, Diethyl ester, 6126
198.100442	$C_9H_{14}N_2O_3$	Metharbital, 5819 Probarbital, 7550
198.104465	$C_{14}H_{14}O$	Benzyl Ether, 1148
198.115699	$C_{13}H_{14}N_2$	p,p'-Diaminodiphenylmethane, 2937 Tacrine, 8812
198.140851	$C_{15}H_{18}$	Cadalene, 1598 Guaiazulene, 4406
198.148061	$C_9H_{18}N_4O$	Amidinomycin, 402
198.161980	$C_{12}H_{22}O_2$	Menthyl Acetate, 5665 Undecylenic Acid, Methyl ester, 9505
198.184447	$C_{10}H_{22}N_4$	Guanethidine, 4413 Spherophysine, 8517
198.961449	$C_6H_6AsNO_2$	Oxophenarsine, 6765
199.022920	$C_6H_5N_3O_5$	Picramic Acid, 7209
199.045571	$C_{12}H_9NS$	Phenothiazine, 7052
199.063329	$C_{12}H_9NO_2$	Dictamnine, 3065 o-Nitrobiphenyl, 6416
199.076217	$C_9H_{14}NO_2P$	Toldimfos, 9211
199.110948	$C_{12}H_{13}N_3$	4,4'-Diaminodiphenylamine, 2936
199.168462	$C_{10}H_{21}N_3O$	Diethylcarbamazine, 3095
199.230000	$C_{13}H_{29}N$	Octamylamine, 6560
199.818720	SeSn	Stannous Selenide, 8569
199.843934	Cl_2Te	Tellurium Dichloride, 8866

199.

Molecular Weight	Empirical Formula	Compound Name, Monograph Number
199.852089	C_2HCl_5	Pentachloroethane, 6900
199.883622	$C_3H_6Br_2$	Propylene Dibromide, 7642 Trimethylene Bromide, 9383
199.947291	$C_7H_5BrO_2$	p-Bromobenzoic Acid, 1407
199.983676	C_8H_9BrO	p-Bromophenol, Ethyl ether, 1432
199.987226	C_4F_8	Octafluorocyclobutane, 6555
199.995703	$C_7H_4O_7$	Meconic Acid, 5608, 6253
199.998870	$C_7H_5ClN_2O_3$	Aklomide, 188
200.006936	$C_6H_4N_2O_6$	2,4-Dinitroresorcinol, 3281
200.008591	$C_4H_9O_7P$	D-Erythrose 4-Phosphate, 3614
200.017498	$C_8H_9ClN_2S$	p-Chlorobenzylpseudothiourea, 2102
200.024022	$C_9H_9ClO_3$	(4-Chloro-o-toloxy)acetic Acid, 2158
200.050716	$C_9H_{12}O_3S$	Ethyl p-Toluenesulfonate, 3808
200.054555	$C_6H_8N_4O_4$	Ronidazole, 8021
200.059721	$C_8H_9FN_2O_3$	Ftorafur, 4126
200.061950	$C_8H_{12}N_2O_2S$	Thiobarbital, 9051
200.083730	$C_{13}H_{12}O_2$	Monobenzone, 6085
200.094963	$C_{12}H_{12}N_2O$	Harmalol, 4469
200.104859	$C_{10}H_{16}O_4$	Camphoric Acid, 1735, 1450
200.131349	$C_{13}H_{16}N_2$	Tetrahydrozoline, 8934
200.177630	$C_{12}H_{24}O_2$	Ethyl Caprate, 3707 Lauric Acid, 5230
200.641549	$[C_{27}H_{35}N_3]^{2+}$	Methyl Green, Cation (half mass) 5949
200.906979	$C_3AlN_3S_3$	Aluminum Thiocyanate, 371
201.009580	$C_7H_7NO_4S$	Carzenide, 1870
201.036069	$C_{10}H_7N_3S$	Thiabendazole, 9017
201.045965	$C_8H_{11}NO_3S$	Penicillanic Acid, 6883
201.078123	$C_7H_{12}ClN_5$	Simazine, 8282
201.078979	$C_{12}H_{11}NO_2$	4-Amino-1-naphthol, N-Acetyl deriv, 462 Carbaryl, 1790
201.847654	$CaMoO_4$	Calcium Molybdate(VI), 1681
201.949082	$C_3H_7IO_2$	Glyceryl Iodide, 4322

Use in conjunction with The Merck Index, Ninth Edition

Molecular Weight	Empirical Formula	Compound Name, Monograph Number
201.958835	$C_8H_4Cl_2O_2$	Phthaloyl Chloride, 7182
201.961115	$C_6H_7AsO_3$	Benzenearsonic Acid, 1070
201.962941	$C_7H_7BrO_2$	Bromosaligenin, 1438
201.978134	$C_6H_3ClN_2O_4$	1-Chloro-2,4-dinitrobenzene, 2111 2-Chloro-1,3-dinitrobenzene, 2112
201.981163	$B_4Na_2O_7$	Sodium Borate, 8334
201.993595	$C_7H_6O_5S$	2-Sulfobenzoic Acid, 7466
202.039672	$C_9H_{11}ClO_3$	Chlorphenesin, 2167
202.056470	$C_{11}H_{10}N_2S$	1-(1-Naphthyl)-2-thiourea, 6244
202.058972	$C_7H_{10}N_2O_5$	Enteromycin, Methyl ester, 3528
202.062994	$C_{12}H_{10}O_3$	2-Naphthoxyacetic Acid, 6224
202.076170	$C_6H_{16}AlNaO_4$	Vitride, 9695
202.078250	$C_{16}H_{10}$	Pyrene, 7746
202.099380	$C_{13}H_{14}O_2$	Tremetone, 9268
202.120509	$C_{10}H_{18}O_4$	Ethyl Adipate, 3689 Hexylene Glycol, Diacetate, 4585 Sebacic Acid, 8167
202.146999	$C_{13}H_{18}N_2$	Medmain, 5615
202.215747	$C_{10}H_{26}N_4$	Spermine, 8515
202.240899	$[C_{12}H_{30}N_2]^{2+}$	Hexamethonium, 4557, 4558, 4580
202.791872	Cu_3N	Cuprous Nitride, 2670
202.927960	H_3O_3Sm	Samarium Hydroxide, 8106
202.988844	$C_6H_5NO_5S$	3-Sulfoisonicotinic Acid, 6274
203.003533	$C_6H_{12}Cl_3N$	2,2',2''-Trichlorotriethylamine, 9328
203.050177	$C_{12}H_{10}ClN$	4-Amino-4'-chlorodiphenyl, 439
203.054221	$C_6H_9N_3O_5$	α-Ketoglutaric Acid, Semicarbazone, 5152
203.058243	$C_{11}H_9NO_3$	Kynurenic Acid, Methyl ester, 5178 Quininic Acid, 7883
203.079373	$C_8H_{13}NO_5$	Oryzacidin, 6722
203.094629	$C_{12}H_{13}NO_2$	Indolebutyric Acid, 4842 Methsuximide, 5880
203.098001	$C_9H_{17}NO_2S$	Lethane® 384, 5300
203.105862	$C_{11}H_{13}N_3O$	Ampyrone, 627

Molecular Weight	Empirical Formula	Compound Name, Monograph Number
203.131014	$C_{13}H_{17}NO$	Crotamiton, 2588
203.134386	$C_{10}H_{21}NOS$	Pebulate, 6859
203.825793	CrO_4Sr	Strontium Chromate(VI), 8631
203.929462	EuH_3O_3	Europic Hydroxide, 3853
203.934139	C_7H_6BrCl	p-Bromobenzyl Chloride, 1409
203.943602	C_6H_5I	Iodobenzene, 4890
204.009245	$C_7H_8O_5S$	Guaiacolsulfonic Acid, 7415
204.027003	$C_7H_8O_7$	Anhydromethylenecitric Acid, 689, 5826, 8318
204.034193	$C_{12}H_9ClO$	2-Phenyl-6-chlorophenol, 7082 4-Phenyl-2-chlorophenol, 7083
204.060703	$C_4H_8N_6O_4$	Oxalenediuramidoxime, 6742
204.072120	$C_{11}H_{12}N_2S$	Tetramisole, 8949
204.089878	$C_{11}H_{12}N_2O_2$	Ethotoin, 3675 5-Ethyl-5-phenylhydantoin, 3781 Fenozolone, 3911 Thozalinone, 9123 L-Tryptophan, 9458
204.111007	$C_8H_{16}N_2O_4$	Ethylidene Diurethane, 3752
204.112344	$C_9H_{12}N_6$	Triethylenemelamine, 9346
204.115030	$C_{13}H_{16}O_2$	Cinnamic Acid, n-Butyl ester, 2288 R-11, 7906 Dihydrotremetone, 9268
204.126263	$C_{12}H_{16}N_2O$	Bufotenine, 1467 Caulophylline, 1904 Psilocin, 7711
204.162649	$C_{13}H_{20}N_2$	4-Dimethylamino-1-phenylpiperidine, 3222
204.183778	$C_{10}H_{24}N_2O_2$	Ethambutol, 3645
204.187801	$C_{15}H_{24}$	Cadinenes, 1599 Caryophyllene, 1868 α-Caryophyllene, 1869 Copaene, 2491 α-Farnesene, 3876 β-Farnesene, 3877 Thujopsene, 9134 Ylangene, 9765
204.896947	CuF_6Si	Cupric Hexafluorosilicate, 2643
204.996428	$C_7H_8ClNO_2S$	N-Chloro-4-methylbenzenesulfonamide, 2038
205.037508	$C_{10}H_7NO_4$	Xanthurenic Acid, 9733
205.079945	$[C_{11}H_{13}N_2S]^+$	3-o-Aminobenzyl-4-methylthiazolium, 431

Use in conjunction with The Merck Index, Ninth Edition

Molecular Weight	Empirical Formula	Compound Name, Monograph Number
205.085127	$C_{10}H_{11}N_3O_2$	Neo-cupferron, 6271 Tryptazan, 9457
205.095023	$C_8H_{15}NO_5$	Chitaric Acid, Dimethylamide, 2020 N-Acetyl-D-fucosamine, 4128 N-Acetylmycosamine, 6153
205.131408	$C_9H_{19}NO_4$	Dexpanthenol, 2903
205.132746	$C_{10}H_{15}N_5$	Phenformin, 7022 Trapidil, 9264
205.146664	$C_{13}H_{19}NO$	Amfepramone, 396
205.775719	Cu_2Se	Cuprous Selenide, 2673
205.803659	Br_2OS	Thionyl Bromide, 9084
205.816014	Cu_2O_3S	Cuprous Sulfite, 2675
205.818027	Cl_5P	Phosphorus Pentachloride, 7161
205.822813	BrI	Iodine Monobromide, 4877
205.843804	$C_2H_2AsCl_3$	Dichloro(2-chlorovinyl)arsine, 3036
205.853606	$AgClO_4$	Silver Perchlorate, 8267
205.869784	$CaCl_2O_6$	Calcium Chlorate, 1651
205.893171	$Na_2O_6S_2$	Sodium Dithionate, 8360
205.896493	F_6SiZn	Zinc Hexafluorosilicate, 9797
205.973111	$C_{10}H_7Br$	1-Bromonaphthalene, 1426 2-Bromonaphthalene, 1427
206.042653	$C_7H_{10}O_7$	D-Glucoascorbic Acid, 4281
206.043523	$C_8H_{14}O_2S_2$	Thioctic Acid, 9061
206.057909	$C_{11}H_{10}O_4$	Limettin, ´5332 Scoparone, 8156
206.079038	$C_8H_{14}O_6$	Ethyl Tartrate, 3803
206.087770	$C_{11}H_{14}N_2S$	Pyrantel, 7738
206.105528	$C_{11}H_{14}N_2O_2$	Ethylphenacemide, 3779 Levulinic Acid, Phenylhydrazone, 5316
206.109551	$C_{16}H_{14}$	1,4-Diphenyl-1,3-butadiene, 3328
206.130680	$C_{13}H_{18}O_2$	Ibuprofen, 4796
206.141913	$C_{12}H_{18}N_2O$	Oxotremorine, 6767 Pivalylbenzhydrazine, 7295
206.167065	$C_{14}H_{22}O$	C_{14}-Aldehyde, 216 Irone, 4943

Molecular Weight	Empirical Formula	Compound Name, Monograph Number
206.178299	$C_{13}H_{22}N_2$	Dicyclohexylcarbodiimide, 3072
206.810906	Cl_3Ru	Ruthenium Trichloride, 8061
206.825778	CuO_4Se	Cupric Selenate, 2653
206.909205	$BeCl_2O_8$	Beryllium Perchlorate, 1200
207.014328	$C_7H_4F_3NO_3$	TFM, 8966
207.025875	$C_7H_{14}BrNO$	Ibrotamide, 4794
207.046634	$C_9H_9N_3OS$	2-Acetamido-6-aminobenzothiazole, 33
207.064391	$C_9H_9N_3O_3$	Isonicotinoyl Hydrazone Pyruvic Acid, 8618 Piperonal, Semicarbazone, 7269
207.089543	$C_{11}H_{13}NO_3$	Corydaldine, 2524 Hydrastinine, 4651 4-Salicyloylmorpholine, 8094 Thurfyl Nicotinate, 9136
207.092915	$C_8H_{17}NO_3S$	Felinine, 3883
207.100777	$C_{10}H_{13}N_3O_2$	Guanoxan, 4420 5-(α-Phenylethyl)semioxamazide, 7094
207.110673	$C_8H_{17}NO_5$	Deanol, Hemisuccinate, 2824
207.113353	$[C_{11}H_{15}N_2O_2]^+$	Methylnicopholinium, 6337
207.125929	$C_{12}H_{17}NO_2$	Butacetin, 1490 o-Butoxyacetanilide, 1514 Promecarb, 7579
207.812061	Cl_3Rh	Rhodium Chloride, 7980
207.825324	O_4SeZn	Zinc Selenate, 9823
207.842941	$BaCl_2$	Barium Chloride, 974
207.864603	$MoNa_2O_4$	Sodium Molybdate(VI), 8403
207.897761	F_4Xe	Xenon Tetrafluoride, 9738
207.933014	$C_6H_2Cl_2O_4$	Chloranilic Acid, 2044
208.019416	$C_{10}H_8O_3S$	1-Naphthalenesulfonic Acid, 6200 2-Naphthalenesulfonic Acid, 6201
208.037173	$C_{10}H_8O_5$	Fraxetin, 4116
208.051779	$C_6H_{12}N_2O_4S$	Lanthionine, 5210
208.052430	$C_{14}H_8O_2$	Anthraquinone, 722 Morphenol, 6105 Phenanthrenequinone, 6997
208.073559	$C_{11}H_{12}O_4$	Ethyl Acetylsalicylate, 3687 Benzyl Succinic Acid, 8660

Use in conjunction with The Merck Index, Ninth Edition

Molecular Weight	Empirical Formula	Compound Name, Monograph Number
208.078556	$C_{10}H_{18}Cl_2$	Terpinene, Dihydrochloride, 8885
208.084792	$C_{10}H_{12}N_2O_3$	Allobarbital, 245 p-(Carbamylmethoxy)acetanilide, 1785 Kynurenine, 5179
208.088815	$C_{15}H_{12}O$	Anthranol, Methyl ether, 721 Chalcone, 1991
208.100048	$C_{14}H_{12}N_2$	Bendazol, 1040 Neocuproine, 6272
208.109945	$C_{12}H_{16}O_3$	Asarone, 847 Guaiacol Valerate, 4404 Isoamyl Salicylate, 4974
208.113088	$C_8H_{20}O_4Si$	Ethyl Silicate, 3795
208.121178	$C_{11}H_{16}N_2O_2$	Pilocarpine, 7224
208.121178	$[C_{22}H_{32}N_4O_4]^{2+}$	Distigmine (half mass), 3380
208.125201	$C_{16}H_{16}$	[2.2]Metacyclophane, 5771
208.932330	GdH_3O_3	Gadolinium Hydroxide, 4174
209.041525	$C_7H_{16}BrNO$	Meprochol, 5691
209.095297	$C_{13}H_{11}N_3$	Acriflavine, 119, 120, 3453 Proflavine, 7567, 7568
209.105193	$C_{11}H_{15}NO_3$	Anhalamine, 686 Hydroxyphenamate, 4750 p-Lactophenetide, 5191 Propoxur, 7625
209.126323	$C_8H_{19}NO_5$	N-Ethylglucamine, 6477
209.141579	$C_{12}H_{19}NO_2$	Bamethan, 956
209.152812	$C_{11}H_{19}N_3O$	Citral, Semicarbazone, 2303
209.777182	Br_2Cr	Chromous Bromide, 2238
209.855091	CdO_4S	Cadmium Sulfate, 1616
209.937091	$C_4H_{12}As_2$	Cacodyl, 1592
209.940598	$C_7H_5Cl_3O$	2,4,6-Trichloroanisole, 9308 Trichlorocresols, 9313-9315
210.016438	$C_9H_6O_6$	Trimellitic Acid, 9369
210.019605	$C_9H_7ClN_2O_2$	2-Acetylzoxazolamine, 9853
210.037567	$C_6H_{10}O_8$	Galactaric Acid, 4176 D-Glucaric Acid, 4279
210.042928	$C_{12}H_6N_2O_2$	Phanquone, 6980
210.046299	$C_9H_{10}N_2O_2S$	1,3-Bis(hydroxymethyl)-2-benzimidazolinethione, 1273

Molecular Weight	Empirical Formula	Compound Name, Monograph Number
210.052824	$C_{10}H_{10}O_5$	Fumigatin, Monoacetate, 4139 Opianic Acid, 6691
210.068080	$C_{14}H_{10}O_2$	Benzil, 1084
210.073953	$C_7H_{14}O_7$	D-manno-Heptulose, 4528
210.079313	$C_{13}H_{10}N_2O$	Pyocyanine, 7735
210.089209	$C_{11}H_{14}O_4$	Dimethyl Carbate, 3230 Homogentisic Acid, Methyl ester dimethyl ether, 4619
210.090546	$C_{12}H_{10}N_4$	2,3-Diaminophenazine, 2939
210.100442	$C_{10}H_{14}N_2O_3$	Aprobarbital, 783 N,N-Bis(2-hydroxyethyl)-p-nitrosoaniline, 1272 Dolcental, 3417
210.104465	$C_{15}H_{14}O$	Lactaroviolin, 5182
210.115699	$C_{14}H_{14}N_2$	Naphazoline, 6192
210.140851	$C_{16}H_{18}$	1,1-Di-p-tolylethane, 3395 1,2-Di-p-tolylethane, 3396
210.654657	$[C_{27}H_{39}N_3O]^{2+}$	Pentacynium (half mass), 6902
211.022920	$C_7H_5N_3O_5$	Nitromide, 6434
211.035809	$C_4H_{10}N_3O_5P$	Phosphocreatine, 7150
211.040006	$C_{10}H_{10}ClNO_2$	Chlorthenoxazin, 2183
211.076392	$C_{11}H_{14}ClNO$	Propachlor, 7588
211.084458	$C_{10}H_{13}NO_4$	Glyceryl p-Aminobenzoate, 4321 Methyldopa, 5925 3-O-Methyldopa, 5926
211.099714	$C_{14}H_{13}NO$	Desylamine, 2895 Diphenylacetamide, 3323 Diphenylacetic Acid, Amide, 3324
211.110948	$C_{13}H_{13}N_3$	1,3-Diphenylguanidine, 3333 Methyl Pyridyl Ketone, Phenylhydrazone, 5986 Nitrin, 6398
211.112777	$C_{12}H_{18}ClN$	Mefenorex, 5622
211.120844	$C_{11}H_{17}NO_3$	Dimetan®, 3199 Dioxethedrine, 3303 Isoproterenol, 5079 Mescaline, 5752 Metaproterenol, 5782 Methoxamine, 5860 Varon, Free base, 9594
211.132077	$C_{10}H_{17}N_3O_2$	Isolan®, 5032
211.136100	$C_{15}H_{17}N$	Benethamine, 1043 Ethylbenzylaniline, 3700

Use in conjunction with The Merck Index, Ninth Edition

Molecular Weight	Empirical Formula	Compound Name, Monograph Number
211.168462	$C_{11}H_{21}N_3O$	Rhodinal, Semicarbazone, 7976
211.816277	IRb	Rubidium Iodide, 8045
211.823769	BrCs	Cesium Bromide, 1965
211.844546	K_3O_4P	Potassium Phosphate, Tribasic, 7447
211.881261	N_2O_6Sr	Strontium Nitrate, 8637
211.895825	F_6Mo	Molybdenum Hexafluoride, 6070
211.929933	$C_4H_4O_6Zn$	Zinc Tartrate, 9831
211.936390	HfO_2	Hafnium Dioxide, 4442
211.957036	$C_4H_8N_2S_4$	Nabam, Free acid, 6169
212.000412	$C_6H_4N_4O_3S$	Triafur, 9278
212.003863	BiH_3	Bismuthine, 1290
212.006936	$C_7H_4N_2O_6$	3,4-Dinitrobenzoic Acid, 3271 3,5-Dinitrobenzoic Acid, 3272
212.029587	$C_{13}H_8OS$	Thioxanthone, 9109
212.035255	$C_9H_9ClN_2O_2$	α-Chloro-α-phenylacetylurea, 2136
212.068474	$C_{10}H_{12}O_5$	Propyl Gallate, 7648
212.083730	$C_{14}H_{12}O_2$	Benzoin, 1102 Benzyl Benzoate, 1141 Diphenylacetic Acid, 3324 Pinosylvine, 7246
212.094963	$C_{13}H_{12}N_2O$	1-Benzoyl-2-phenylhydrazine, 1127 Carbanilide, 1788 Harmine, 4471 2-(1-Naphthylamino)-2-oxazoline, 6235
212.098335	$C_{10}H_{16}N_2OS$	Albutoin, 207
212.106196	$C_{12}H_{12}N_4$	Chrysoidine, 2256-2258 p-Diaminoazobenzene, 2933
212.116093	$C_{10}H_{16}N_2O_3$	Butabarbital, 1487 Butethal, 1504
212.131349	$C_{14}H_{16}N_2$	1,2-Dianilinoethane, 2951 o-Tolidine, 9212
212.774719	Br_2Mn	Manganese Bromide, 5553
212.944995	AlN_3O_9	Aluminum Nitrate, 351
212.974819	$C_9H_5Cl_2NO$	Chloroxine, 2164, 4457
212.978925	C_8H_8BrNO	p-Bromoacetanilide, 1397 p-Bromoacetophenone, Oxime, 1400

213.

Molecular Weight	Empirical Formula	Compound Name, Monograph Number
213.002185	$C_6H_3N_3O_6$	sym-Trinitrobenzene, 9393
213.007182	$C_5H_9Cl_2N_3O_2$	Carmustine, 1845
213.009580	$C_8H_7NO_4S$	Indican (Metabolic Indican), 4825
213.055656	$C_{10}H_{12}ClNO_2$	Baclofen, 944 Chlorpropham, 2177
213.078979	$C_{13}H_{11}NO_2$	Nicotinic Acid Benzyl Ester, 6344 N-Phenylanthranilic Acid, 7073 Salicylanilide, 8090
213.092042	$C_{11}H_{16}ClNO$	Clorprenaline, 2365
213.100108	$C_{10}H_{15}NO_4$	Kainic Acid, 5127
213.101445	$C_{11}H_{11}N_5$	Phenazopyridine, 7004
213.104817	$C_8H_{15}N_5S$	Simetryne®, 8285
213.115364	$C_{14}H_{15}NO$	2-Amino-1,2-diphenylethanol, 443
213.771612	Br_2Fe	Ferrous Bromide, 3965
213.796590	Cl_4Ge	Germanium Tetrachloride, 4243
213.852929	IKO_3	Potassium Iodate, 7425
213.856422	$Cl_2O_5S_2$	Pyrosulfuryl Chloride, 7796
213.866482	Na_2O_3Sn	Sodium Stannate(IV), 8445
213.873905	$INaO_4$	Sodium Metaperiodate, 8398
213.899272	$C_4H_8Br_2$	α-Butylene Dibromide, 1554
213.962941	$C_8H_7BrO_2$	p-Bromobenzoic Acid, Methyl ester, 1407
213.987071	$C_7H_6N_2O_2S_2$	Holomycin, 4603
214.016062	$C_6H_6N_4O_3S$	Niridazole, 6382
214.025753	$C_7H_7ClN_4O_2$	8-Chlorotheophylline, 2154, 3193, 7280
214.036987	$C_6H_7ClN_6O$	N-Amidino-3-amino-6-chloropyrazinecarboxamide, 401
214.039672	$C_{10}H_{11}ClO_3$	Clofibric Acid, 2343
214.041214	$C_8H_{10}N_2O_3S$	N^1-Acetylsulfanilamide, 100 N^4-Acetylsulfanilamide, 101 p-Tolylsulfonylmethylnitrosamide, 9244
214.052447	$C_7H_{10}N_4O_2S$	Sulfaguanidine, 8700
214.062994	$C_{13}H_{10}O_3$	Benzoresorcinol, 1117 Phenyl Carbonate, 7081 Phenyl Salicylate, 7116
214.076058	$C_{11}H_{15}ClO_2$	Phenaglycodol, 6992

Molecular Weight	Empirical Formula	Compound Name, Monograph Number
214.081622	$C_{14}H_{14}S$	Benzyl Sulfide, 1171
214.088833	$C_8H_{14}N_4OS$	Metribuzin, 6027
214.099380	$C_{14}H_{14}O_2$	Hydrobenzoin, 4664
214.110613	$C_{13}H_{14}N_2O$	Harmaline, 4468 10-Methoxyharmalan, 5867 N-(p-Methoxyphenyl)-p-phenylenediamine, 5873 Phenyramidol, 7128
214.131743	$C_{10}H_{18}N_2O_3$	Desthiobiotin, 2893
214.229666	$C_{14}H_{30}O$	Myristyl Alcohol, 6161
214.937402	DyH_3O_3	Dysprosium Hydroxide, 3457
214.958190	$C_7H_6BrNO_2$	5-Bromoanthranilic Acid, 1404
214.975213	$C_5H_7Cl_2NO_4$	N-(Dichloroacetyl)-DL-serine, 3025
215.025230	$C_8H_9NO_4S$	N-Acetylsulfanilic Acid, 102
215.036463	$C_7H_9N_3O_3S$	Sulfanilylurea, 8725
215.069477	$C_{11}H_9N_3O_2$	1,2-Naphthoquinone, 2-Semicarbazone, 6221
215.080710	$C_{10}H_9N_5O$	Kinetin, 5159
215.093773	$C_8H_{14}ClN_5$	Atrazine, 887
215.094629	$C_{13}H_{13}NO_2$	4-Acetamido-2-methyl-1-naphthol, 35 3-Amino-2-naphthoic Acid, Ethyl ester, 461
215.106349	$[C_9H_{16}ClN_4]^+$	N-(3-Chloroallyl)hexaminium, 2089
215.118751	$C_9H_{18}N_3OP$	Metepa, 5790
215.126992	$C_9H_{17}N_3O_3$	Dopastin, 3424
215.142248	$C_{13}H_{17}N_3$	Tramazoline, 9258
215.167400	$C_{15}H_{21}N$	Fencamfamine, 3893
215.772019	Br_2Ni	Nickel Bromide, 6316
215.853929	O_4SSn	Stannous Sulfate, 8570
215.895840	F_5Sb	Antimony Pentafluoride, 734
215.899055	CeF_4	Ceric Fluoride, 1947
215.984683	$C_6H_4FeN_6$	Hydroferrocyanic Acid, 4680
216.031712	$C_6H_8N_4O_3S$	Nithiazide, 6385
216.042259	$C_{12}H_8O_4$	Methoxsalen, 5861 Naphthalic Acid, 6204
216.056864	$C_8H_{12}N_2O_3S$	6-Aminopenicillanic Acid, 468

Molecular Weight	Empirical Formula	Compound Name, Monograph Number
216.066556	$C_9H_{13}ClN_2O_2$	Terbacil, 8876
216.078644	$C_{13}H_{12}O_3$	Allenolic Acid, 241 Euparin, 3846 2-Naphthyl Lactate, 6240
216.089878	$C_{12}H_{12}N_2O_2$	Tetrantoin, 8952
216.097088	$C_6H_{12}N_6O_3$	Trimethylolmelamine, 9386
216.126263	$C_{13}H_{16}N_2O$	Adrenoglomerulotropin, 160
216.136159	$C_{11}H_{20}O_4$	Azelaic Acid, Dimethyl ester, 918 n-Butylmalonic Acid, Diethyl ester, 1564 Di-tert-butyl Malonate, 3008 Ethyl Diethylmalonate, 3725 Pimelic Acid, Diethyl ester, 7231
216.151415	$C_{15}H_{20}O$	DL-ar-Turmerone, 9477
216.769870	Br_2Co	Cobaltous Bromide, 2387
216.938524	ErH_3O_3	Erbium Hydroxide, 3556
216.972014	$C_6H_8AsNO_3$	Arsanilic Acid, 817, 8319
217.004494	$C_7H_7NO_5S$	4-Amino-2-sulfobenzoic Acid, 493 Methyl p-Nitrobenzenesulfonate, 5968
217.006119	$C_9H_9Cl_2NO$	Propanil, 7595
217.019183	$C_7H_{14}Cl_3N$	Novembichin, 6529
217.042505	$C_{10}H_{13}Cl_2N$	Aniline Mustard, 693
217.056136	$C_{12}H_{11}NOS$	Thionalide, 9080
217.077265	$C_9H_{15}NO_3S$	Mycobacidin, 6148
217.098015	$C_8H_{16}N_3O_2P$	ODEPA, 6574
217.110279	$C_{13}H_{15}NO_2$	Fenimide, 3903 Glutethimide, 4309 Methastyridone, 5821 Securinine, 8174
217.131408	$C_{10}H_{19}NO_4$	α-Aminoadipic Acid, Diethyl ester, 418
217.146664	$C_{14}H_{19}NO$	Ethoxyquin, 3681
217.183050	$C_{15}H_{23}N$	Prolintane, 7575
217.821757	BaSe	Barium Selenide, 999
217.862052	BaO_3S	Barium Sulfite, 1004
217.871568	$BaHO_3P$	Barium Phosphite, 996
217.904156	$H_2MgO_8S_2$	Magnesium Bisulfate, 5479
217.923189	$H_4MgO_8P_2$	Magnesium Phosphate, Monobasic, 5506

Use in conjunction with The Merck Index, Ninth Edition

Molecular Weight	Empirical Formula	Compound Name, Monograph Number
217.956029	$C_6H_7AsO_4$	p-Hydroxybenzenearsonic Acid, 4711
217.965547	HgO	Mercuric Oxide, Red, 5719 Mercuric Oxide, Yellow, 5720
217.988510	$C_7H_6O_6S$	Salicylsulfuric Acid, 8097 Sulfosalicylic Acid, 8761, 409
218.032757	$C_{10}H_6N_2O_4$	5-Nitroquinaldic Acid, 6454
218.040151	$C_{12}H_{10}SO_2$	Diphenyl Sulfone, 3348
218.049843	$C_{13}H_{11}ClO$	Clorophene, 2364
218.057909	$C_{12}H_{10}O_4$	Diresorcinol, 3372 Piperic Acid, 7260
218.065990	$C_7H_{14}N_4S_2$	Methallibure, 5803
218.069142	$C_{11}H_{10}N_2O_3$	Phenylmethylbarbituric Acid, 7109
218.079038	$C_9H_{14}O_6$	Triacetin, 9275 Tricarballylic Acid, Trimethyl ester, 9301
218.105528	$C_{12}H_{14}N_2O_2$	Abrine, 4 Mephenytoin, 5682 Methetoin, 5838 Primaclone, 7542
218.109551	$C_{17}H_{14}$	1,2-Cyclopentenophenanthrene, 2748
218.126657	$C_9H_{18}N_2O_4$	Meprobamate, 5690
218.141913	$C_{13}H_{18}N_2O$	Fenoxazoline, 3909
218.167065	$C_{15}H_{22}O$	Vetivones, 9625
218.913450	C_6O_6V	Vanadium Carbonyl, 9576
218.954501	C_6H_6NI	p-Iodoaniline, 4888
219.053158	$C_{11}H_9NO_4$	Xanthurenic Acid, Methyl ester, 9733
219.077280	$C_7H_{14}N_3O_3P$	Uredepa, 9533
219.100777	$C_{11}H_{13}N_3O_2$	Senecialdehyde, p-Nitrophenylhydrazone, 8195
219.110673	$C_9H_{17}NO_5$	Pantothenic Acid, 6820, 3158
219.149738	$[C_{13}H_{19}N_2O]^+$	Methylbufoteninium, 1467
219.162314	$C_{14}H_{21}NO$	Fabianine, 3860
219.791931	Cl_4Se	Selenium Tetrachloride, 8189
219.793584	N_2Zn_3	Zinc Nitride, 9807
219.810433	Cl_3In	Indium Trichloride, 4837
219.811269	P_4S_3	Tetraphosphorus Trisulfide, 8955

Molecular Weight	Empirical Formula	Compound Name, Monograph Number
219.814878	O$_4$SiZn$_2$	Zinc Silicate, 9825
219.849422	CH$_2$Cu$_2$O$_5$	Cupric Carbonate, Basic, 2631
219.894764	F$_6$K$_2$Si	Potassium Hexafluorosilicate, 7420
219.909998	C$_6$CrO$_6$	Chromium Carbonyl, 2230
219.938517	C$_6$H$_5$IO	o-Iodophenol, 4901 p-Iodophenol, 4902 Iodosobenzene, 4909
219.942444	H$_3$O$_3$Tm	Thulium Hydroxide, 9135
219.945903	C$_4$H$_7$Cl$_2$O$_4$P	Dichlorvos, 3058
219.957311	C$_4$H$_7$Cl$_3$N$_2$O$_2$	Mecloralurea, 5606
219.969400	C$_8$H$_6$Cl$_2$O$_3$	3,6-Dichloro-2-methoxybenzoic Acid, 3045 (2,4-Dichlorophenoxy)acetic Acid, 3049
220.012892	C$_{10}$H$_8$N$_2$S$_2$	3,3-Dithiodipyridine, 3391
220.029107	C$_{12}$H$_9$ClO$_2$	Sulphenone, 8783
220.037173	C$_{11}$H$_8$O$_5$	Purpurogallin, 7733
220.048407	C$_{10}$H$_8$N$_2$O$_4$	α-Furildioxime, 4159
220.084792	C$_{11}$H$_{12}$N$_2$O$_3$	5-Hydroxytryptophan, 4769
220.103420	C$_{12}$H$_{16}$N$_2$S	Morantel, 6099 Xylazine, 9742
220.157563	C$_{13}$H$_{20}$N$_2$O	Prilocaine, 7541
220.182716	C$_{15}$H$_{24}$O	Butylated Hydroxytoluene, 1532 DBMC, 2819 α-Santalol, 8113 β-Santalol, 8114
220.766271	Br$_2$Cu	Cupric Bromide, 2629
221.024356	C$_{11}$H$_8$ClNO$_2$	5-Chloro-8-quinolinol, Acetate (ester), 2147
221.046823	C$_9$H$_8$ClN$_5$	Chlorazanil, 2045
221.089937	C$_8$H$_{15}$NO$_6$	N-Acetylglucosamine, 4289
221.098669	C$_{11}$H$_{15}$N$_3$S	Cuminaldehyde Thiosemicarbazone, 2621
221.105193	C$_{12}$H$_{15}$NO$_3$	Anhalonine, 688 Carbofuran, 1808 N-Ethyl-p-hydroxyacetanilide Acetate, 3748 Hydrocotarnine, 4679 Metaxalone, 5786
221.116427	C$_{11}$H$_{15}$N$_3$O$_2$	Pivalizid®, 7294

Use in conjunction with The Merck Index, Ninth Edition

Molecular Weight	Empirical Formula	Compound Name, Monograph Number
221.141579	$C_{13}H_{19}NO_2$	Aklonine, Free base, 189 Bufencarb, 1460 Carnegine, 1847 Dioscorine, 3294
221.152812	$C_{12}H_{19}N_3O$	Ambonestyl®, 387 Procarbazine, 7558
221.177964	$C_{14}H_{23}NO$	Affinin, 167 Tastromine, 8849
221.765818	Br_2Zn	Zinc Bromide, 9786
221.807886	P_2S_5	Phosphorus Pentasulfide, 7164
221.823595	K_2O_4Se	Potassium Selenate, 7454
221.846133	$K_2O_5S_2$	Potassium Metabisulfite, 7428
221.857446	Cl_2Sm	Samarium Dichloride, 8106
221.882019	$Mg_2O_7P_2$	Magnesium Pyrophosphate, 5509
221.882068	Cl_2MgO_8	Magnesium Perchlorate, 5502
221.884768	CH_3IO_3S	Methiodal, 5842
221.896493	F_5I	Iodine Pentafluoride, 4879
221.907119	$H_2Na_2O_7P_2$	Sodium Acid Pyrophosphate, 8311
221.919270	$AlH_6O_6P_3$	Aluminum Hypophosphite, 343
221.977150	$HOTl$	Thallium Hydroxide, 8976
221.988134	$C_4H_6N_4O_3S_2$	Acetazolamide, 41
222.000389	$C_6H_{11}BrN_2O_2$	Bromisovalum, 1395
222.025541	$C_8H_{15}BrO_2$	α-Bromo-n-caproic Acid, Ethyl ester, 1415
222.034882	$C_3H_6N_6O_6$	Cyclonite, 2741
222.052824	$C_{11}H_{10}O_5$	Gladiolic Acid, 4263
222.054698	$C_6H_{11}BO_8$	Gluconic Acid Cyclic 4,5-Borate, 1644
222.056380	$C_6H_{18}O_3Si_3$	Hexamethylcyclotrisiloxane, 4559
222.067429	$C_7H_{14}N_2O_4S$	L-Cystathionine, 2776
222.068080	$C_{15}H_{10}O_2$	Flavone, 4008 Isoflavone, 5028 2-Methylanthraquinone, 5894 Morphenol, Methyl ether, 6105 Phenindione, 7027
222.085186	$C_7H_{14}N_2O_6$	Nitral, 6387

Molecular Weight	Empirical Formula	Compound Name, Monograph Number
222.089209	$C_{12}H_{14}O_4$	Apiol, 767 Dimecrotic Acid, 3189 Ethyl Phthalate, 3783 Ferulic Acid, Ethyl ester, 3986
222.104465	$C_{16}H_{14}O$	Anthranol, Ethyl ether, 721 Dypnone, 3456
222.111676	$C_{10}H_{14}N_4O_2$	Morphazinamide, 6104
222.125595	$C_{13}H_{18}O_3$	Mandelic Acid Isoamyl Ester, 5546
222.136828	$C_{12}H_{18}N_2O_2$	Mexacarbate, 6036 Nicametate, 6308
222.146724	$C_{10}H_{22}O_5$	Tetraglyme, 8926
222.173213	$C_{13}H_{22}N_2O$	Noruron, 6526
222.198366	$C_{15}H_{26}O$	Carotol, 1856 Cedrol, 1910 Farnesol, 3878 Guaiol, 4407 Ledol, 5288 Nerolidol, 6300 Patchouli Alcohol, 6854
222.209599	$[C_{14}H_{26}N_2]^{2+}$	N,N'-Dimethyltremorinium, 9270
222.969634	CAuN	Gold Monocyanide, 4358
223.025621	$C_8H_{14}ClNS_2$	CDEC, 1907
223.030315	$C_{10}H_9NO_3S$	Badische Acid, 945 Cassella's Acid F, 1887 1,6-Cleve's Acid, 2321, 2322 1-Naphthylamine-4-sulfonic Acid, 6230 1-Naphthylamine-5-sulfonic Acid, 6231 1-Naphthylamine-8-sulfonic Acid, 6232 2-Naphthylamine-1-sulfonic Acid, 6233 2-Naphthylamine-5-sulfonic Acid, 6234
223.041548	$C_9H_9N_3O_2S$	p-Carboxybenzaldehyde thiosemicarbazone, 8487
223.060961	$C_7H_{14}NO_5P$	Monocrotophos, 6087
223.063329	$C_{14}H_9NO_2$	1-Aminoanthraquinone, 419
223.074562	$C_{13}H_9N_3O$	Phenazine-α-carboxamide, 2148
223.084458	$C_{11}H_{13}NO_4$	Bendiocarb, 1041 Benzacetin, 1054 Mephenoxalone, 5680 Threonine, Monobenzoyl deriv, 9124
223.099714	$C_{15}H_{13}NO$	N-2-Fluorenylacetamide, 4039
223.103086	$C_{12}H_{17}NOS$	Tiletamine, 9165
223.114319	$C_{11}H_{17}N_3S$	Safranal, Thiosemicarbazone, 8075

Molecular Weight	Empirical Formula	Compound Name, Monograph Number
223.120844	$C_{12}H_{17}NO_3$	Anhalonidine, 687 Deacetylanisomycin, 705 Bufexamac, 1463 Cerulenin, 1959 Ethamivan, 3646 Norbutrine, 6497 Phenisonone, 7030 Rimiterol, 8011
223.144653	$[C_{12}H_{19}N_2O_2]^+$	Neostigmine, 6290, 6291
223.157229	$C_{13}H_{21}NO_2$	Gaiactamine, 4175 Tigloidine, 9162 Toliprolol, 9214
223.193615	$C_{14}H_{25}NO$	N-Isobutyldeca-trans-2:trans-4-dienamide, 6868
223.870512	Na_2O_3Te	Sodium Tellurate(IV), 8455
223.971556	OPb	Lead Monoxide, 5267
223.972813	FTl	Thallium Fluoride, 8975
223.995703	$C_9H_4O_7$	Puberulonic Acid, 7722
224.014331	$C_{10}H_8O_4S$	Asaprol, Free acid, 845 Cassella's Acid, 1886 Croceic Acid, 2582 1-Naphthol-2-sulfonic Acid, 6217 1-Naphthol-4-sulfonic Acid, 6218 2-Naphthol-6-sulfonic Acid, 6219
224.044977	$C_7H_{13}O_6P$	Mevinphos, 6035
224.054555	$C_8H_8N_4O_4$	Nifuradene, 6356
224.068474	$C_{11}H_{12}O_5$	Elenolide, 3496
224.083730	$C_{15}H_{12}O_2$	Dibenzoylmethane, 2973 α-Phenylcinnamic Acid (cis-Form), 7085 α-Phenylcinnamic Acid (trans-Form), 7086
224.090940	$C_9H_{12}N_4O_3$	Etofylline, 3828 Methoxycaffeine, 5864 Pyridoxal, Semicarbazone, 7762
224.104859	$C_{12}H_{16}O_4$	Aspidinol, 870
224.116093	$C_{11}H_{16}N_2O_3$	5-Allyl-5-butylbarbituric Acid, 280 Butalbital, 1493 Butylvinal, 1582 Enallylpropymal, 3514 Nifenalol, 6353 Talbutal, 8817 Vinbarbital, 9636
224.118773	$[C_{14}H_{14}N_3]^+$	Acriflavine, Cation, 119, 120, 3453
224.139902	$[C_{11}H_{18}N_3O_2]^+$	Murexine, Cation, 6133
224.152478	$C_{12}H_{20}N_2O_2$	Aspergillic Acid, 865

Molecular Weight	Empirical Formula	Compound Name, Monograph Number
224.177630	$C_{14}H_{24}O_2$	Geraniol, Butyrate, 4235
224.188864	$C_{13}H_{24}N_2O$	Cuscohygrine, 2686
224.928680	$C_4H_4INO_2$	N-Iodosuccinimide, 4911
224.947092	H_3O_3Yb	Ytterbium Hydroxide, 9772
225.009580	$C_9H_7NO_4S$	8-Hydroxyquinoline-5-sulfonic Acid, 4764
225.038570	$C_8H_7N_3O_5$	Dinitolmide, 3265 Furazolidone, 4151
225.053827	$C_{12}H_7N_3O_2$	Phanquone, Monoxime, 6980
225.078979	$C_{14}H_{11}NO_2$	α-Benzilmonoxime, 1084 β-Benzilmonoxime, 1084 9-Carbazoleacetic Acid, 1794 10-Acetylphenoxazine, 7053
225.090212	$C_{13}H_{11}N_3O$	Dihydrophenazine-α-carboxamide, 2148 Isonicotinic Acid Benzylidenehydrazide, 5045 Tinuvin® P, 9175
225.100108	$C_{11}H_{15}NO_4$	Mephenesin Carbamate, 5677
225.115364	$C_{15}H_{15}NO$	p-Dimethylaminobenzophenone, 3220
225.126598	$C_{14}H_{15}N_3$	o-Aminoazotoluene, 422 p-Dimethylaminoazobenzene, 3216
225.136494	$C_{12}H_{19}NO_3$	Macromerine, 5466 N-Methylmescaline, 5752 6-(Diethylcarbamoyl)-3-cyclohexene-1-carboxylic Acid, 7846 Terbutaline, 8879
225.796456	O_3Y_2	Yttrium Oxide, 9773
225.810382	Cl_3Sb	Antimony Trichloride, 745
225.822800	$AgMnO_4$	Silver Permanganate, 8268
225.928204	$C_4H_4K_2O_6$	Potassium Tartrate, 7467
225.933683	$C_8Cl_2N_2O_2$	2,3-Dichloro-5,6-dicyanobenzoquinone, 3037
225.958835	$C_{10}H_4Cl_2O_2$	Dichlone, 3016
226.030670	$C_{12}H_{10}CaO_2$	Calcium Phenoxide, 1692
226.033819	$C_7H_6N_4O_5$	Nifuraldezone, 6357
226.039672	$C_{11}H_{11}ClO_3$	Alclofenac, 209
226.047738	$C_{10}H_{10}O_6$	Chorismic Acid, 2215 Prephenic Acid, 7537
226.050905	$C_{10}H_{11}ClN_2O_2$	Aminochlorthenoxazin, 440

Molecular Weight	Empirical Formula	Compound Name, Monograph Number
226.062994	$C_{14}H_{10}O_3$	Anthrarobin, 723 Benzoic Anhydride, 1101 Dithranol, 3394
226.068868	$C_7H_{14}O_8$	Glucoheptonic Acid, 4286
226.070205	$C_8H_{10}N_4O_4$	Divicine, Diacetate, 3401
226.077600	$C_{10}H_{14}N_2O_2S$	Methallatal, 5801
226.095357	$C_{10}H_{14}N_2O_4$	5-Allyl-5-(2-hydroxypropyl)barbituric Acid, 287 Carbidopa, Anhydrous form, 1800 Porphobilinogen, 7373
226.099380	$C_{15}H_{14}O_2$	Benzoin, Methyl ether, 1102 Diphenylacetic Acid, Methyl ester, 3324 Pinosylvine, Monomethyl ether, 7246
226.106590	$C_9H_{14}N_4O_3$	Carnosine, 1850 Nimorazole, 6372
226.110613	$C_{14}H_{14}N_2O$	p-Anisaldehyde, Phenylhydrazone, 696 Metyrapone, 6032
226.120509	$C_{12}H_{18}O_4$	Butopyronoxyl, 1512
226.131743	$C_{11}H_{18}N_2O_3$	Amobarbital, 605 Pentobarbital, 6928
226.168128	$C_{12}H_{22}N_2O_2$	Crotethamide, 2589
226.793876	$CoCr_2O_4$	Cobaltous Chromate(III), 2390
226.954614	O_2Pt	Platinic Oxide, 7306
226.987526	$CuH_{12}N_4O_4S$	Tetraamminecopper Sulfate, 8897
226.990469	$C_{10}H_7Cl_2NO$	Chlorquinaldol, 2180
227.002579	$C_3H_5N_3O_9$	Nitroglycerin, 6429
227.017835	$C_7H_5N_3O_6$	2,4,6-Trinitrotoluene, 9397
227.051719	$C_{12}H_9N_3S$	Thionine, 9083
227.071307	$C_{11}H_{14}ClNO_2$	Buclosamide, 1457
227.094629	$C_{14}H_{13}NO_2$	Benzoin Oxime, 1103 Diphenane, 3313 Flindersine, 4013
227.115758	$C_{11}H_{17}NO_4$	Dimethophrine, 3208
227.120467	$C_9H_{17}N_5S$	Ametryne®, 395
227.167400	$C_{16}H_{21}N$	Morphinan, 6107
227.236148	$C_{13}H_{29}N_3$	Dodecylguanidine, 3414
227.891280	$C_2H_4Cl_3O_4P$	Triclofos, 9336

227.

Molecular Weight	Empirical Formula	Compound Name, Monograph Number
227.913090	H_5IO_6	Periodic Acid, 6958
227.972209	$H_8N_2O_8S_2$	Ammonium Peroxydisulfate, 571
227.978591	$C_9H_9BrO_2$	p-Bromobenzoic Acid, Ethyl ester, 1407
228.002721	$C_8H_8N_2O_2S_2$	Thiolutin®, 9076
228.049003	$C_7H_{16}O_4S_2$	Sulfonmethane, 8756
228.072120	$C_{13}H_{12}N_2S$	sym-Diphenylthiourea, 3350
228.074622	$C_9H_{12}N_2O_5$	Deoxyuridine, 2870
228.078644	$C_{14}H_{12}O_3$	Benzilic Acid, 1086 Benzyl Salicylate, 1168 Guaiacol Benzoate, 4400 Oxybenzone, 6770 Trioxsalen, 9401 Xanthyletin, 9735
228.091708	$C_{12}H_{17}ClO_2$	Fenpentadiol, 3912
228.093250	$C_{10}H_{16}N_2O_2S$	5-sec-Butyl-5-ethyl-2-thiobarbituric Acid, 1558
228.093900	$C_{18}H_{12}$	1,2-Benzanthracene, 1063 3,4-Benzphenanthrene, 1129 Chrysene, 2252 Naphthacene, 6193 Triphenylene, 9407
228.111007	$C_{10}H_{16}N_2O_4$	3-Carene, Nitrosate, 1838
228.115030	$C_{15}H_{16}O_2$	Bisphenol A, 1324
228.136159	$C_{12}H_{20}O_4$	Ethyl Camphorate, 1287 Traumatic Acid, 9265
228.147393	$C_{11}H_{20}N_2O_3$	Desthiobiotin, Methyl ester, 2893
228.162649	$[C_{30}H_{40}N_4]^{2+}$	Dequalinium (half mass), 2873, 2874
228.208930	$C_{14}H_{28}O_2$	Ethyl Laurate, 3758 Myristic Acid, 6160
228.793797	Mn_3O_4	Manganese Oxide, 5564
228.836792	Cl_2CuO_6	Cupric Chlorate, 2632
228.849976	MnO_7P_2	Manganese Pyrophosphate, 5566
228.901960	$C_6H_3Cl_4N$	N-Serve®, 6536
228.997100	$C_6H_3N_3O_7$	Picric Acid, 7210
228.999625	$C_5H_{12}NO_3PS_2$	Dimethoate, 3206
229.006119	$C_{10}H_9Cl_2NO$	Dicryl, 3064
229.017353	$C_9H_9Cl_2N_3$	Clonidine, 2353

Use in conjunction with The Merck Index, Ninth Edition

Molecular Weight	Empirical Formula	Compound Name, Monograph Number
229.037508	$C_{12}H_7NO_4$	Resazurin, 7933
229.047886	$C_6H_8ClN_7O$	Amiloride, 407
229.058637	$C_9H_{11}NO_6$	Showdomycin, 8224
229.073893	$C_{13}H_{11}NO_3$	Fagarine, 3871 Osalmid, 6724 Phenyl Aminosalicylate, 7072
229.085127	$C_{12}H_{11}N_3O_2$	INF, 4848
229.088499	$C_9H_{15}N_3O_2S$	Ergothioneine, 3585
229.109423	$C_9H_{16}ClN_5$	Propazin®, 7602 Trietazine, 9340
229.110279	$C_{14}H_{15}NO_2$	Dihydroflindersine, 4013
229.146664	$C_{15}H_{19}NO$	Furfurylmethylamphetamine, 4158 Pronethalol, 7587
229.759712	As_2Se	Arsenic Hemiselenide, 823
229.770746	Cl_2Se_2	Selenium Chloride, 8181
229.780119	Cl_4Zr	Zirconium Chloride, 9840
229.805105	Ag_2O	Silver Oxide, 8265
229.808343	CO_3Rb_2	Rubidium Carbonate, 8042
229.817764	As_2O_5	Arsenic Pentoxide, 827
229.847843	IKO_4	Potassium Periodate, 7440
229.860377	$C_3FeN_3S_3$	Ferric Thiocyanate, 3955
229.864988	$CsHO_4S$	Cesium Bisulfate, 1963
229.877431	$Cr_2Li_2O_7$	Lithium Dichromate(VI), 5367
229.879111	N_2O_6Pd	Palladium Nitrate, 6801
229.891043	BaN_2O_4	Barium Nitrite, 989
229.957855	$C_8H_7BrO_3$	p-Bromomandelic Acid, 1424
229.973049	$C_7H_3ClN_2O_5$	3,5-Dinitrobenzoyl Chloride, 3273
229.981115	$C_6H_2N_2O_8$	Nitranilic Acid, 6390
229.991677	$C_8H_7ClN_2O_2S$	Diazoxide, 2968
230.019156	$C_5H_{11}O_8P$	D-Ribose-5-phosphoric Acid, 7998
230.020026	$C_6H_{15}O_3PS_2$	Methyl Demeton, 5923
230.030626	$C_{10}H_{15}BrO$	3-Bromo-d-camphor, 1414
230.049843	$C_{14}H_{11}ClO$	Diphenylacetic Acid, Chloride, 3324

230.

Molecular Weight	Empirical Formula	Compound Name, Monograph Number
230.057909	$C_{13}H_{10}O_4$	Isomaltol, Benzoate, 5036 Maltol, Benzoate, 5540 Phthiocol, Acetate, 7185 Salicylresorcinol, 8095 Visnagin, 9664
230.061281	$C_{10}H_{14}O_4S$	Thymolsulfonic Acid, 9146
230.073165	$C_{17}H_{10}O$	Benzanthrone, 1064
230.079038	$C_{10}H_{14}O_6$	Ketipic Acid, Diethyl ester, 5150
230.080376	$C_{11}H_{10}N_4O_2$	1,4-Naphthoquinone, Semicarbazone, 6222
230.087770	$C_{13}H_{14}N_2S$	Metizoline, 6015
230.094294	$C_{14}H_{14}O_3$	Euparin, O-Methyl deriv, 3846 2-Isovaleryl-1,3-indanedione, 5098 Kawain, 5139 2-Naphthoxyacetic Acid, Ethyl ester, 6224 Naproxen, 6245 Pindone, 7241
230.105528	$C_{13}H_{14}N_2O_2$	Acetylvasicine, 9595
230.115424	$C_{11}H_{18}O_5$	Apiose, Di-O-isopropylidene, 768
230.130680	$C_{15}H_{18}O_2$	Procerin, 7559
230.141913	$C_{14}H_{18}N_2O$	Propyphenazone, 7660
230.151809	$C_{12}H_{22}O_4$	Dibutyl Succinate, 3010 Di-tert-butyl Succinate, 3011
230.224580	$C_{14}H_{30}O_2$	Decamethylene Glycol, Diethyl ether, 2832
230.790277	Cr_2CuO_4	Cupric Chromite, 2635
230.900524	$C_3Cl_3N_3O_3$	Symclosene, 8802
230.919367	N_3O_9Sc	Scandium Nitrate, 8145
230.953104	$C_7H_6BrNO_3$	5-Bromosalicylhydroxamic Acid, 1436
231.013620	$C_7H_9N_3O_2S_2$	Sulfathiourea, 8741, 8742
231.031378	$C_7H_9N_3O_4S$	Isoniazid Methanesulfonate, 5042
231.056530	$C_9H_{13}NO_4S$	p-Phenetidinomethanesulfonic Acid, 7019
231.058155	$C_{11}H_{15}Cl_2N$	N,N-Bis(2-chloroethyl)benzylamine, 1263
231.089543	$C_{13}H_{13}NO_3$	Euparin, Oxime, 3846
231.092915	$C_{10}H_{17}NO_3S$	Mycobacidin, Methyl ester, 6148
231.100777	$C_{12}H_{13}N_3O_2$	Isocarboxazid, 5012 Triaziquone, 9284
231.123484	$C_{12}H_{16}F_3N$	Fenfluramine, 3902

Use in conjunction with The Merck Index, Ninth Edition

Molecular Weight	Empirical Formula	Compound Name, Monograph Number
231.125929	$C_{14}H_{17}NO_2$	3-Quinuclidinol, dl-Form benzoate (ester), 7905
231.137162	$C_{13}H_{17}N_3O$	Aminopyrine, 487
231.162314	$C_{15}H_{21}NO$	Metazocine, 5787
231.267448	$[C_{13}H_{33}N_3]^{2+}$	Azamethonium, 911
231.784476	Fe_3O_4	Ferrosoferric Oxide, 3964
231.801804	O_4SeSr	Strontium Selenate, 8642
231.811035	Cl_3I	Iodine Trichloride, 4881
231.922667	H_6O_6Te	Telluric(VI) Acid, 8863
231.929054	C_8H_6BrClO	p-Chlorophenacyl Bromide, 2132
231.935413	$AuCl$	Gold Monochloride, 4357
231.935697	O_3W	Tungsten Trioxide, 9470
232.013441	$C_5H_4F_8O$	2,2,3,3,4,4,5,5,-Octafluoro-1-pentanol, 6556
232.017018	$C_9H_{10}Cl_2N_2O$	Diuron, 3400
232.017582	C_2H_6Hg	Dimethylmercury, 3239
232.059640	$C_{10}H_8N_4O_3$	Nifurprazine, 6362
232.067035	$C_{12}H_{12}N_2OS$	4,4'-Sulfinyldianiline, 8745
232.073559	$C_{13}H_{12}O_4$	Piperic Acid, Methyl ester, 7260
232.076931	$C_{10}H_{16}O_4S$	d-Camphorsulfonic Acid, 1736, 9374
232.082348	$C_{10}H_{11}F_3N_2O$	Fluometuron, 4033
232.084792	$C_{12}H_{12}N_2O_3$	Anhydro-L-kynurenine Monoacetate, 5179 Nalidixic Acid, 6179 Phenobarbital, 7032, 7033, 7649
232.109945	$C_{14}H_{16}O_3$	Dihydrokawain, 5139
232.121178	$C_{13}H_{16}N_2O_2$	Aminoglutethimide, 446 Melatonin, 5635 Mofebutazone, 6066
232.125201	$C_{18}H_{16}$	3'-Methyl-1,2-cyclopentenophenanthrene, 5921
232.142307	$C_{10}H_{20}N_2O_4$	Mebutamate, 5598
232.146330	$C_{15}H_{20}O_2$	Alantolactone, 196
232.157563	$C_{14}H_{20}N_2O$	Pyrrocaine, 7800 Siduron, 8227 Tymazoline, 9488
232.163460	$C_{14}H_{21}BO_2$	Tolboxane, 9208
232.251464	$[C_{13}H_{32}N_2O]^{2+}$	Plegatil, Cation, 7312

Molecular Weight	Empirical Formula	Compound Name, Monograph Number
232.808593	Ag$_2$F	Silver Subfluoride, 8274
232.991343	C$_8$H$_8$ClNO$_3$S	N-Acetylsulfanilyl Chloride, 103
233.001034	C$_9$H$_9$Cl$_2$NO$_2$	Diloxanide, 3186
233.068808	C$_{12}$H$_{11}$NO$_4$	Casimiroin, 1882
233.080041	C$_{11}$H$_{11}$N$_3$O$_3$	Erythrocentaurin, Semicarbazone, 3599
233.095297	C$_{15}$H$_{11}$N$_3$	Nicotelline, 6338
233.116427	C$_{12}$H$_{15}$N$_3$O$_2$	6-Methoxy-α-tetralone, Semicarbazone, 5877
233.141579	C$_{14}$H$_{19}$NO$_2$	Levophacetoperane, 5312 Methylphenidate, 5977 Piperoxan, 7272
233.201774	[C$_{30}$H$_{50}$N$_4$]$^{2+}$	Isocurine, Cation (half mass), 5020
233.808175	AgBrO$_3$	Silver Bromate, 8246
233.809572	AgI	Silver Iodide, 8259
233.813116	C$_2$Cl$_6$	Hexachloroethane, 4545
233.837927	BaO$_3$Ti	Barium Titanate(IV), 1008
233.844650	C$_3$H$_5$Br$_2$Cl	1,2-Dibromo-3-chloropropane, 2988
233.856966	BaO$_4$S	Barium Sulfate, 1001
233.866483	BaHO$_4$P	Barium Phosphate, Dibasic, 995
233.867972	C$_6$H$_4$Br$_2$	p-Dibromobenzene, 2986
233.900735	CaH$_4$O$_8$P$_2$	Calcium Phosphate, Monobasic, 1694
233.929039	B$_4$K$_2$O$_7$	Potassium Tetraborate, 7470
233.942704	HgS	Mercuric Sulfide, Black, 5728 Mercuric Sulfide, Red, 5729
233.950206	C$_6$H$_3$ClN$_2$O$_4$S	2,4-Dinitrobenzenesulfenyl Chloride, 3270
233.954167	C$_7$H$_7$IO	o-Iodoanisole, 4889
233.986592	C$_7$H$_7$ClN$_2$O$_3$S	4-Chloro-3-sulfamoylbenzamide, 2151
233.992156	C$_{10}$H$_6$N$_2$OS$_2$	Quinomethionate, 7897
234.037567	C$_8$H$_{10}$O$_8$	Succinyl Peroxide, 8677
234.057533	C$_{10}$H$_{10}$N$_4$OS	Methisazone, 5849
234.076946	C$_8$H$_{15}$N$_2$O$_4$P	O,O-Diethyl O-(3-Methyl-5-pyrazolyl) Phosphate, 3105
234.081143	C$_{14}$H$_{15}$ClO	6-tert-Butyl-1-chloro-2-naphthol, 1550
234.100442	C$_{12}$H$_{14}$N$_2$O$_3$	5-Allyl-5-(2-cyclopenten-1-yl)barbituric Acid, 283

Use in conjunction with The Merck Index, Ninth Edition

Molecular Weight	Empirical Formula	Compound Name, Monograph Number
234.104465	$C_{17}H_{14}O$	Dibenzalacetone, 2970
234.111676	$C_{11}H_{14}N_4O_2$	Mepirizole, 5687
234.136828	$C_{13}H_{18}N_2O_2$	Lenacil, 5292
234.140851	$C_{18}H_{18}$	Retene, 7957
234.161980	$C_{15}H_{22}O_2$	Drimenin, 3433 Helminthosporal, 4488
234.173213	$C_{14}H_{22}N_2O$	3-Diethylaminobutyranilide, 3082 Lidocaine, 5323
234.209599	$C_{15}H_{26}N_2$	Genisteine-Alkaloid, 4222 β-Isosparteine, 5090 Sparteine, 8510
234.940509	O_3Re	Rhenium Trioxide, 7971
234.956976	$C_5H_8Cl_3NO_3$	Carbocloral, 1806
235.066701	$C_{12}H_{13}NO_2S$	Carboxin, 1830 5-Ethyldihydro-6-phenyl-2H-1,3-thiazine-2,4(3H)-dione, 3726 Phenythilone, 7129
235.105587	$C_9H_{17}NO_6$	N-Methyl-α-L-glucosamine, N-Acetyl deriv, 5946
235.120844	$C_{13}H_{17}NO_3$	Lophophorine, 5405
235.136100	$C_{17}H_{17}N$	Azapetine, 913 Berbine, 1180
235.157229	$C_{14}H_{21}NO_2$	Amylocaine, 652 4-(Dimethylamino)benzoic Acid, Pentyl ester, 3219 Meprylcaine, 5693
235.168462	$C_{13}H_{21}N_3O$	Procainamide, 7553
235.807699	InSb	Indium Antimonide, 4830
235.885458	$C_2H_2MoO_7$	Oxalomolybdic Acid, 6744
235.953664	$C_9H_4N_2S_3$	Thioquinox, 9096
236.003784	$C_5H_8N_4O_3S_2$	Methazolamide, 5822 Propazolamide, 7603
236.016039	$C_7H_{13}BrN_2O_2$	Carbromal, 1833
236.028065	$C_6H_8N_2O_8$	Isosorbide Dinitrate, 5089
236.065321	$C_8H_{16}N_2O_2S_2$	4,4'-Dithiodimorpholine, 3390
236.068474	$C_{12}H_{12}O_5$	Fraxetin, Dimethyl ether, 4116 Radicinin, 7909
236.073183	$C_{10}H_{12}N_4OS$	Thiacetazone, 9018
236.083730	$C_{16}H_{12}O_2$	Anthranol, Acetate, 721

236.

Molecular Weight	Empirical Formula	Compound Name, Monograph Number
236.090940	$C_{10}H_{12}N_4O_3$	Adrenochrome, Monosemicarbazone, 159, 1791
236.094963	$C_{15}H_{12}N_2O$	Carbamazepine, 1781
236.108415	$C_8H_{24}O_2Si_3$	Octamethyltrisiloxane, 6559
236.116093	$C_{12}H_{16}N_2O_3$	Cyclobarbital, 2717 Hexobarbital, 4576, 4577
236.128669	$[C_{13}H_{18}NO_3]^+$	Methylhydrocotarninium, 4679
236.152478	$C_{13}H_{20}N_2O_2$	Butethamine, 1506 Metabutethamine, 5767 Procaine, Base, 7554, 7555-7557
236.165054	$[C_{14}H_{22}NO_2]^+$	Methylcarneginium, 1847
236.177630	$C_{15}H_{24}O_2$	Helminthosporol, 4489 Tiglic Acid, Geranyl ester, 9161
236.214016	$C_{16}H_{28}O$	Guaiol, Methyl ether, 4407
236.919376	$AlH_4NO_8S_2$	Aluminum Ammonium Sulfate, 319
237.019123	$F_6GaH_{12}N_3$	Ammonium Hexafluorogallate, 548
237.020813	$C_9H_7N_3O_3S$	Furazolium Chloride, Free base, 4152
237.030504	$C_{10}H_8ClN_3O_2$	Drazoxolon, 3432
237.076611	$C_8H_{16}NO_5P$	Dicrotophos, 3063
237.078979	$C_{15}H_{11}NO_2$	Viridicatin, 9659
237.079665	$C_8H_{11}N_7S$	Ambazone, 382
237.086189	$C_9H_{11}N_5O_3$	Biopterin, 1241
237.092042	$C_{13}H_{16}ClNO$	Ketamine, 5147
237.100108	$C_{12}H_{15}NO_4$	Cotarnine, 2533-2535 Ethopabate, 3672 Flavipucine, A6
237.115364	$C_{16}H_{15}NO$	Cyheptamide, 2762
237.136494	$C_{13}H_{19}NO_3$	Anacardiol, 661 Gigantine, 4252 N-(3-Methoxypropyl)-1,4-benzodioxan-2-methylamine, 5876 Pellotine, 6869 Viloxazine, 9634
237.151750	$C_{17}H_{19}N$	Etifelmin, 3820
237.878997	CdN_2O_6	Cadmium Nitrate, 1609
237.883000	$Na_2O_8S_2$	Sodium Persulfate, 8417
237.903963	CrN_3O_9	Chromic Nitrate, 2223

Use in conjunction with The Merck Index, Ninth Edition

Molecular Weight	Empirical Formula	Compound Name, Monograph Number
237.935513	$C_8H_5Cl_3O_2$	Fenac®, 3886
237.993595	$C_{10}H_6O_5S$	β-Naphthoquinone-4-sulfonic Acid, 8405
238.033819	$C_8H_6N_4O_5$	Nitrofurantoin, 6421
238.047738	$C_{11}H_{10}O_6$	o-Pyrocatechuic Acid Diacetate, 7781
238.070205	$C_9H_{10}N_4O_4$	1-Theobromineacetic Acid, 8996, 8997-9003 7-Theophyllineacetic Acid, 9005, 17, 1230, 6951
238.084124	$C_{12}H_{14}O_5$	Cinametic Acid, 2270 3,5-Diacetylfilicinic acid, 3997
238.095357	$C_{11}H_{14}N_2O_4$	5'-Nitro-2'-propoxyacetanilide, 6451
238.099380	$C_{16}H_{14}O_2$	Benzyl Cinnamate, 1145
238.106590	$C_{10}H_{14}N_4O_3$	8-Ethoxycaffeine, 3677 1-(β-Hydroxypropyl)theobromine, 4760 Proxyphylline, 7691
238.110613	$C_{15}H_{14}N_2O$	5,5-Diphenyl-4-imidazolidinone, 3336
238.120509	$C_{13}H_{18}O_4$	Felogen, 3884
238.131743	$C_{12}H_{18}N_2O_3$	5-Allyl-5-neopentylbarbituric Acid, 290 Secobarbital, 8172
238.146999	$C_{16}H_{18}N_2$	Agroclavine, 177
238.193280	$C_{15}H_{26}O_2$	d-Bornyl Isovalerate, 1354 Daucol, 2814
238.229666	$C_{16}H_{30}O$	Muscone, 6138
238.873758	$Cl_4CuH_8N_2$	Ammonium Cupric Chloride, 534
238.957456	$C_7H_7Cl_2NO_2S$	Dichloramine T, 3018
239.007472	$C_{10}H_9NO_2S_2$	Anethole Trithione, Oxime, 679
239.025230	$C_{10}H_9NO_4S$	1-Amino-2-naphthol-4-sulfonic Acid, 463 1-Amino-2-naphthol-6-sulfonic Acid, 464
239.094629	$C_{15}H_{13}NO_2$	o-Aminoacetophenone, N-Benzoyl derivative, 414 Dibenzoylmethane, Monoxime, 2973
239.107692	$C_{13}H_{18}ClNO$	Solan, 8477
239.115758	$C_{12}H_{17}NO_4$	Methyldopa, Ethyl ester, 5925
239.126992	$C_{11}H_{17}N_3O_3$	5-Ethyl-5-(1-piperidyl)barbituric Acid, 3788
239.131014	$C_{16}H_{17}NO$	Diphenamid, 3312
239.152144	$C_{13}H_{21}NO_3$	Albuterol, 206 Isoetharine, 5026
239.167400	$C_{17}H_{21}N$	Benzphetamine, 1130 N,N-Diethylbenzhydrylamine, 3092

Use in conjunction with The Merck Index, Ninth Edition

239.

Molecular Weight	Empirical Formula	Compound Name, Monograph Number
239.199763	$C_{13}H_{25}N_3O$	Cuscohygrine, Oxime, 2686
239.812479	K_2MoO_4	Potassium Molybdate(VI), 7431
239.865427	Na_2O_4Te	Sodium Tellurate(VI), 8456
239.873305	$Cl_4H_8N_2Zn$	Ammonium Tetrachlorozincate, 593
239.873394	$Mg_3O_7Si_2$	Magnesium Silicates, Serpentine (mineral), 5514
239.875059	$C_4K_2N_4Ni$	Potassium Tetracyanonickelate(II), 7475
239.926011	$C_6H_3Cl_3N_2O_2$	4-Amino-3,5,6-trichloropicolinic Acid, 497, 6978
239.943263	ClTl	Thallium Chloride, 8974
239.948713	PbS	Lead Sulfide, 5279
239.966470	O_2Pb	Lead Dioxide, 5257
239.967439	F_2Hg	Mercuric Fluoride, 5714
239.973730	$C_{10}H_8OS_3$	Anethole Trithione, 679
239.988336	$C_6H_{12}N_2S_4$	Thiram, 9111
239.998698	$C_4H_8N_4O_4S_2$	Tetramethylenedisulfotetramine, 8944
240.023851	$C_6H_{12}N_2O_4S_2$	Cystine, 2780
240.031712	$C_8H_8N_4O_3S$	Thiofuradene, 9067
240.037503	$C_8H_{17}I$	sec-Octyl Iodide, 6573
240.042259	$C_{14}H_8O_4$	Alizarin, 231 Anthrarufin, 724 Danthron, 2802 Quinizarin, 7886
240.058658	$C_{15}H_9FO_2$	Previscan®, 7538
240.078644	$C_{15}H_{12}O_3$	Chrysarobin, Pure substance, 2250 2-(p-Toluyl)benzoic Acid, 9237
240.089878	$C_{14}H_{12}N_2O_2$	Benzil Dioxime, 1085
240.093250	$C_{11}H_{16}N_2O_2S$	Buthalital, 1507
240.101111	$C_{13}H_{12}N_4O$	Diphenylcarbazone, 3330
240.115030	$C_{16}H_{16}O_2$	Benzoin, Ethyl ether, 1102 γ-(4-Biphenylyl)butyric Acid, 1250 Diphenylacetic Acid, Ethyl ester, 3324 2-(2-Hydroxy-1-naphthyl)cyclohexanone, 4746 Xenbucin, 9737
240.122241	$C_{10}H_{16}N_4O_3$	Anserine, 714 Dimetilan, 3255

Use in conjunction with The Merck Index, Ninth Edition

Molecular Weight	Empirical Formula	Compound Name, Monograph Number
240.126263	$C_{15}H_{16}N_2O$	Benmoxine, 1046 p-Dimethylaminobenzophenone, syn-Oxime, 3220 Phenatine, 7001
240.147393	$C_{12}H_{20}N_2O_3$	Hexethal, 4574 Nerol, Allophanate, 6299 Tetrabarbital, 8898
240.162649	$C_{16}H_{20}N_2$	N,N'-Dibenzylethylenediamine, 2979, 1066, 1067 Pheniramine, 7029
240.183778	$C_{13}H_{24}N_2O_2$	Cropropamide, 2586
240.208930	$C_{15}H_{28}O_2$	Exaltolide®, 3857 Menthyl Valerate, 5669
240.231397	$C_{13}H_{28}N_4$	Hendecanediamidine, 4506
240.256549	$[C_{15}H_{32}N_2]^{2+}$	Pentolinium, 6930
240.779252	Co_3O_4	Cobaltic-Cobaltous Oxide, 2381
240.910026	$C_6H_2Cl_3NO_3$	3,4,6-Trichloro-2-nitrophenol, 9321
240.972251	O_2Po	Polonium Dioxide, 7339
241.061804	$C_{10}H_{12}ClN_3O_2$	Tranid, 9260
241.067369	$C_{13}H_{11}N_3S$	Azure C, Free base, 935
241.085127	$C_{13}H_{11}N_3O_2$	Acroteben®, 124 Flavoteben®, 4010 1-Isonicotinoyl-2-salicylidenehydrazine, 5047
241.086957	$C_{12}H_{16}ClNO_2$	Lodal®, 5401
241.095023	$C_{11}H_{15}NO_5$	Methocarbamol, 5852
241.110279	$C_{15}H_{15}NO_2$	Apo-β-erythroidine, 774 N-Methylflindersine, 4013 Isoapo-β-erythroidine, 4977 Mefenamic Acid, 5621
241.136118	$C_{10}H_{19}N_5S$	Prometryne®, 7585
241.183050	$C_{17}H_{23}N$	N-Methylmorphinan, 6107
241.183050	$[C_{34}H_{46}N_2]^{2+}$	Hedaquinium (half mass), 4478
241.874772	$AlNaO_8S_2$	Aluminum Sodium Sulfate, 365
241.898393	FeN_3O_9	Ferric Nitrate, 3945
241.956367	$C_6H_{10}O_2S_4$	Dixanthogen, 3402
242.018371	$C_9H_{10}N_2O_2S_2$	Aureothricin, 900
242.054436	$[C_{11}H_{17}BrN]^+$	Bretylium, 1376
242.057053	$C_9H_{11}ClN_4O_2$	7-(β-Chloroethyl)theophylline, 2117

242.

Molecular Weight	Empirical Formula	Compound Name, Monograph Number
242.057909	$C_{14}H_{10}O_4$	Benzoyl Peroxide, 1126 Diphenic Acid, 3317 Salicil, 8084
242.061281	$C_{11}H_{14}O_4S$	2-(Ethylsulfonyl)ethanol, Benzoate, 3802
242.064653	$C_8H_{18}O_4S_2$	Sulfonethylmethane, 8754
242.065120	$C_8H_{10}N_4O_5$	Nidroxyzone, 6350
242.070972	$C_{12}H_{15}ClO_3$	Clofibrate, 2342
242.080376	$C_{12}H_{10}N_4O_2$	N,N'-Bis[isonicotinic acid] Hydrazide, 1274 N,N'-Bis[nicotinic acid] Hydrazide, 1320 Lumichrome, 5415
242.082206	$C_{11}H_{15}ClN_2O_2$	Iproclozide, 4931
242.090272	$C_{10}H_{14}N_2O_5$	Thymidine, 9138
242.094294	$C_{15}H_{14}O_3$	Benzilic Acid, Methyl ester, 1086 Fenoprofen, 3907 Lapachol, 5211 Mexenone, 6037 Phenylacetic Acid, o-Methoxyphenyl ester, 7069
242.101505	$C_9H_{14}N_4O_4$	Morsydomine, 6120
242.108900	$C_{11}H_{18}N_2O_2S$	Thiopental, 9087
242.109551	$C_{19}H_{14}$	2-Methyl-3,4-benzphenanthrene, 5901
242.116761	$C_{13}H_{14}N_4O$	sym-Diphenylcarbazide, 3329
242.130680	$C_{16}H_{18}O_2$	Bisphenol B, 1325
242.141913	$C_{15}H_{18}N_2O$	Selagine, 8176
242.188195	$C_{14}H_{26}O_3$	Menthyl Ethoxyacetate, 5667
242.260966	$C_{16}H_{34}O$	Cetyl Alcohol, 1983
243.011903	$C_9H_{10}NO_3PS$	Ciafos, 2263
243.056530	$C_{10}H_{13}NO_4S$	N-Acetylsulfanilic Acid, Ethyl ester, 102
243.085521	$C_9H_{13}N_3O_5$	Cytarabine, 2782 Cytidine, 2783
243.086858	$C_{10}H_9N_7O$	Furterene, 4164
243.104799	$C_{18}H_{13}N$	6-Chrysenamine, 2251
243.160972	$[C_{14}H_{19}N_4]^+$	Amprolium, Cation, 625
243.162314	$C_{16}H_{21}NO$	Norlevorphanol, 6514
243.198700	$C_{17}H_{25}N$	Phencyclidine, 7008
243.795554	Mo_2O_3	Molybdenum Sesquioxide, 6071

Use in conjunction with The Merck Index, Ninth Edition

Molecular Weight	Empirical Formula	Compound Name, Monograph Number
243.809590	CdTe	Cadmium Telluride, 1618
243.812913	$LaCl_3$	Lanthanum Chloride, 5209
243.865240	$C_6O_2Cl_4$	Chloranil, 2043
243.896649	F_6Te	Tellurium Hexafluoride, 8868
244.038044	$C_{14}H_{12}S_2$	Mesulphen, 5765
244.048407	$C_{12}H_8N_2O_4$	Iodinin, 4882
244.063012	$C_8H_{12}N_4O_3S$	Carnidazole, 1848
244.069536	$C_9H_{12}N_2O_6$	Uridine, 9540
244.073559	$C_{14}H_{12}O_4$	Cotoin, 2537 Dioxybenzone, 3304 Fulvoplumierin, 4135
244.080770	$C_8H_{12}N_4O_5$	Ribavirin, 7991
244.084792	$C_{13}H_{12}N_2O_3$	5-Allyl-5-phenylbarbituric Acid, 291
244.088164	$C_{10}H_{16}N_2O_3S$	Amidephrine, 400 Biotin, 1244
244.089958	$C_{15}H_{13}FO_2$	Flurbiprofen, A8
244.109945	$C_{15}H_{16}O_3$	Osthole, 6730
244.121178	$C_{14}H_{16}N_2O_2$	Dianisidine, 2952 Rolicyprine, 8019
244.125201	$C_{19}H_{16}$	Triphenylmethane, 9408
244.157563	$C_{15}H_{20}N_2O$	Anagyrine, 663
244.193949	$C_{16}H_{24}N_2$	Isoaminile, 4957 Xylometazoline, 9747
244.251464	$[C_{14}H_{32}N_2O]^{2+}$	Plegarol, Cation, 7311
244.812000	$CeCl_3$	Cerous Chloride, 1953
244.992014	$C_6H_3N_3O_8$	Styphnic Acid, 8655
245.045486	$C_{10}H_{12}ClNO_4$	Chlorphenesin Carbamate, 2168
245.064785	$C_8H_{11}N_3O_6$	6-Azauridine, 917
245.066363	$C_{11}H_{10}F_3NO_2$	Flumetramide, 4027
245.068808	$C_{13}H_{11}NO_4$	Menadoxime, Free acid, 5656
245.116427	$C_{13}H_{15}N_3O_2$	Pyrolan, 7788
245.141579	$C_{15}H_{19}NO_2$	Tropacocaine, 9437

245.

Molecular Weight	Empirical Formula	Compound Name, Monograph Number
245.152812	$C_{14}H_{19}N_3O$	4-[(2-Dimethylaminoethyl)amino]-6-methoxyquinoline, 7846 Isopyrin, 5080 Methafurylene, 5800 Oxolamine, 6761
245.177964	$C_{16}H_{23}NO$	Pyrovalerone, 7797 Tolperisone, 9219
245.742297	Br_2Sr	Strontium Bromide, 8627
245.759406	As_2S_3	Arsenic Trisulfide, 834
245.782262	Ag_2S	Silver Sulfide, 8276
245.814215	$PrCl_3$	Praseodymium Chloride, 7506
245.814359	K_2O_3Sn	Potassium Stannate(IV), 7460
245.868857	$C_4K_2N_4Zn$	Potassium Tetracyanozincate, 7477
245.894568	F_6Xe	Xenon Hexafluoride, 9738
245.973448	F_2Pb	Lead Fluoride, 5258
246.000389	$C_8H_{11}BrN_2O_2$	Isocil, 5014
246.023182	$C_6H_{14}O_6S_2$	Busulfan, 1486
246.030197	$C_{10}H_{15}OPS_2$	Fonofos, 4091
246.052824	$C_{13}H_{10}O_5$	Pimpinellin, 7237
246.053694	$C_{14}H_{14}S_2$	Dibenzyl Disulfide, 2978
246.064057	$C_{12}H_{10}N_2O_4$	Nifurpirinol, 6361
246.065200	$C_9H_{11}FN_2O_5$	Floxuridine, 4018
246.073953	$C_{10}H_{14}O_7$	D-Threose, Triacetate, 9125
246.089209	$C_{14}H_{14}O_4$	Phthalofyne, 7181 Piperic Acid, Ethyl ester, 7260
246.100442	$C_{13}H_{14}N_2O_3$	N-Acetyltryptophan, 1624 Mephobarbital, 5683
246.115699	$C_{17}H_{14}N_2$	Ellipticine, 3498 Flavopereirine, 4009 Olivacine, 6681
246.125595	$C_{15}H_{18}O_3$	Ambrosin, 388 α-Santonin, 8118 Xanthatin, 9720
246.136828	$C_{14}H_{18}N_2O_2$	Hypaphorine, 4784
246.161980	$C_{16}H_{22}O_2$	Geranylhydroquinone, 4237
246.173213	$C_{15}H_{22}N_2O$	Mepivacaine, 5688

Use in conjunction with The Merck Index, Ninth Edition

Molecular Weight	Empirical Formula	Compound Name, Monograph Number
246.209599	$C_{16}H_{26}N_2$	Pimetine, 7233
246.814289	$NdCl_3$	Neodymium Chloride, 6273
246.855882	F_6K_2Mn	Potassium Hexafluoromanganate(IV), 7419
246.903567	$C_7H_3BrClNO_2$	Bromchlorenone, 1385
246.946193	$C_6H_6AsNO_5$	Nitarsone, 6384
246.963213	$C_6H_2ClN_3O_6$	Picryl Chloride, 7220
246.978673	$C_7H_5NO_7S$	2-Nitro-4-sulfobenzoic Acid, 6465
247.024576	$C_8H_{10}NO_6P$	Pyridoxal 5-Phosphate, 7763
247.053070	$C_{11}H_{15}Cl_2NO$	Dichlorisoproterenol, 3020
247.062678	$C_8H_{13}N_3O_4S$	Tinidazole, 9172
247.105587	$C_{10}H_{17}NO_6$	Linamarin, 5337
247.110948	$C_{16}H_{13}N_3$	Yellow AB, 9761
247.120844	$C_{14}H_{17}NO_3$	EEDQ, 3478 Fagaramide, 3870
247.132077	$C_{13}H_{17}N_3O_2$	Parbendazole, 6840
247.157229	$C_{15}H_{21}NO_2$	β-Eucaine, 3837 Hexahydrodesmethoxy-β-erythroidine, 4554 Ketobemidone, 5151 Meperidine, 5675 Prodilidine, 7564 Tolpronine, 9221
247.193615	$C_{16}H_{25}NO$	Butidrine, 1509 Lycopodine, 5440
247.764314	BBr_3	Boron Tribromide, 1361
247.836739	Cl_2F_3Sb	Antimony Dichlorotrifluoride, 732
247.883622	$C_7H_6Br_2$	p-Bromobenzyl Bromide, 1408
247.923968	$C_8H_6BrClO_2$	p-Bromobenzyl Chloroformate, 1410
247.933431	$C_7H_5IO_2$	o-Iodobenzoic Acid, 4891
247.943343	$C_6H_7O_3Sb$	Benzenestibonic Acid, 1072
247.963903	$C_{10}H_{10}Cl_2Ti$	Titanocene Dichloride, 9191
247.974779	AuH_3O_3	Gold Trihydroxide, 4373
248.011933	$C_9H_{10}Cl_2N_2O_2$	Linuron, 5348
248.038453	$C_8H_{13}N_2O_3PS$	Thionazin, 9082
248.047344	$C_{16}H_8O_3$	β-Brazanquinone, 1374

Molecular Weight	Empirical Formula	Compound Name, Monograph Number
248.056210	$C_8H_{13}N_2O_5P$	Pyridoxamine Phosphate, 7764
248.061950	$C_{12}H_{12}N_2O_2S$	Dapsone, 2808 Sulfabenz, 8688
248.082874	$C_{12}H_{13}ClN_4$	Pyrimethamine, 7768
248.083730	$C_{17}H_{12}O_2$	2-Naphthyl Benzoate, 6236
248.113962	$C_{12}H_{25}Br$	Lauryl Bromide, 5235
248.116093	$C_{13}H_{16}N_2O_3$	Acetylpheneturide, 96
248.141245	$C_{15}H_{20}O_3$	Arborescin, 799 Illudin M, 4806 Perezone, 6950
248.148455	$C_9H_{20}N_4O_4$	Negamycin, 6266
248.152478	$C_{14}H_{20}N_2O_2$	Etryptamine Acetate, 3835 Pindolol, 7240 Piridocaine, 7283
248.188864	$C_{15}H_{24}N_2O$	Aphylline, 762 Lupanine, 5423 Matrine, 5589 Trimecaine, 9368
248.872973	$EuSO_4$	Europous Sulfate, 3853
249.045965	$C_{12}H_{11}NO_3S$	N-Phenylsulfanilic Acid, 7118
249.057199	$C_{11}H_{11}N_3O_2S$	Sulfapyridine, 8733
249.078979	$C_{16}H_{11}NO_2$	Benzoxiquine, 1122 Cinchophen, 2277, 2278
249.151750	$C_{18}H_{19}N$	Benzoctamine, 1094
249.152893	$C_{15}H_{20}FNO$	4'-Fluoro-4-piperidinobutyrophenone, 4064
249.172879	$C_{15}H_{23}NO_2$	Alprenolol, 307 Nupharidine, 6540
249.233074	$[C_{16}H_{29}N_2]^+$	N-Methyl-β-isosparteinium, 5090
249.762833	$CHBr_3$	Bromoform, 1418
249.834124	BaO_3S_2	Barium Thiosulfate, 1007
249.876093	$Na_4O_6P_2$	Sodium Hypophosphate, 8384
249.946262	H_2O_4W	Tungstic(VI) Acid, 9471
250.019434	$C_6H_{10}N_4O_3S_2$	Butazolamide, 1501 Isobutamide, 4982
250.044151	$C_{13}H_8F_2O_3$	Diflunisal, A4
250.047738	$C_{12}H_{10}O_6$	Methylescutol, 5837

Use in conjunction with The Merck Index, Ninth Edition

Molecular Weight	Empirical Formula	Compound Name, Monograph Number
250.052447	$C_{10}H_{10}N_4O_2S$	Sulfadiazine, 8693, 8694 Sulfapyrazine, 8731, 8732
250.054103	$C_8H_{15}N_2O_3PS$	O,O-Diethyl O-(3-Methyl-5-pyrazolyl) Phosphorothioate, 3106
250.084124	$C_{13}H_{14}O_5$	Citrinin, 2308
250.095357	$C_{12}H_{14}N_2O_4$	5-Furfuryl-5-isopropylbarbituric Acid, 4157
250.110613	$C_{16}H_{14}N_2O$	Glycosine, 4338 Methaqualone, 5818
250.120509	$C_{14}H_{18}O_4$	Cinoxate, 2298 o-Thymotic Acid, Acetonyl ester, 9150
250.131743	$C_{13}H_{18}N_2O_3$	Heptabarbital, 4513 2-(3,4,5-Trimethoxybenzyl)-2-imidazoline, 9378
250.146999	$C_{17}H_{18}N_2$	Amphetaminil, 620
250.168128	$C_{14}H_{22}N_2O_2$	Naepaine, 6173 Tutocaine®, 9481
250.168128	$[C_{28}H_{44}N_4O_4]^{2+}$	Hexadistigmine, Cation (half mass), 4551
250.193280	$C_{16}H_{26}O_2$	2,5-Di-tert-pentylhydroquinone, 3306
250.204514	$C_{15}H_{26}N_2O$	Retamine, 7956
250.229666	$C_{17}H_{30}O$	Civetone, 2315
250.872668	$FeH_6O_6P_3$	Ferric Hypophosphite, 3943
250.965522	$C_7H_6ClNO_5S$	6-Chloro-5-nitrotoluene-3-sulfonic Acid, 8345
251.022832	$C_8H_{11}Cl_2N_3O_2$	Uracil Mustard, 9507
251.093773	$C_{11}H_{14}ClN_5$	Cycloguanil, 2726
251.100502	$C_9H_{17}NO_7$	Muramic Acid, 6130
251.101839	$C_{10}H_{13}N_5O_3$	Cordycepin, 2506
251.138225	$C_{11}H_{17}N_5O_2$	Dimethazan, 3201
251.167400	$C_{18}H_{21}N$	2-Diphenylmethylpiperidine, 3341
251.167400	$[C_{36}H_{42}N_2]^{2+}$	Hexafluorenium (half mass), 4552
251.199763	$C_{14}H_{25}N_3O$	Solanone, Semicarbazone, 8481
251.261300	$C_{17}H_{33}N$	2,2'-Dicyclohexyl-N-methyldiethylamine, 3073
251.914170	$Cr_2H_8N_2O_7$	Ammonium Dichromate(VI), 535
251.957254	$H_8MgN_2O_8S_2$	Ammonium Magnesium Sulfate, 558
252.017046	$C_5H_8FeN_8O$	Ammonium Nitroferricyanide, 564

252.

Molecular Weight	Empirical Formula	Compound Name, Monograph Number
252.022104	$C_{12}H_{10}Cl_2N_2$	2,2'-Dichlorobenzidine, 3031 3,3'-Dichlorobenzidine, 3032
252.063388	$C_{12}H_{12}O_6$	1,2,4-Benzenetriol, Triacetate, 1076 Pyrogallol Triacetate, 7785
252.078644	$C_{16}H_{12}O_3$	Anisindione, 703
252.085855	$C_{10}H_{12}N_4O_4$	Nebularine, 6263
252.089878	$C_{15}H_{12}N_2O_2$	Phenopyrazone, 7049 Phenytoin, 7130
252.093900	$C_{20}H_{12}$	Benzo[a]pyrene, 1113 Benzo[e]pyrene, 1114 Perylene, 6969
252.095751	$C_8H_{16}N_2O_7$	Cycasin, 2705
252.099774	$C_{13}H_{16}O_5$	Dihydrocitrinin, 2308
252.115716	$C_{10}H_{16}N_6S$	Cimetidine, A3
252.129635	$C_{13}H_{20}N_2OS$	Thiocaine, 9053
252.136159	$C_{14}H_{20}O_4$	Frequentin, 4119
252.137497	$C_{15}H_{16}N_4$	Neutral Red, Free base, 6304
252.147393	$C_{13}H_{20}N_2O_3$	Hydroxyprocaine, 4755, 6850
252.208930	$C_{16}H_{28}O_2$	Hydnocarpic Acid, 4639
252.904240	$C_6H_6AsCl_2NO$	Dichlorophenarsine, 3048
252.958209	ClH_2HgN	Mercuric Chloride, Ammoniated, 5711
252.996428	$C_{11}H_8ClNO_2S$	Fenclozic Acid, 3898
253.040880	$C_{11}H_{11}NO_4S$	Actinoquinol, 137 Woodward's Reagent K, 9715
253.052113	$C_{10}H_{11}N_3O_3S$	Sulfamethoxazole, 8709
253.069871	$C_{10}H_{11}N_3O_5$	Suosan, Free acid, 8793
253.077265	$C_{12}H_{15}NO_3S$	Vanitiolide, 9592
253.081104	$C_9H_{11}N_5O_4$	Eritadenine, 3593 Ichthyopterin, 4799 Neopterin, 6287
253.106256	$C_{11}H_{15}N_3O_4$	Pyridinol Carbamate, 7758
253.107593	$C_{12}H_{11}N_7$	Triamterene, 9282
253.109423	$C_{11}H_{16}ClN_5$	Chlorguanide, 2058
253.110279	$C_{16}H_{15}NO_2$	9-Carbazoleacetic Acid, Ethyl ester, 1794
253.113651	$C_{13}H_{19}NO_2S$	Isobornyl Thiocyanoacetate, Technical, 4981

Use in conjunction with The Merck Index, Ninth Edition

Molecular Weight	Empirical Formula	Compound Name, Monograph Number
253.121512	$C_{15}H_{15}N_3O$	Ethacridine, 3642
253.131408	$C_{13}H_{19}NO_4$	N-Acetylmescaline, 5752
253.146664	$C_{17}H_{19}NO$	Nefopam, 6265
253.183050	$C_{18}H_{23}N$	Tolpropamine, 9222 N,N-1-Trimethyl-3,3-diphenylpropylamine, 9382
253.201494	$C_{11}H_{23}N_7$	Hydramitrazine, 4647
253.734963	P_2Zn_3	Zinc Phosphide, 9818
253.748107	Br_2OSe	Selenium Oxybromide, 8184
253.812818	Cl_2O_6Sr	Strontium Chlorate, 8629
253.825404	$BaCrO_4$	Barium Chromate(VI), 975
253.835962	$K_2O_7S_2$	Potassium Pyrosulfate, 7452
253.837111	$Ca_2O_7P_2$	Calcium Pyrophosphate, 1700
253.855528	$C_2BaN_2S_2$	Barium Thiocyanate, 1006
253.880390	$C_6H_4BrClO_2S$	p-Bromobenzenesulfonyl Chloride, 1406
253.901087	$C_4H_3IN_2OS$	Iothiouracil, 4924
253.930427	$C_8H_5Cl_3O_3$	(2,4,5-Trichlorophenoxy)acetic Acid, 9324
253.955496	$C_6H_6O_7S_2$	Phenoldisulfonic Acid, 7039
253.976780	C_2HgN_2	Mercuric Cyanide, 5712
254.006267	$C_{10}H_6O_8$	Pyromellitic Acid, 7790
254.024895	$C_{11}H_{10}O_5S$	Menadione Sulfonic Acid, 5655
254.039501	$C_7H_{14}N_2O_4S_2$	Djenkolic Acid, 3404
254.043523	$C_{12}H_{14}O_2S_2$	Diethyl Dithiolisophthalate, 3099
254.057909	$C_{15}H_{10}O_4$	Alizarin, 1-Methyl ether, 231 Alizarin, 2-Methyl ether, 231 Chrysin, 2254 Chrysophanic Acid, 2260 Daidzein, 2796 Rubiadin, 8039
254.072514	$C_{11}H_{14}N_2O_3S$	Sulfadicramide, 8695
254.094294	$C_{16}H_{14}O_3$	Ketoprofen, 5154 3-(o-Methoxyphenyl)-2-phenylacrylic Acid, 5872 2-(p-Toluyl)benzoic Acid, Methyl ester, 9237
254.101505	$C_{10}H_{14}N_4O_4$	Dyphylline, 3455
254.108900	$C_{12}H_{18}N_2O_2S$	Spirothiobarbital, 8538 Thiamylal, 9035

Molecular Weight	Empirical Formula	Compound Name, Monograph Number
254.109551	$C_{20}H_{14}$	Cholanthrene, 2188 Triptycene, 9415
254.115424	$C_{13}H_{18}O_5$	Meprophendiol, 5692
254.116761	$C_{14}H_{14}N_4O$	Phenamidine, 6994
254.118591	$C_{13}H_{19}ClN_2O$	Butanilicaine, 1498
254.126657	$C_{12}H_{18}N_2O_4$	D-Fucose, Phenylhydrazone, 4129 L-Fucose, Phenylhydrazone, 4130
254.141913	$C_{16}H_{18}N_2O$	Amphenone B, 615 Elymoclavine, 3500
254.151809	$C_{14}H_{22}O_4$	Palitantin, 6797
254.178299	$C_{17}H_{22}N_2$	Phenbenzamine, 7006 4,4'-Tetramethyldiaminodiphenylmethane, 8942
254.977235	$C_8H_5N_3O_3S_2$	Atrican, 888
255.013620	$C_9H_9N_3O_2S_2$	Sulfathiazole, 8739, 8740 Thiazolsulfone, 9041
255.083019	$C_{14}H_{13}N_3S$	Azure A, Free base, 933
255.089543	$C_{15}H_{13}NO_3$	p-Hydroxyacetanilide Benzoate, 4707 3'-Methylphthalanilic Acid, 5979
255.102607	$C_{13}H_{18}ClNO_2$	Cloforex, 2345
255.125929	$C_{16}H_{17}NO_2$	p,α-Dimethylbenzyl Alcohol, Phenylurethan, 3226
255.137162	$C_{15}H_{17}N_3O$	Cetoxime, 1978
255.138992	$C_{14}H_{22}ClNO$	Clobutinol, 2334
255.147058	$C_{13}H_{21}NO_4$	Meteloidine, 5789
255.160972	$[C_{15}H_{19}N_4]^+$	Toluylene Blue, Cation, 9238
255.162314	$C_{17}H_{21}NO$	Diphenhydramine, 3316, 1230, 3193 Phenyltoloxamine, 7121, 7122 Tofenacin, 9204
255.173548	$C_{16}H_{21}N_3$	Tripelennamine, 9404, 9750
255.198700	$C_{18}H_{25}N$	Dimemorfan, 3192
255.808013	$CaCr_2O_7$	Calcium Dichromate(VI), 1659
255.818389	K_2O_3Te	Potassium Tellurate(IV), 7468
255.922580	$C_4H_8Cl_3O_4P$	Trichlorfon, 9303
255.941146	O_4Os	Osmium Tetroxide, 6729
255.980633	$C_{10}H_6Cl_2N_2O_2$	Pyrrolnitrin, 7806
255.984743	$C_7H_{14}AsClO_3$	Chloroarsenol, 2093

Use in conjunction with The Merck Index, Ninth Edition

Molecular Weight	Empirical Formula	Compound Name, Monograph Number
255.997636	$C_9H_8N_2O_3S_2$	Phenosulfazole, 7051
256.024556	$C_{10}H_{14}NiO_4$	Nickel Acetylacetonate, 6314
256.029107	$C_{15}H_9ClO_2$	Clorindione, 2363
256.037173	$C_{14}H_8O_5$	Anthragallol, 719 Purpurin, 7732
256.058303	$C_{11}H_{12}O_7$	Piscidic Acid, 7287
256.073559	$C_{15}H_{12}O_4$	o-(p-Anisoyl)benzoic Acid, 707 Methyl Benzoylsalicylate, 5900 Phenyl Acetylsalicylate, 7070
256.076726	$C_{15}H_{13}ClN_2$	Chlormidazole, 2076
256.078268	$C_{13}H_{12}N_4S$	Diphenylthiocarbazone, 3349
256.088164	$C_{11}H_{16}N_2O_3S$	Phenbutamide, 7007
256.096026	$C_{13}H_{12}N_4O_2$	Lumiflavine, 5416
256.109945	$C_{16}H_{16}O_3$	Lapachol, Methyl ether, 5211
256.113629	$[C_9H_{23}NO_3PS]^+$	Echothiopate, 3470
256.125201	$C_{20}H_{16}$	9,10-Dimethyl-1,2-benzanthracene, 3224
256.132411	$C_{14}H_{16}N_4O$	Ethoxazene, 3676
256.157563	$C_{16}H_{20}N_2O$	Chanoclavine, 1996
256.168797	$C_{15}H_{20}N_4$	2-[Benzyl(2-dimethylaminoethyl)amino]pyrimidine, 1147
256.170139	$[C_{17}H_{22}NO]^+$	Bephenium, 1175
256.240230	$C_{16}H_{32}O_2$	Ethyl Myristate, 3769 2-Hexyldecanoic Acid, 4584 Palmitic Acid, 6805
256.822941	$BaMnO_4$	Barium Manganate(VI), 986
256.826299	$SmCl_3$	Samarium Trichloride, 8106
256.992014	$C_7H_3N_3O_8$	2,4,6-Trinitrobenzoic Acid, 9394
256.994540	$C_6H_{12}NO_4PS_2$	Formothion, 4104
257.001034	$C_{11}H_9Cl_2NO_2$	Barban, 962
257.068808	$C_{14}H_{11}NO_4$	4-Benzamidosalicylic Acid, 1061 Diphenylamine-2,2'-dicarboxylic acid, 3326
257.072180	$C_{11}H_{15}NO_4S$	Ethebenecid, 3657
257.081871	$C_{12}H_{16}ClNO_3$	Meclofenoxate, 5604
257.101171	$C_{10}H_{15}N_3O_5$	Benserazide, 1052 MADU, 5471

257.

Molecular Weight	Empirical Formula	Compound Name, Monograph Number
257.105193	$C_{15}H_{15}NO_3$	Tolmetin, 9215
257.116427	$C_{14}H_{15}N_3O_2$	Indolmycin, 4843
257.129003	$[C_{15}H_{17}N_2O_2]^+$	Benzpyrinium, 1132
257.140724	$C_{11}H_{20}ClN_5$	Chlorazine, 2046
257.177964	$C_{17}H_{23}NO$	3-Hydroxy-N-methylmorphinan, 4739
257.799540	CdO_4Se	Cadmium Selenate, 1613
257.827801	$EuCl_3$	Europic Chloride, 3853
257.848710	$AlKO_8S_2$	Aluminum Potassium Sulfate, 360
257.950144	$C_4H_7IN_2O_3$	2-Iodoethyl Allophanate, 4895
257.952770	$C_9H_7BrO_4$	5-Bromosalicylic Acid Acetate, 1437
257.972837	$C_6H_5Na_3O_7$	Sodium Citrate, 8349
257.975237	$C_7H_7BrN_4O_2$	8-Bromotheophylline, 6809, 7736
257.975296	$C_6H_{11}IO_3$	Iodinated Glycerol, 4873
258.000389	$C_9H_{11}BrN_2O_2$	Metobromuron, 6016
258.013286	$C_9H_{10}N_2O_3S_2$	Ethoxzolamide, 3682
258.037567	$C_{10}H_{10}O_8$	Aceglatone, 18
258.044757	$C_{15}H_{11}ClO_2$	Clobenfurol, 2328
258.046299	$C_{13}H_{10}N_2O_2S$	Phethenylic Acid, 7132
258.051327	$C_8H_{19}O_3PS_2$	Demeton, 2859
258.052824	$C_{14}H_{10}O_5$	Gentisin, 4231 Salicylsalicylic acid, 8096
258.064057	$C_{13}H_{10}N_2O_4$	Myxin, 6168 Thalidomide, 8968
258.089209	$C_{15}H_{14}O_4$	Euparin, Acetate, 3846 Menadiol Diacetate, 5649 3-(4-Methoxy-1-naphthoyl)propionic Acid, 5870 Peucedanin, 6976 Xanthoxyletin, 9730 Yangonin, 9759
258.103814	$C_{11}H_{18}N_2O_3S$	Biotin, Methyl ester, 1244
258.159295	$C_{13}H_{18}N_6$	Zolertine, 9852
258.161980	$C_{17}H_{22}O_2$	Cicutoxin, 2266 Enanthotoxin, 3516
258.183110	$C_{14}H_{26}O_4$	Ethyl Sebacate, 3794
258.194343	$C_{13}H_{26}N_2O_3$	Elaiomycin, 3486

Use in conjunction with The Merck Index, Ninth Edition

Molecular Weight	Empirical Formula	Compound Name, Monograph Number
258.198366	$C_{18}H_{26}O$	Versalide®, 9621 Xibornol, 9740
258.219495	$C_{15}H_{30}O_3$	Trethocanic Acid, 9273
258.303499	$[C_{16}H_{38}N_2]^{2+}$	Decamethonium, 2830
258.982578	$C_8H_{10}AsNO_4$	Arsacetin, 816
259.053070	$C_{12}H_{15}Cl_2NO$	Karsil, 5136
259.054612	$C_{10}H_{14}ClN_3OS$	Azintamide, 923
259.059306	$C_{12}H_9N_3O_4$	Magneson, 5523
259.084458	$C_{14}H_{13}NO_4$	Skimmianine, 8298
259.112777	$C_{16}H_{18}ClN$	N-(2-Chloroethyl)dibenzylamine, 2115 Clobenzorex, 2330
259.116821	$C_{10}H_{17}N_3O_5$	Linatine, 5339 Orotic Acid, Choline ester, 6711
259.132077	$C_{14}H_{17}N_3O_2$	Tremetone, Semicarbazide, 9268
259.155886	$[C_{14}H_{19}N_4O]^+$	Pyrithiamine, Free base, Cation, 7774
259.157229	$C_{16}H_{21}NO_2$	Propranolol, 7628
259.168462	$C_{15}H_{21}N_3O$	Primaquine, 7543 Quinocide, 7887
259.214744	$[C_{28}H_{58}N_2O_6]^{2+}$	Prodeconium (half mass), 7562
259.777610	Cl_4Sn	Stannic Chloride, 8552
259.809910	CsI	Cesium Iodide, 1969
259.860192	$Mg_2O_8Si_3$	Magnesium trisilicate, 5514
259.893300	F_7I	Iodine Heptafluoride, 4876
259.896540	$C_7H_4Cl_4O_2$	Drosophilin A, 3439
259.944155	BiClO	Bismuth Chloride Oxide, 1285
259.977827	$C_7H_9AsN_2O_4$	Carbarsone, 1789
259.988607	BiH_3O_3	Bismuth Hydroxide, 1289
260.012834	$C_7H_{17}O_2PS_3$	Phorate, 7142
260.016039	$C_9H_{13}BrN_2O_2$	Bromacil, 1380
260.024822	$C_7H_{15}Cl_2N_2O_2P$	Cyclophosphamide, 2751
260.029721	$C_6H_{13}O_9P$	Fructose-6-phosphate, 4124 Glucose-6-phosphate, 4293 α-Glucose-1-phosphate, 4294 Inositol Monophosphate, 4855

Molecular Weight	Empirical Formula	Compound Name, Monograph Number
260.046693	$C_9H_{12}N_2O_5S$	Sulocarbilate, 8780
260.047831	$C_{14}H_{11}ClNO_2$	Diclofenac Sodium, Free acid, 3059
260.068474	$C_{14}H_{12}O_5$	Khellin, 5156
260.083079	$C_{10}H_{16}N_2O_4S$	Biotin l-Sulfoxide, 1245
260.089603	$C_{11}H_{16}O_7$	D-2-Deoxyribose, 1,3,4-Triacetate, 2869
260.104859	$C_{15}H_{16}O_4$	Dihydroyangonin, 9759
260.116093	$C_{14}H_{16}N_2O_3$	Abrine, Acetyl deriv, 4 Nadoxolol, 6171 Phenetharbital, 7013
260.120115	$C_{19}H_{16}O$	Triphenylcarbinol, 9406
260.125989	$C_{12}H_{20}O_6$	Diacetoneglucose, 2922
260.127326	$C_{13}H_{16}N_4O_2$	Diaveridine, 2960
260.131349	$C_{18}H_{16}N_2$	N,N'-Diphenyl-p-phenylenediamine, 3342
260.134721	$C_{15}H_{20}N_2S$	Methaphenilene, 5816
260.152478	$C_{15}H_{20}N_2O_2$	Fenspiride, 3916
260.163711	$C_{14}H_{20}N_4O$	Imolamine, 4816
260.173607	$C_{12}H_{24}N_2O_4$	Carisoprodol, 1841
260.188864	$C_{16}H_{24}N_2O$	Oxymetazoline, 6779
260.250401	$C_{19}H_{32}$	Androstane, 671 Etiocholane, 3822 Tridecylbenzene, 9338
260.826622	Cl_2CuO_8	Cupric Perchlorate, 2649
261.063723	$C_{13}H_{11}NO_5$	Oxolinic Acid, 6762
261.074956	$C_{12}H_{11}N_3O_4$	Isomaltol, p-Nitrophenylhydrazone, 5036
261.126598	$C_{17}H_{15}N_3$	Yellow OB, 9762
261.129970	$C_{14}H_{19}N_3S$	Methapyrilene, 5817 Thenyldiamine, 8993
261.139174	$[C_{18}H_{17}N_2]^+$	Methylellipticinium, 3498
261.151750	$C_{19}H_{19}N$	Phenindamine, 7026
261.172879	$C_{16}H_{23}NO_2$	Alphaprodine, 306 Betaprodine®, 1212 γ-Diethylaminopropyl Cinnamate, 3089 Ethoheptazine, 3668 Etoxadrol, 3833 Hexylcaine, 4583 Piperocaine, 7268 Properidine, 7606

Use in conjunction with The Merck Index, Ninth Edition

Molecular Weight	Empirical Formula	Compound Name, Monograph Number
261.209265	$C_{17}H_{27}NO$	Amixetrine, 505
261.820455	$MgMn_2O_8$	Magnesium Permanganate, 5503
261.824961	$Cr_2Na_2O_7$	Sodium Dichromate(VI), 8356
261.826168	Cl_2O_8Zn	Zinc Perchlorate, 9813
261.861302	$Al_2MgO_8Si_2$	Aluminum Magnesium Silicate, 348
261.861979	$Mg_3O_8P_2$	Magnesium Phosphate, Tribasic, 5507
261.878345	$CrH_4NO_8S_2$	Ammonium Chromic Sulfate, 530
261.880872	BaN_2O_6	Barium Nitrate, 988
261.927837	$C_5FeN_6Na_2O$	Sodium Nitroferricyanide, 8408
261.969620	F_3Tl	Thallium Trifluoride, 8985
262.016350	$C_{11}H_{12}Cl_2O_3$	(2,4-Dichlorophenoxy)acetic Acid, Isopropyl ester, 3049
262.017892	$C_9H_{11}ClN_2O_3S$	Diapamide, 2953
262.035475	$C_8H_{11}N_2O_6P$	Pyridoxal 5-Phosphate, Oxime, 7763
262.038817	$C_9H_{12}Cl_2N_4O$	Guanochlor, 4418
262.047738	$C_{13}H_{10}O_6$	Maclurin, 5465
262.070205	$C_{11}H_{10}N_4O_4$	Carbadox, 1780
262.074228	$C_{16}H_{10}N_2O_2$	Indigo, 4827
262.075883	$C_{14}H_{15}O_3P$	Dibenzyl Phosphite, 2980
262.077600	$C_{13}H_{14}N_2O_2S$	Benzylsulfamide, 1170 p-Sulfanilylbenzylamine, 8722
262.088833	$C_{12}H_{14}N_4OS$	Thiochrome, 9058
262.091139	$C_{18}H_{15}P$	Triphenylphosphine, 9410
262.120509	$C_{15}H_{18}O_4$	Artemisin, 842 Desacetylmatricarin, 5588 Helenalin, 4481 Parthenin, 6849
262.126561	$[C_{15}H_{20}NOS]^+$	Thenium, 8991
262.131743	$C_{14}H_{18}N_2O_3$	Methohexital, 5853 Physovenine, 7190 Reposal, 7931
262.138953	$C_8H_{18}N_6O_4$	Streptidine, 8606
262.141639	$C_{12}H_{22}O_6$	Etoglucid, 3830
262.156895	$C_{16}H_{22}O_3$	Fencibutirol, 3897 Homosalate, 4621

Molecular Weight	Empirical Formula	Compound Name, Monograph Number
262.180704	$[C_{16}H_{24}NO_2]^+$	Hexahydrodesmethoxy-N-methyl-β-erythroidinium, 4554
262.266051	$C_{19}H_{34}$	Fichtelite, 3995
262.821137	BeI_2	Beryllium Iodide, 1196
262.830669	$GdCl_3$	Gadolinium Chloride, 4174
262.941108	$C_6H_6AsNO_6$	Roxarsone, 8034
262.995993	$C_4H_{11}NO_8P_2$	Glyphosine, 4349
263.001733	$C_8H_{10}NO_5PS$	Methyl Parathion, 5973
263.003533	$C_{11}H_{12}Cl_3N$	Amphecloral, 613
263.071307	$C_{14}H_{14}ClNO_2$	Clopirac, 2360
263.080243	$C_{14}H_{17}NS_2$	Dimethylthiambutene, 3250
263.120467	$C_{12}H_{17}N_5S$	Methiotriazamine, 5848
263.126992	$C_{13}H_{17}N_3O_3$	Benzyl 4-Carbamyl-1-piperazinecarboxylate, 1143 Dioxypyramidon, 3305
263.152144	$C_{15}H_{21}NO_3$	Hydroxypethidine, 4749
263.167400	$C_{19}H_{21}N$	Nortriptyline, 6525 Protriptyline, 7689
263.188529	$C_{16}H_{25}NO_2$	Butacarb, 1489 Butethamate, 1505 Cyclexanone, 2712 Gravitol(e), Free base, 4387 Thymyl N-Isoamylcarbamate, 9151
263.199763	$C_{15}H_{25}N_3O$	β-Irone, Semicarbazone, 4945
263.736550	$AlBr_3$	Aluminum Bromide, 328
263.831908	Cl_3Tb	Terbium Chloride, 8877
263.847003	C_6HCl_5O	Pentachlorophenol, 6901, 1981, 6554
263.881559	$C_8Cl_4N_2$	Tetrachloroisophthalonitrile, 8908
263.928346	$C_7H_5IO_3$	m-Iodosalicylic Acid, 4908 o-Iodosobenzoic Acid, 4910
263.962067	$C_6H_9NO_3Sb$	4-Aminobenzenestibonic Acid, 423
264.008678	$C_8H_{15}Cl_3O_3$	Chlorhexadol, 2059
264.027883	O_2Th	Thorium Oxide, 9120
264.039107	$C_{12}H_{12}N_2OS_2$	p-Dimethylaminobenzalrhodanine, 3218
264.049470	$C_{10}H_8N_4O_5$	Picrolonic Acid, 7213
264.068098	$C_{11}H_{12}N_4O_2S$	Sulfamerazine, 8703, 8704 Sulfaperine, 8728

Use in conjunction with The Merck Index, Ninth Edition

Molecular Weight	Empirical Formula	Compound Name, Monograph Number
264.078644	$C_{17}H_{12}O_3$	1-Naphthyl Salicylate, 6242 2-Naphthyl Salicylate, 6243
264.078864	$C_{10}H_{20}N_2S_3$	Sulfirame, 8746
264.093250	$C_{13}H_{16}N_2O_2S$	Thialbarbital, 9019
264.099774	$C_{14}H_{16}O_5$	Methyl Citrinin, 2308
264.115030	$C_{18}H_{16}O_2$	Cinnamyl Cinnamate, 2294
264.126263	$C_{17}H_{16}N_2O$	Etaqualone, 3641
264.136159	$C_{15}H_{20}O_4$	Abscisic Acid, 6 Hirsutic Acid C, 4592 Illudin S, 4806 Santonic Acid, 8117 Tanacetin, 8824
264.137497	$C_{16}H_{16}N_4$	Stilbamidine, 8597
264.158626	$C_{13}H_{20}N_4O_2$	Pentifylline, 6927 Tibicor, 9158
264.162649	$C_{18}H_{20}N_2$	Mianserin, 6041
264.183778	$C_{15}H_{24}N_2O_2$	2-Diisopropylaminoethyl p-Aminobenzoate, 3182 Helminthosporal, Bis-oxime, 4488 Hydroxylupanine, 4733 Tetracaine, 8904
264.879432	$C_4H_6MnN_2S_4$	Maneb, 5549
264.902490	Cl_2Pt	Platinous Chloride, 7307
265.038482	$C_9H_{13}Cl_2N_3O_2$	Dopan, 3423
265.044718	$C_9H_7N_5O_5$	Erythropterin, 3611
265.073893	$C_{16}H_{11}NO_3$	3-Hydroxy-2-phenylcinchoninic Acid, 4751
265.091000	$C_8H_{15}N_3O_7$	Streptozocin, 8620
265.099732	$C_{11}H_{15}N_5OS$	Thiethazone, 9042
265.110279	$C_{17}H_{15}NO_2$	O,N-Dimethylviridicatin, 9659 O,O-Dimethylviridicatin, 9659
265.112308	$[C_{12}H_{17}N_4OS]^+$	Thiamine, Cation, 9024, 9022, 9025, 9028, 9029, 9031
265.131408	$C_{14}H_{19}NO_4$	Anisomycin, 705
265.157898	$C_{17}H_{19}N_3$	Antazoline, 715
265.167794	$C_{15}H_{23}NO_3$	Ethomoxane, 3671 Oxprenolol, 6768 Parethoxycaine, 6842
265.240565	$C_{17}H_{31}NO$	Myrtecaine, 6165
265.739762	Br_3HSi	Tribromosilane, 9295

265.

Molecular Weight	Empirical Formula	Compound Name, Monograph Number
265.775330	O_4Rb_2S	Rubidium Sulfate, 8047
265.787329	Nb_2O_5	Niobium Pentoxide, 6378
265.871008	$Na_4O_7P_2$	Tetrasodium Pyrophosphate, 8957
265.872774	$FeH_4NO_8S_2$	Ammonium Ferric Sulfate, 540
265.975598	BiF_3	Bismuth Fluoride, 1288
265.978610	C_2H_5ClHg	Ethylmercuric Chloride, 3764
265.999051	$Cl_3CoH_{18}N_6$	Hexaaminecobalt Trichloride, 4541
266.003115	$C_7H_{10}N_2O_5S_2$	Sulfanilamidomethanesulfonic Acid, 8718 Sulphocidine Album, Free acid, 8784
266.026521	$C_{14}H_{12}Cl_2O$	DMC, 3405
266.042653	$C_{12}H_{10}O_7$	Echinochrome A, 3465
266.056506	$C_6H_{12}FeN_9$	Ammonium Ferricyanide, 542
266.090272	$C_{12}H_{14}N_2O_5$	N-(p-Aminobenzoyl)glutamic Acid, 430 2-Cyclohexyl-4,6-dinitrophenol, 2738
266.094294	$C_{17}H_{14}O_3$	Benzarone, 1065
266.096324	$[C_{12}H_{16}N_3O_2S]^+$	Oxythiamine, Cation, 6792
266.101505	$C_{11}H_{14}N_4O_4$	Tubercidin, 9463
266.130680	$C_{18}H_{18}O_2$	Dienestrol, 3076 Diphenesenic Acid, 3315 Equilenin, 3551
266.141913	$C_{17}H_{18}N_2O$	Diphenazoline, 3314
266.155181	$C_{12}H_{26}O_4S$	Lauryl Sulfuric Acid, 8392
266.163043	$C_{14}H_{22}N_2O_3$	Atenolol, 880 Bucolome, 1459 Practolol, 7500 Trimetazidine, 9372
266.164698	$C_{12}H_{27}O_4P$	Tributyl Phosphate, 9298
266.178299	$C_{18}H_{22}N_2$	Cyclizine, 2716 Desipramine, 2881
266.962228	NO_3Tl	Thallium Nitrate, 8978
266.985384	$C_{12}H_7Cl_2NO_2$	2,6-Dichloroindophenol, 3042
267.023983	$C_8H_5N_5O_6$	Murexide, Free acid, 6132
267.053158	$C_{15}H_9NO_4$	1-Aminoanthraquinone-2-carboxylic Acid, 420
267.058155	$C_{14}H_{15}Cl_2N$	Chlornaphazine, 2077

Use in conjunction with The Merck Index, Ninth Edition

Molecular Weight	Empirical Formula	Compound Name, Monograph Number
267.067763	$C_{11}H_{13}N_3O_3S$	Sulfamoxole, 8716 Sulfisoxazole, 8748
267.095417	$C_9H_{17}NO_8$	Neuraminic Acid, 6301
267.096754	$C_{10}H_{13}N_5O_4$	Adenosine, 143 Vidarabine, Anhydrous, A15
267.121906	$C_{12}H_{17}N_3O_4$	Agaritine, 175
267.125929	$C_{17}H_{17}NO_2$	Apomorphine, 775
267.137162	$C_{16}H_{17}N_3O$	Lysergamide, 5451
267.149738	$[C_{17}H_{19}N_2O]^+$	Pyronine® Y, Cation, 7792
267.162314	$C_{18}H_{21}NO$	Azacyclonol, 909 Pipradrol, 7279
267.183444	$C_{15}H_{25}NO_3$	Butoxamine, 1513 Metoprolol, 6024
267.292601	$[C_{36}H_{74}N_2]^{2+}$	Triclobisonium (half mass), 9332
267.728772	Br_3P	Phosphorus Tribromide, 7167
267.750642	Cl_5Nb	Niobium Pentachloride, 6376
267.824604	CH_2I_2	Methylene Iodide, 5935
267.863722	$BaH_4O_4P_2$	Barium Hypophosphite, 983
267.869800	$C_4H_6As_2O_4$	Arsenoacetic Acid, 836
267.946077	$C_9H_7Cl_3O_3$	Silvex, 8278
267.988028	$C_{13}H_{10}Cl_2S$	Chlorbenside, 2047
268.005785	$C_{13}H_{10}Cl_2O_2$	Bis(p-chlorophenoxy)methane, 1266 Dichlorophen(e), 3047
268.037173	$C_{15}H_8O_5$	Coumestrol, 2549
268.044384	$C_9H_8N_4O_6$	Urfadyn, 9537
268.055151	$C_8H_{16}N_2O_4S_2$	Homocystine, 4616 Penicillamine Cysteine Disulfide, 6881
268.073559	$C_{16}H_{12}O_4$	Alizarin, Dimethyl ether, 231 Formononetin, 4102 Methylchrysin, 2254 Quinizarin, Dimethyl ether, 7886
268.080770	$C_{10}H_{12}N_4O_5$	Inosine, 4851, 4853
268.084792	$C_{15}H_{12}N_2O_3$	Hydrofuramide, 4683
268.109945	$C_{17}H_{16}O_3$	Eugenol Benzoate, 3843

Molecular Weight	Empirical Formula	Compound Name, Monograph Number
268.121178	$C_{16}H_{16}N_2O_2$	Isolysergic Acid, 5035 Lysergic Acid, 5452 Rugulovasines, 8053
268.125201	$C_{21}H_{16}$	3-Methylcholanthrene, 5919
268.134241	$[C_{28}H_{42}Cl_2N_4O_2]^{2+}$	Ambenonium (half mass), 383
268.146330	$C_{18}H_{20}O_2$	Diethylstilbestrol, 3113 Equilin, 3552
268.157563	$C_{17}H_{20}N_2O$	N,N'-Diethylcarbanilide, 3096 Michler's Ketone, 6042
268.167459	$C_{15}H_{24}O_4$	Acoric Acid, 116
268.276616	$C_{18}H_{36}O$	Oleyl Alcohol, 6678
268.287849	$[C_{17}H_{36}N_2]^{2+}$	Trimethidinium, 9375
268.313001	$C_{19}H_{40}$	Pristane, 7548
268.835741	$DyCl_3$	Dysprosium Chloride, 3457
268.931635	$C_7H_5Cl_2NO_4S$	Halazone, 4444
269.027728	$C_{12}H_{12}ClNO_2S$	Dansyl Chloride, 2801
269.029270	$C_{10}H_{11}N_3O_2S_2$	Sulfamethylthiazole, 8711 Sulfasomizole, 8737
269.040792	$C_{10}H_{17}Cl_2NOS$	Di-allate, 2929
269.049529	$C_6H_{11}N_3O_9$	Propatyl Nitrate, 7601
269.098669	$C_{15}H_{15}N_3S$	Azure B, Free base, 934
269.101171	$C_{11}H_{15}N_3O_5$	D-2-Deoxyribose, p-Nitrophenylhydrazone, 2869
269.116427	$C_{15}H_{15}N_3O_2$	Methyl Red, 5989
269.120450	$C_{20}H_{15}N$	β,β'-Dinaphthylamine, 3264
269.121454	$C_{10}H_{24}NO_3PS$	Tetram®, 8938
269.127660	$C_{14}H_{15}N_5O$	Benzodet, 1096
269.154642	$C_{15}H_{24}ClNO$	Caryophyllene, Nitrosochloride, 1868 α-Caryophyllene, Nitrosochloride, 1869
269.162708	$C_{14}H_{23}NO_4$	4,4'-Iminodicyclohexanecarboxylic Acid, 4812
269.177964	$C_{18}H_{23}NO$	2-(2-Biphenylyloxy)triethylamine, 1251 Mephenhydramine, 5679 p-Methylbenzhydryl 2-Dimethylaminoethyl Ether, 5898 Orphenadrine, 6714 Racefemine, 7907
269.781640	Cl_4Te	Tellurium Tetrachloride, 8870
269.830877	$K_2O_8S_2$	Potassium Persulfate, 7442

Use in conjunction with The Merck Index, Ninth Edition

Molecular Weight	Empirical Formula	Compound Name, Monograph Number
269.836890	Cl_3Ho	Holmium Chloride, 4602
269.837314	$Na_2O_6S_4$	Sodium Tetrathionate, 8462
269.909362	ClO_3Re	Rhenium Oxychloride, 7970
269.950411	$C_6H_6O_8S_2$	Tiron®, Free acid, 9178
269.953578	$C_6H_7ClN_2O_4S_2$	Clofenamide, 2339
269.986530	$C_6H_{11}IN_2O_2$	Iodival, 4884
270.024519	$C_9H_{10}N_4O_2S_2$	Sulfamethizole, 8707
270.036774	$C_{11}H_{15}BrN_2O$	Bromanylpromide, 1383
270.040615	O_2U	Uranium Dioxide, 9512
270.052824	$C_{15}H_{10}O_5$	Aloe-Emodin, 301 Apigenin, 764 Baicalein, 947 Emodin, 3512 Genistein, 4221 Purpurin, 2-Methyl ether, 7732 Sulfuretin, 8767
270.055991	$C_{15}H_{11}ClN_2O$	Mecloqualone, 5605
270.073953	$C_{12}H_{14}O_7$	Piscidic Acid, p-O-Methyl derivative, 7287
270.089209	$C_{16}H_{14}O_4$	Diphenic Acid, Dimethyl ester, 3317 Imperatorin, 4817
270.090183	$C_{10}H_{20}Cl_2N_2O_2$	N^6,N^6-Bis(2-chloroethyl)lysine, 1264
270.092376	$C_{16}H_{15}ClN_2$	Medazepam, 5612
270.103814	$C_{12}H_{18}N_2O_3S$	Tolbutamide, 9209
270.106494	$[C_{15}H_{16}N_3S]^+$	Tolonium, 9217
270.113506	$C_{13}H_{19}ClN_2O_2$	2-Chloroprocaine, 2140
270.119070	$C_{16}H_{18}N_2S$	Fenethazine, 3900
270.121572	$C_{12}H_{18}N_2O_5$	Bacilysin, 940 D-Gulose, Phenylhydrazone, 4429 Hypoglycine B, 4788 D-Mannose, Phenylhydrazone, 5578
270.125595	$C_{17}H_{18}O_3$	4-tert-Butylphenyl Salicylate, 1574
270.161980	$C_{18}H_{22}O_2$	Dihydroequilin, 3150 5,6,8-Estratrien-3β-ol-17-one, 3634 Estrone, 3636 Hexestrol, 4571 8-Isoestrone, 5025 Trenbolone, 9271
270.173213	$C_{17}H_{22}N_2O$	Doxylamine, 3430

270.

Molecular Weight	Empirical Formula	Compound Name, Monograph Number
270.255881	$C_{17}H_{34}O_2$	Isopropyl Myristate, 5075 Margaric Acid, 5581
270.292266	$C_{18}H_{38}O$	Stearyl Alcohol, 8583
270.836863	Cl_3Er	Erbium Chloride, 3556
271.044920	$C_{10}H_{13}N_3O_2S_2$	Subathizone, 8659
271.060649	$[C_{15}H_{11}O_5]^+$	Pelargonidin, 6865
271.066701	$C_{15}H_{13}NO_2S$	Metiazinic Acid, 6013
271.084458	$C_{15}H_{13}NO_4$	p-Nitrophenylacetic Acid, Benzyl ester, 6444 Phenetsal, 7021
271.099063	$C_{11}H_{17}N_3O_3S$	1-Butyl-3-metanilylurea, 1569 Carbutamide, 1835
271.116821	$C_{11}H_{17}N_3O_5$	Carbubarbital, 1834
271.120844	$C_{16}H_{17}NO_3$	Normorphine, 6520
271.133907	$C_{14}H_{22}ClNO_2$	Bupranolol, 1481
271.157229	$C_{17}H_{21}NO_2$	Apoatropine, 769 Desomorphine, 2885
271.193615	$C_{18}H_{25}NO$	Cyclazocine, 2710 Racemethorphan, 7908
271.214744	$C_{15}H_{29}NO_3$	Gardol®, Free acid, 4205
271.740033	Br_2Cd	Cadmium Bromide, 1602
271.813303	K_2O_4Te	Potassium Tellurate(VI), 7469
271.908337	Cl_2Hg	Mercuric Chloride, 5710
271.915657	MgO_4W	Magnesium Tungstate(VI), 5522
271.962478	C_2AuN_2Na	Sodium Dicyanoaurate(I), 8357
272.013680	$C_6H_{12}N_2O_6S_2$	Cystine S-Dioxides, 2781
272.025564	$C_{13}H_8N_2O_3S$	Nitroscanate, A11
272.032088	$C_{14}H_8O_6$	Quinalizarin, 7836
272.046693	$C_{10}H_{12}N_2O_5S$	7-Aminocephalosporanic Acid, 438
272.068474	$C_{15}H_{12}O_5$	Naringenin, 6251
272.089603	$C_{12}H_{16}O_7$	Arbutin, 800
272.090940	$C_{13}H_{12}N_4O_3$	Vacor, 9564
272.119464	$C_{12}H_{20}N_2O_3S$	Sotalol, 8499
272.131349	$C_{19}H_{16}N_2$	Sempervirine, 8194

Use in conjunction with The Merck Index, Ninth Edition

Molecular Weight	Empirical Formula	Compound Name, Monograph Number
272.144617	$C_{14}H_{24}O_3S$	3,6-Bis(1,1-dimethylethyl)-1-naphthalenesulfonic Acid, 3001 3,7-Bis(1,1-dimethylethyl)-1-naphthalenesulfonic Acid, 3001
272.177630	$C_{18}H_{24}O_2$	3-Epihexahydroequilenin, 3542 Estradiol, 3626 α-Estradiol, 3627 Hexahydroequilenin, 4555 Isoestradiol, 5024 Nimbiol, 6369
272.188864	$C_{17}H_{24}N_2O$	Dimethisoquin, 3204
272.998229	$C_9H_{12}AsNO_4$	3-Methylarsacetin, 5896
273.019435	$C_3H_8HgO_2$	Merisoprol Hg 197, 5746
273.022643	$C_{11}H_{12}ClNO_3S$	Chlormezanone, 2075
273.111341	$C_{14}H_{15}N_3O_3$	Euparin, Semicarbazone, 3846
273.136494	$C_{16}H_{19}NO_3$	α-Erythroidine, 3600 β-Erythroidine, 3601 Evadol, 3854 Lunacrine, 5420
273.172879	$C_{17}H_{23}NO_2$	α-Terpineol, Phenylurethan, 8886 Tilidine, 9167
273.184113	$C_{16}H_{23}N_3O$	DL-ar-Turmerone, Semicarbazone, 9477
273.196689	$[C_{34}H_{50}N_4O_2]^{2+}$	Benzoquinonium (half mass), 1116
273.254204	$[C_{15}H_{33}N_2O_2]^+$	Dibutoline, 3004, 3005
273.794934	CAg_2O_3	Silver Carbonate, 8248
273.806167	$Ag_2N_2O_2$	Silver Hyponitrite, 8257
273.840783	Cl_3Tm	Thulium Chloride, 9135
273.847631	$C_4H_4Br_2O_4$	2,3-Dibromosuccinic Acid, 2997
273.870531	$C_4H_6N_2S_4Zn$	Zineb, 9835
274.028484	$C_8H_{19}O_2PS_3$	Disulfoton, 3383
274.047738	$C_{14}H_{10}O_6$	Phenicin, 7025
274.063969	$C_{12}H_{16}Cl_2N_2O$	Neburon, 6264
274.080972	$C_{11}H_{18}N_2O_2S_2$	Thionarcon, 9081
274.084124	$C_{15}H_{14}O_5$	Guaiacol Carbonate, 4401 Methysticin, 6012 Phloretin, 7136 Uvinul® D49, 9560
274.123676	$C_{16}H_{19}ClN_2$	Dexchlorpheniramine, 2901 Chlorpheniramine, 2169

274.

Molecular Weight	Empirical Formula	Compound Name, Monograph Number
274.156895	$C_{17}H_{22}O_3$	Bornyl Salicylate, 1355 Podocarpic Acid, 7321
274.189258	$C_{13}H_{26}N_2O_4$	Tybamate, 9483
274.193280	$C_{18}H_{26}O_2$	Isanic Acid, 4950 Nandrolone, 6185
274.204514	$C_{17}H_{26}N_2O$	Phenampromid(e), 6995
274.869310	N_3O_9Y	Yttrium Nitrate, 9773
274.975782	C_3AuN_3	Gold Tricyanide, 4372
274.977493	$C_8H_{10}AsNO_5$	Acetarsone, 39, 40, 3434 Glycarsamide, 4315
275.028602	$C_{10}H_{13}NO_4S_2$	Meticrane, 6014
275.054221	$C_{12}H_9N_3O_5$	Nifuroxazide, 6359
275.055876	$C_{10}H_{14}NO_6P$	Diethyl p-Nitrophenyl Phosphate, 3107
275.065454	$C_{11}H_9N_5O_4$	Acetylfuratrizine, 87
275.080243	$C_{15}H_{17}NS_2$	Tipepidine, 9177
275.094629	$C_{18}H_{13}NO_2$	Tetrophan, 8962
275.115758	$C_{15}H_{17}NO_4$	Actiphenol, 139
275.139880	$C_{11}H_{22}N_3O_3P$	Meturedepa, 6031
275.152144	$C_{16}H_{21}NO_3$	Annotinine, 711 Dihydro-β-erythroidine, 3152 Homatropine, 4606 Norhyoscyamine, 6512
275.163377	$C_{15}H_{21}N_3O_2$	Mexolamine, 6039 Physostigmine, 7189
275.167400	$C_{20}H_{21}N$	Cyclobenzaprine, 2718
275.188529	$C_{17}H_{25}NO_2$	Falicain, Free base, 3872 Proheptazine, 7571 Promedol, 7580
275.199763	$C_{16}H_{25}N_3O$	Propiram, 7621
275.842331	$AlCl_3O_9$	Aluminum Chlorate, 333
275.878537	$C_8H_6Br_2O$	p-Bromophenacyl Bromide, 1429
275.884896	$AuBr$	Gold Monobromide, 4356
275.940030	F_5Ta	Tantalum Pentafluoride, 8833
275.970927	$C_4H_8N_2O_9Ti$	Ammonium Titanium Oxalate, 596
275.978591	$C_{13}H_9BrO_2$	p-Bromobenzoic Acid, Phenyl ester, 1407

Use in conjunction with The Merck Index, Ninth Edition

Molecular Weight	Empirical Formula	Compound Name, Monograph Number
276.032000	$C_{12}H_{14}Cl_2O_3$	(2,4-Dichlorophenoxy)acetic Acid, Butyl ester, 3049
276.033542	$C_{10}H_{13}ClN_2O_3S$	Chlorpropamide, 2176
276.045631	$C_{14}H_{12}O_4S$	Vitamin K-S(II), 9690
276.051125	$C_9H_{13}N_2O_6P$	Pyridoxal 5-Phosphate, O-Methyloxime, 7763
276.056864	$C_{13}H_{12}N_2O_3S$	Sulfabenzamide, 8689
276.072177	$C_{11}H_{11}F_3N_2O_3$	Flutamide, 4083
276.082703	$C_8H_{16}N_6OS_2$	Kethoxal, Bis(thiosemicarbazone), 5149
276.099774	$C_{15}H_{16}O_5$	Fuscin, 4170 Lactucin, 5195 Dihydromethysticin, 6012
276.120903	$C_{12}H_{20}O_7$	Ethyl Citrate, 3719
276.126263	$C_{18}H_{16}N_2O$	1-Xylylazo-2-naphthol, 9751
276.132137	$C_{11}H_{20}N_2O_6$	L-Saccharopine, 8072
276.141156	$C_{15}H_{26}Cl_2$	Caryophyllene, Dihydrochloride, 1868
276.147393	$C_{15}H_{20}N_2O_3$	Pixifenide, 7297
276.162649	$C_{19}H_{20}N_2$	Mebhydroline, 5596
276.172545	$C_{17}H_{24}O_3$	Cyclandelate, 2708 Menthyl Salicylate, 5668
276.183778	$C_{16}H_{24}N_2O_2$	Molindone, 6067 N,N,N',N'-Tetraethylphthalamide, 8922
276.196354	$[C_{17}H_{26}NO_2]^+$	Methylethoheptazinium, 3668
276.208930	$C_{18}H_{28}O_2$	Bolandiol, 1336
276.220164	$C_{17}H_{28}N_2O$	Etidocaine, 3819
276.256549	$[C_{18}H_{32}N_2]^{2+}$	N,N'-Dimethylpimetinium, 7233
276.963948	$C_{12}H_9AsClN$	Phenarsazine Chloride, 6999
277.006119	$C_{14}H_9Cl_2NO$	3,9-Dichloro-7-methoxyacridine, 3044
277.017383	$C_9H_{12}NO_5PS$	Fenitrothion, 3905
277.038194	$C_9H_7N_7O_2S$	Azathioprine, 915
277.091000	$C_9H_{15}N_3O_7$	Lycomarasmine, 5437
277.095023	$C_{14}H_{15}NO_5$	Folescutol, 4084
277.095893	$C_{15}H_{19}NS_2$	Ethylmethylthiambutene, 3767
277.167794	$C_{16}H_{23}NO_3$	Estil, 3625

Molecular Weight	Empirical Formula	Compound Name, Monograph Number
277.183050	$C_{20}H_{23}N$	Amitriptyline, 504 Maprotiline, 5579
277.204179	$C_{17}H_{27}NO_2$	4-(Dimethylamino)benzoic Acid, 2-Ethylhexyl ester, 3219
277.276950	$C_{19}H_{35}N$	Perhexiline, 6954
277.738871	Br_2Sn	Stannous Bromide, 8561
277.751995	As_2O_4Zn	Zinc Meta-arsenite, 9805
277.757748	C_2HBr_3O	Bromal, 1381
277.783938	$CrH_2O_6Zn_2$	Zinc Chromate(VI) Hydroxide, 9790
277.791205	Br_2MgO_6	Magnesium Bromate, 5481
277.792494	In_2O_3	Indium Oxide, 4832
277.793999	I_2Mg	Magnesium Iodide, 5494
277.794133	$C_3H_5Br_3$	1,2,3-Tribromopropane, 9294
277.798000	$CaMn_2O_8$	Calcium Permanganate, 1689
277.857802	$C_7H_4Br_2O_2$	3,5-Dibromosalicylaldehyde, 2995
277.914346	Cl_2Pb	Lead Chloride, 5254
278.020026	$C_{10}H_{15}O_3PS_2$	Fenthion, 3919
278.023062	$C_3H_8HgO_2$	Merisoprol Hg 197 (Hg 202), 5746
278.026604	$C_9H_{15}BrN_2O_3$	Acetylcarbromal, 76
278.065120	$C_{11}H_{10}N_4O_5$	α-Methylene Butyrolactone, 2,4-Dinitrophenylhydrazine deriv, 5931
278.079038	$C_{14}H_{14}O_6$	Radicinin, Monoacetate, 7909
278.083748	$C_{12}H_{14}N_4O_2S$	Sulfamethazine, 8706 Sulfisomidine, 8747
278.094294	$C_{18}H_{14}O_3$	Cinnamic Acid, Anhydride, 2288
278.109551	$C_{22}H_{14}$	1,2:5,6-Dibenzanthracene, 2971 2,3:6,7-Dibenzphenanthrene, 2974 Pentacene, 6899 Picene, 7202
278.111401	$C_{10}H_{18}N_2O_7$	Versen-Ol®, Free acid, 9622
278.115424	$C_{15}H_{18}O_5$	Coriamyrtin, 2507
278.126657	$C_{14}H_{18}N_2O_4$	α-Ribazole, 7992
278.130029	$C_{11}H_{22}N_2O_4S$	Pantetheine, 6816
278.137891	$C_{13}H_{18}N_4O_3$	Pentoxifylline, 6931

Use in conjunction with The Merck Index, Ninth Edition

Molecular Weight	Empirical Formula	Compound Name, Monograph Number
278.151809	$C_{16}H_{22}O_4$	n-Butyl Phthalate, 1575 Hirsutic Acid C, Methyl ester, 4592
278.163043	$C_{15}H_{22}N_2O_3$	Tolycaine, 9239
278.178299	$C_{19}H_{22}N_2$	Triprolidine, 9414
278.188195	$C_{17}H_{26}O_3$	(p-Nonylphenoxy)acetic Acid, 6491
278.190875	$[C_{20}H_{24}N]^+$	Diphemanil, 3309
278.199428	$C_{16}H_{26}N_2O_2$	Amydricaine, 629 Dimethocaine, 3207
278.199428	$[C_{32}H_{52}N_4O_4]^{2+}$	Demecarium (half mass), 2855
278.212004	$[C_{17}H_{28}NO_2]^+$	Methylcyclexanonium, 2712
278.224580	$C_{18}H_{30}O_2$	Estrane-3α,17α-diol, 3633 Linolenic Acid, 5344
278.775021	Cr_2CuO_7	Cupric Dichromate(VI), 2637
278.843059	$BeK_2O_8S_2$	Beryllium Potassium Sulfate, 1202
278.845431	Cl_3Yb	Ytterbium Chloride, 9772
278.939245	$C_7H_6INO_3$	Iopax®, Free acid, 4917
279.089543	$C_{17}H_{13}NO_3$	Isatophan, 4953 Monoacetylviridicatin, 9659
279.125929	$C_{18}H_{17}NO_2$	Aporeine, 779
279.133140	$C_{12}H_{17}N_5O_3$	7-Morpholinomethyltheophylline, 6118
279.162314	$C_{19}H_{21}NO$	Cidoxepin, 2267 Doxepin, 3427
279.169525	$C_{13}H_{21}N_5O_2$	Etamiphyllin, 3639
279.176920	$C_{15}H_{25}N_3S$	β-Irone, Thiosemicarbazone, 4945 γ-Irone, Thiosemicarbazone, 4946
279.183444	$C_{16}H_{25}NO_3$	Moxisylyte, 6122 Propinal, 7608
279.198700	$C_{20}H_{25}N$	Fenpiprane, 3914
279.207253	$[C_{16}H_{27}N_2O_2]^+$	N-Methylhydroxylupaninium, 4733
279.735240	Se_2Sn	Stannic Selenide, 8557
279.773398	$C_2H_3Br_3O$	Tribromoethanol, 9291
279.774567	Cr_2O_7Zn	Zinc Dichromate(VI), 9793
279.847343	Cl_3Lu	Lutetium Chloride, 5429
279.872584	BaF_6Si	Barium Hexafluorosilicate, 981

279.

Molecular Weight	Empirical Formula	Compound Name, Monograph Number
279.923261	$C_7H_5IO_4$	Amidoxyl®, Free acid, 406
279.954902	$BiNaO_3$	Sodium Bismuthate(V), 8329
279.997636	$C_{11}H_8N_2O_3S_2$	Firefly Luciferin, 4002
280.063012	$C_{11}H_{12}N_4O_3S$	Sulfalene, 8702 Sulfameter, 8705 Sulfamethoxypyridazine, 8710
280.109945	$C_{18}H_{16}O_3$	Phenprocoumon, 7060
280.132411	$C_{16}H_{16}N_4O$	Hydroxystilbamidine, 4765
280.146330	$C_{19}H_{20}O_2$	Cyclofenil, Free diol, 2725 Equilenin, Methyl ether, 3551
280.178693	$C_{15}H_{24}N_2O_3$	Hydroxytetracaine, 4767 Mefexamide, 5623
280.181373	$[C_{18}H_{22}N_3]^+$	Flavicid, Cation, 4006
280.193949	$C_{19}H_{24}N_2$	Bamipine, 958 Histapyrrodine, 4596 Imipramine, 4813
280.203845	$C_{17}H_{28}O_3$	Juvenile Hormones, C-17 JH, 5124
280.215078	$C_{16}H_{28}N_2O_2$	Epipropidine, 3545 Tromantadine, 9434
280.240230	$C_{18}H_{32}O_2$	Chaulmoogric Acid, 2000 Linoleic Acid, 5343 Propylure, 7659
280.922692	$C_6FeN_6Na_3$	Sodium Ferricyanide, 8365
280.998841	$C_7H_{14}Cl_3NO_4$	Chloral Betaine, 2030
281.017007	$C_6H_{12}N_5O_2PS_2$	Menazon, 5657
281.064785	$C_{11}H_{11}N_3O_6$	Nitractin, 6386
281.066363	$C_{14}H_{10}F_3NO_2$	Flufenamic Acid, 4023
281.068808	$C_{16}H_{11}NO_4$	1-Aminoanthraquinone-2-carboxylic Acid, Methyl ester, 420
281.080041	$C_{15}H_{11}N_3O_3$	Nitrazepam, 6391
281.092930	$C_{12}H_{16}N_3O_3P$	Benzodepa, 1095
281.105193	$C_{17}H_{15}NO_3$	Indoprofen, A10
281.126323	$C_{14}H_{19}NO_5$	Trimetozine, 9390
281.137556	$C_{13}H_{19}N_3O_4$	Dipropalin®, 3359
281.138894	$C_{14}H_{15}N_7$	Diminazene, 3258
281.141579	$C_{18}H_{19}NO_2$	Apocodeine, 770

Use in conjunction with The Merck Index, Ninth Edition

Molecular Weight	Empirical Formula	Compound Name, Monograph Number
281.152812	$C_{17}H_{19}N_3O$	Phentolamine, 7064
281.162708	$C_{15}H_{23}NO_4$	Cycloheximide, 2733
281.177964	$C_{19}H_{23}NO$	Cinnamedrine, 2287 Diphenylpyraline, 3346, 7280
281.214350	$C_{20}H_{27}N$	Alverine, 374
281.271865	$C_{18}H_{35}NO$	Petroselinic Acid, Amide, 6974
281.794316	$AgIO_3$	Silver Iodate, 8258
281.801415	BaO_4Se	Barium Selenate, 998
281.808169	O_8S_2Zr	Zirconium Sulfate, 9848
281.813116	C_6Cl_6	Hexachlorobenzene, 4544
281.822543	F_6K_2Zr	Potassium Hexafluorozirconate(IV), 7421
282.049671	$C_{12}H_{14}N_2O_2S_2$	Ujothion, 9499
282.050456	$C_9H_{15}O_8P$	Bomyl, 1344
282.052824	$C_{16}H_{10}O_5$	Pseudobaptigenin, 7695
282.061612	$C_{13}H_9F_3N_2O_2$	Niflumic Acid, 6355
282.089209	$C_{17}H_{14}O_4$	Rubiadin, Dimethyl ether, 8039
282.100442	$C_{16}H_{14}N_2O_3$	Bendazac, 1039 Elbon®, 3490
282.136828	$C_{17}H_{18}N_2O_2$	Isolysergic Acid, Methyl ester, 5035
282.149404	$[C_{18}H_{20}NO_2]^+$	Methylapomorphinium, 776
282.161980	$C_{19}H_{22}O_2$	Equilin, Methyl ether, 3552 Mestilbol, 5763
282.183110	$C_{16}H_{26}O_4$	Acoric Acid, Methyl ester, 116
282.209599	$C_{19}H_{26}N_2$	Quebrachamine, 7817
282.222175	$[C_{20}H_{28}N]^+$	Emepronium, 3506
282.243304	$[C_{17}H_{32}NO_2]^+$	Methylanisotropinium, 706
282.255881	$C_{18}H_{34}O_2$	Elaidic Acid, 3485 Oleic Acid, 6674 Petroselinic Acid, 6974 Vaccenic Acid, 9562
282.807678	$CrKO_8S_2$	Chromic Potassium Sulfate, 2227
283.078421	$[C_{12}H_{16}ClN_4S]^+$	Beclotiamine, Cation, 1026
283.091669	$C_{10}H_{13}N_5O_5$	Guanosine, 4419
283.120844	$C_{17}H_{17}NO_3$	Sparsiflorine, 8509

283.

Molecular Weight	Empirical Formula	Compound Name, Monograph Number
283.193615	$C_{19}H_{25}NO$	2-(Benzhydryloxy)triethylamine, 1082 Levallorphan, 5309
283.723686	Br_3OP	Phosphorus Oxybromide, 7158
283.843214	$C_4H_2Br_2N_2O_3$	Dibromin®, 2984
283.892746	BrTl	Thallium Bromide, 8972
283.907148	$FeH_8N_2O_8S_2$	Ammonium Ferrous Sulfate, 544
283.969228	$C_7H_9ClN_2O_4S_2$	Disulfamide, 3381
283.970254	F_4Pb	Lead Tetrafluoride, 5282
284.013475	$C_{10}H_9ClN_4O_2S$	Metachloridine, 5769 Sulfachloropyridazine, 8691
284.032088	$C_{15}H_8O_6$	Munjistin, 6129 Rhein, 7966
284.040169	$C_{10}H_{12}N_4O_2S_2$	Sulfaethidole, 8698
284.060428	$C_6H_{12}N_4O_9$	Trolnitrate, 9433
284.068474	$C_{16}H_{12}O_5$	Acacetin, 9 Emodin, 3-Methyl ether, 3512 Genistein, 4′-Methyl ether, 4221 Oroxylin A, 6713 Prunetin, 7693 Purpurin, 2,4-Dimethyl ether, 7732
284.071641	$C_{16}H_{13}ClN_2O$	Diazepam, 2961 Mazindol, 5591
284.075684	$C_{10}H_{12}N_4O_6$	Xanthosine, 9729
284.089603	$C_{13}H_{16}O_7$	Helicin, 4484 Vacciniin, 9563
284.090880	$C_6H_{16}FeN_{10}$	Ammonium Ferrocyanide, 543
284.092596	$C_{12}H_{17}N_2O_4P$	Psilocybin, 7712
284.098335	$C_{16}H_{16}N_2OS$	Ahistan, 179
284.104859	$C_{17}H_{16}O_4$	Lapachol, Acetate, 5211
284.119464	$C_{13}H_{20}N_2O_3S$	Etozolin, 3834
284.122144	$[C_{16}H_{18}N_3S]^+$	Methylene Blue, Cation, 5929
284.134721	$C_{17}H_{20}N_2S$	Isopromethazine, 5062 Promazine, 7578 Promethazine, 7581
284.152478	$C_{17}H_{20}N_2O_2$	Tropicamide, 9445
284.177630	$C_{19}H_{24}O_2$	Estrone, Methyl ether, 3636 8-Isoestrone, dl-Methyl ether, 5025 Methyltrienolone, 6003

Use in conjunction with The Merck Index, Ninth Edition

Molecular Weight	Empirical Formula	Compound Name, Monograph Number
284.201440	$[C_{19}H_{26}NO]^+$	Bibenzonium, 1224
284.214016	$C_{20}H_{28}O$	Lynestrenol, 5447 Vitamin A_2, 9667 Vitamin A Aldehyde, 9668
284.225249	$C_{19}H_{28}N_2$	Eurazyl, Free base, 3852 Iprindole, 4930
284.271531	$C_{18}H_{36}O_2$	Stearic Acid, 8582
284.331726	$[C_{19}H_{42}N]^+$	Cetrimonium, 1980-1982
284.964477	$C_6H_8ClN_3O_4S_2$	Chloraminophenamide, 2039
284.987799	$C_9H_7N_3O_4S_2$	p-Nitrosulfathiazole, 6464
285.027337	$C_{14}H_7NO_6$	Alizarine Orange, 233
285.041942	$C_{10}H_{11}N_3O_5S$	Nifuratel, 6358
285.063723	$C_{15}H_{11}NO_5$	Bostrycoidin, 1365
285.096085	$C_{11}H_{15}N_3O_6$	D-Ribulose, o-Nitrophenylhydrazone, 8001
285.103480	$C_{13}H_{19}NO_4S$	Probenecid, 7551
285.111341	$C_{15}H_{15}N_3O_3$	Verazide, 9612
285.129970	$C_{16}H_{19}N_3S$	Isothipendyl, 5092 Prothipendyl, 7674
285.136494	$C_{17}H_{19}NO_3$	Chavicine, 2001 Coclaurine, 2416 Hydromorphone, 4700 Morphine, 6108, 6109-6113, 6115, 6253 Norcodeine, 6500 Piperine, 7266
285.172879	$C_{18}H_{23}NO_2$	5,6,8-Estratrien-3β-ol-17-one, Oxime, 3634 Medrylamine, 5619
285.176251	$C_{15}H_{27}NO_2S$	Lethane® 60, 5299
285.184113	$C_{17}H_{23}N_3O$	Piperylon, 7273 Pyrilamine, 7767, 7736
285.190009	$C_{17}H_{24}BNO_2$	8-Quinolineboronic Acid, Butyl diester, 7890
285.209265	$C_{19}H_{27}NO$	Pentazocine, 6921
285.907556	$H_8N_2NiO_8S_2$	Ammonium Nickel Sulfate, 562
285.946951	$C_8H_8Cl_2O_5S$	Sesone®, Free acid, 8218
285.962207	$C_{12}H_8Cl_2O_2S$	Fenticlor, 3920
285.995303	$C_{10}H_{11}BrN_2O_3$	Brallobarbital, 1370
286.012762	$C_{14}H_{10}Cl_2F_2$	DFDD, 2913

286.

Molecular Weight	Empirical Formula	Compound Name, Monograph Number
286.018563	$C_8H_6N_4O_8$	Alloxantin, 275
286.035530	O_3U	Uranium Trioxide, 9518
286.047738	$C_{15}H_{10}O_6$	Baptigenin, 959 Datiscetin, 2811 Fisetin, 4003 Kaempferol, 5126 Luteolin, 5428 Scutellarein, 8165
286.050905	$C_{15}H_{11}ClN_2O_2$	Oxazepam, 6751
286.058972	$C_{14}H_{10}N_2O_5$	Cinnabarine, 2285
286.084124	$C_{16}H_{14}O_5$	Brazilin, 1375 Sakuranetin, 8078
286.087496	$C_{13}H_{18}O_5S$	Ethofumesate, 3667
286.098729	$C_{12}H_{18}N_2O_4S$	Anisylbutamide, 709
286.105253	$C_{13}H_{18}O_7$	Methylarbutin, 5895 Salicin, 8085
286.120509	$C_{17}H_{18}O_4$	Diphenolic Acid, 3320
286.120922	$C_{16}H_{22}OSi_2$	1,1,3,3-Tetramethyl-1,3-diphenyldisiloxane, 8943
286.131743	$C_{16}H_{18}N_2O_3$	Pilosine, 7228
286.132368	$C_8H_{24}N_4O_3P_2$	Octamethyl Pyrophosphoramide, 6558
286.150371	$C_{17}H_{22}N_2S$	Thenaldine, 8990
286.154209	$C_{14}H_{18}N_6O$	2,6-Diamino-2'-butyloxy-3,5'-azopyridine, 2935
286.156895	$C_{18}H_{22}O_3$	Methallenestril, 5802
286.168128	$C_{17}H_{22}N_2O_2$	Naphthocaine, 6207
286.179361	$C_{16}H_{22}N_4O$	Thonzylamine, 9114
286.193280	$C_{19}H_{26}O_2$	4-Androstene-3,17-dione, 675 Boldenone, 1338 Estradiol, 3-Methyl ether, 3626 Nimbiol, (+)-Methyl ether, 6369
286.227655	$[C_{35}H_{60}N_2O_4]^{2+}$	Pancuronium (half mass), 6814
286.229666	$C_{20}H_{30}O$	Neovitamin A, 6292 Vitamin A, 9666
286.262029	$[C_{16}H_{34}N_2O_2]^{2+}$	Dicolin, Cation, 3062
286.905407	$CoH_8N_2O_8S_2$	Ammonium Cobaltous Sulfate, 539
287.013812	$C_7H_5N_5O_8$	Nitramine, 6389
287.054221	$C_{13}H_9N_3O_5$	Alizarine Yellow R, 234 Metachrome Yellow, Free acid, 5770

Use in conjunction with The Merck Index, Ninth Edition

Molecular Weight	Empirical Formula	Compound Name, Monograph Number
287.055563	$[C_{15}H_{11}O_6]^+$	Cyanidin, 2695
287.057593	$C_{10}H_{13}N_3O_5S$	Nifurtimox, 6363
287.059248	$C_8H_{18}NO_6PS$	(−)-Felinine Phosphate, 3883
287.070451	$C_{11}H_{15}Cl_2N_5$	Chlorproguanil, 2174
287.105862	$C_{18}H_{13}N_3O$	Rutecarpine, 8057
287.115758	$C_{16}H_{17}NO_4$	Lunine, 5422 Lycorine, 5444 Mephenesin Nicotinate, 5678
287.129672	$[C_{18}H_{15}N_4]^+$	Phenosafranin, Cation, 7050
287.152144	$C_{17}H_{21}NO_3$	Dihydromorphine, 3155 Galanthamine, 4182 Ritodrine, 8015
287.154824	$[C_{20}H_{19}N_2]^+$	Methylsempervirinium, 8194
287.167400	$C_{21}H_{21}N$	Cyproheptadine, 2771
287.188529	$C_{18}H_{25}NO_2$	Allylprodine, 293
287.199763	$C_{17}H_{25}N_3O$	8-(3-Diethylaminopropylamino)-6-methoxyquinoline, 3088 Proxazole, 7690
287.209659	$C_{15}H_{29}NO_4$	Dioxamate, 3299
287.224915	$C_{19}H_{29}NO$	Cycrimine, 2761 Procyclidine, 7561
287.742901	Br_2Te	Tellurium Dibromide, 8865
287.860067	$C_6H_6Cl_6$	Lindane, 5341
287.886181	$C_5H_8N_2S_4Zn$	Propineb, 7609
287.893202	CaO_4W	Calcium Tungstate(VI), 1717
287.936416	C_2AuKN_2	Potassium Dicyanoaurate(I), 7407
287.951936	$C_2H_3AuO_2S$	Myoral, Free acid, 6155
287.976232	$C_{10}H_8O_6S_2$	Armstrong's Acid, 814 1,6-Naphthalenedisulfonic Acid, 6197 2,6-Naphthalenedisulfonic Acid, 6198 2,7-Naphthalenedisulfonic Acid, 6199
288.010954	$C_{10}H_{13}BrN_2O_3$	Propallylonal, 7589
288.036852	$C_8H_{14}N_4NiO_4$	Nickel Dimethylglyoxime, 6321
288.044134	$C_9H_{21}O_2PS_3$	Terbufos, 8878
288.059366	$C_{10}H_{12}N_2O_8$	Orotidine, 6712

288.

Molecular Weight	Empirical Formula	Compound Name, Monograph Number
288.063388	$C_{15}H_{12}O_6$	Eriodictyol, 3591 Fustin, 4173 5,5'-Methylenedisalicylic Acid, 5934
288.089878	$C_{18}H_{12}N_2O_2$	Xanthocillin X, 9724
288.096622	$C_{12}H_{20}N_2O_2S_2$	Methitural, Free acid, 5850
288.099774	$C_{16}H_{16}O_5$	Alkannin, 236 Asebogenin, 858
288.102941	$C_{16}H_{17}ClN_2O$	Tetrazepam, 8959
288.114379	$C_{12}H_{20}N_2O_4S$	Soterenol, 8500
288.122241	$[C_{14}H_{16}N_4O_3]^{2+}$	Obidoxime, 6548 Piromidic Acid, 7286
288.159969	$[C_{17}H_{22}NO_3]^+$	Methylerythroidinium, 3600, 3601 Lunasine, 5421
288.172545	$C_{18}H_{24}O_3$	Doisynolic Acid, 3416 16-Epiestriol, 3541 Estriol, 3635 Podocarpic Acid, Methyl ester, 7321
288.183778	$C_{17}H_{24}N_2O_2$	Phenglutarimide, 7023
288.208930	$C_{19}H_{28}O_2$	Dehydroepiandrosterone, 2846 Normethandrone, 6518 Testosterone, 8890
288.220164	$C_{18}H_{28}N_2O$	Bupivacaine, 1480
288.230060	$C_{16}H_{32}O_4$	Diethylene Glycol Monolaurate, 3101
288.245316	$C_{20}H_{32}O$	Ethylestrenol, 3741
288.281701	$C_{21}H_{36}$	Allopregnane, 250 Pregnane, 7522
288.695518	Br_3Cr	Chromic Bromide, 2217
289.028791	$C_{10}H_{12}ClN_3O_3S$	Quinethazone, 7845
289.086957	$C_{16}H_{16}ClNO_2$	Nicoclonate, 6333
289.110279	$C_{19}H_{15}NO_2$	Allyl 2-Phenylcinchoninate, 292
289.116152	$C_{12}H_{19}NO_7$	Triacetylmycosamine, 6153
289.123342	$C_{17}H_{20}ClNO$	Chlophedianol, 2022
289.127386	$C_{11}H_{19}N_3O_6$	Wildfire Toxin, 9711
289.131408	$C_{16}H_{19}NO_4$	Benzoylecgonine, 1124
289.134575	$C_{16}H_{20}ClN_3$	Chloropyramine, 2145

Use in conjunction with The Merck Index, Ninth Edition

Molecular Weight	Empirical Formula	Compound Name, Monograph Number
289.167794	$C_{17}H_{23}NO_3$	Atropine, 892, 894 Hyoscyamine, 4780 Lycoramine, 5443 Mesembrine, 5753
289.204179	$C_{18}H_{27}NO_2$	Aminohexan, 448 Butaverine, 1500 Caramiphen, 1774, 1775 Dyclonine, 3450 Pentapiperide, 6919
289.240565	$C_{19}H_{31}NO$	Bencyclane, 1038
289.792391	O_3Sb_2	Antimony Trioxide, 749
289.918844	$C_7H_3IN_2O_3$	Nitroxynil, 6477
290.039501	$C_{10}H_{14}N_2O_4S_2$	Sulthiame, 8787
290.042653	$C_{14}H_{10}O_7$	Citromycetin, 2309
290.068430	$C_8H_{20}O_7P_2$	Tetraethyl Pyrophosphate, 8923
290.079038	$C_{15}H_{14}O_6$	Catechin, 1898 Javanicin, 5112 Plumericin, 7317
290.094294	$C_{19}H_{14}O_3$	Aurin, 901
290.105528	$C_{18}H_{14}N_2O_2$	5-Methyl-5-(3-phenanthryl)hydantoin, 5975
290.118591	$C_{16}H_{19}ClN_2O$	Carbinoxamine, 1802 Rotoxamine, 8032
290.137891	$C_{14}H_{18}N_4O_3$	Benomyl, 1047 Trimethoprim, 9377
290.151809	$C_{17}H_{22}O_4$	Illudin M, Monoacetate, 4806
290.175619	$[C_{17}H_{24}NO_3]^+$	Methylhomatropinium, 4607
290.188195	$C_{18}H_{26}O_3$	4-Hydroxy-19-nortestosterone, 4748
290.220558	$[C_{14}H_{30}N_2O_4]^{2+}$	Succinylcholine, 8674-8676
290.224580	$C_{19}H_{30}O_2$	Androstenediol, 673 5-Androstene-3β,16α-diol, 674 Androsterone, 676 Epiandrosterone, 3535 Stanolone, 8573
290.248390	$[C_{19}H_{32}NO]^+$	Mepiperphenidol, Cation, 5686
291.033033	$C_{10}H_{14}NO_5PS$	Parathion, 6838
291.053158	$C_{17}H_9NO_4$	Alizarine Blue, 232
291.056530	$C_{14}H_{13}NO_4S$	p-(Benzylsulfonamido)benzoic Acid, 1172
291.078997	$C_{12}H_{13}N_5O_2S$	Sulfamidochrysoidine, 8714

Molecular Weight	Empirical Formula	Compound Name, Monograph Number
291.079110	$C_{12}H_{19}ClNO_3P$	Crufomate, 2598
291.089543	$C_{18}H_{13}NO_3$	Naptalam, 6246
291.096754	$C_{12}H_{13}N_5O_4$	Toyocamycin, 9257
291.111543	$C_{16}H_{21}NS_2$	Thiambutene, 9020
291.125929	$C_{19}H_{17}NO_2$	α-(p-Methoxyphenyl)-α-2-pyridyl-p-cresol, 5874 Neocinchophen, 6270
291.140534	$C_{15}H_{21}N_3OS$	Zolamine, 9851
291.147058	$C_{16}H_{21}NO_4$	Convolvine, 2488
291.158292	$C_{15}H_{21}N_3O_3$	Geneserine®, 4220
291.173548	$C_{19}H_{21}N_3$	Perlapine, 6964
291.183444	$C_{17}H_{25}NO_3$	Cyclopentolate, 2750 Eucatropine, 3840 Pecilocin, 6860
291.198700	$C_{21}H_{25}N$	Melitracene, 5643
291.933172	$C_7H_7SbO_5$	p-Stibonobenzoic Acid, 805
291.995085	N_6Pb	Lead Azide, 5247
292.049093	$C_9H_8N_8O_2S$	Thiamiprine, 9032
292.073559	$C_{18}H_{12}O_4$	Karanjin, 5134
292.088164	$C_{14}H_{16}N_2O_3S$	2-p-Sulfanilylanilinoethanol, 8721
292.090666	$C_{10}H_{16}N_2O_8$	Edetic Acid, 3476, 1660, 3375, 6090, 7256, 8362, 9421, 9622
292.094688	$C_{15}H_{16}O_6$	Picrotoxinin, 7219, 7218
292.105922	$C_{14}H_{16}N_2O_5$	L-Kynurenine Diacetate, 5179
292.131074	$C_{16}H_{20}O_5$	Curvularin, 2685
292.147304	$C_{14}H_{26}Cl_2N_2$	Bipiperidyl Mustard, 1252
292.157563	$C_{19}H_{20}N_2O$	Vellosimine, 9599
292.167459	$C_{17}H_{24}O_4$	Trichodermin, 9330
292.193949	$C_{20}H_{24}N_2$	Dimethindene, 3202
292.215078	$C_{17}H_{28}N_2O_2$	Ambucetamide, 390
292.236208	$C_{14}H_{32}N_2O_4$	1,1',1'',1'''-(Ethylenedinitrilo)tetra-2-propanol, 3734
292.689948	Br_3Fe	Ferric Bromide, 3932
292.837167	$C_6Cl_5NO_2$	Quintozene, 7903
292.876532	$Cl_5H_{12}N_3Zn$	Ammonium Pentachlorozincate, 569

Use in conjunction with The Merck Index, Ninth Edition

Molecular Weight	Empirical Formula	Compound Name, Monograph Number
293.141579	$C_{19}H_{19}NO_2$	Sorbic Alcohol, Diphenylurethan, 8496
293.156184	$C_{15}H_{23}N_3OS$	Diamthazol, 2949
293.199094	$C_{17}H_{27}NO_3$	Embutramide, 3505 Pramocaine, 7505 Stadacain, 8550
293.214350	$C_{21}H_{27}N$	Butriptyline, 1517 Prozapine, 7692
293.712644	Al_2Se_3	Aluminum Selenide, 363
293.726711	Ag_2Se	Silver Selenide, 8272
293.752663	$C_2HBr_3O_2$	Tribromoacetic Acid, 9287
293.771545	CaI_2	Calcium Iodide, 1674
293.772838	$Cr_2K_2O_7$	Potassium Dichromate(VI), 7406
293.852716	$C_7H_4Br_2O_3$	3,5-Dibromosalicylic Acid, 2996
293.875714	$C_5FeK_2N_6O$	Potassium Nitroprusside, 7434
293.910151	Na_2O_4W	Sodium Tungstate(VI), 8471
294.009910	C_4H_9ClHg	n-Butylmercuric Chloride, 1568
294.067224	$C_{16}H_{11}ClN_4$	Estazolam, 3623
294.078662	$C_{12}H_{14}N_4O_3S$	4'-Formylsuccinanilic Acid Thiosemicarbazone, 4107 Sulfacytine, 8692 Sulfaethoxypyridazine, 8699 Sulfamethomidine, 8708
294.089209	$C_{18}H_{14}O_4$	3-(2-Hydroxy-1-naphthylmethyl)salicylic Acid, 4747
294.102272	$C_{16}H_{19}ClO_3$	Bucloxic Acid, 1458
294.110339	$C_{15}H_{18}O_6$	Cedrin, 1909 Tutin, 9480
294.118420	$C_{10}H_{22}N_4O_2S_2$	β-Alethine, 221
294.121572	$C_{14}H_{18}N_2O_5$	Aspartame, 863
294.136828	$C_{18}H_{18}N_2O_2$	N^1,N^4-Dibenzoylpiperazine, 7254
294.149404	$[C_{19}H_{20}NO_2]^+$	Methylaporeinium, 779
294.161980	$C_{20}H_{22}O_2$	Norgestrienone, 6511
294.168724	$C_{14}H_{30}O_2S_2$	Tiadenol, 9156
294.173213	$C_{19}H_{22}N_2O$	Benanserin, Free base, 1035 Cinchonidine, 2275 Cinchonine, 2276 Cinchotoxine, 2279 Noxiptilin, 6533

294.

Molecular Weight	Empirical Formula	Compound Name, Monograph Number
294.183110	$C_{17}H_{26}O_4$	Embelin, 3502 Gingerol, 4255
294.184447	$C_{18}H_{22}N_4$	Acetonylacetone, Bis(phenylhydrazone), 57
294.186481	$C_{14}H_{30}O_4S$	Tetradecyl Sulfuric Acid, 8460
294.194343	$C_{16}H_{26}N_2O_3$	Proparacaine, 7598 Propoxycaine, 7626
294.209599	$C_{20}H_{26}N_2$	Dimetacrine, 3198 Trimipramine, 9391
294.219495	$C_{18}H_{30}O_3$	Helenynolic Acid, 4483 Juvenile Hormones, C-18 JH, 5124
294.255881	$C_{19}H_{34}O_2$	Chaulmoogric Acid, Methyl ester, 2000 Methyl Linoleate, 5960
294.902833	$C_9H_4Cl_3NO_2S$	Folpet, 4087
294.948827	$C_7H_6ClN_3O_4S_2$	Chlorothiazide, 2155
295.016684	$C_{14}H_{11}Cl_2NO_2$	Diclofenac, 3059 Meclofenamic Acid, 5603
295.056938	$C_8H_{14}N_3O_7P$	AIR, 181
295.080435	$C_{12}H_{13}N_3O_6$	D-Glucuronolactone Isonicotinoylhydrazone, 4301
295.084458	$C_{17}H_{13}NO_4$	1-Aminoanthraquinone-2-carboxylic Acid, Ethyl ester, 420
295.090997	$C_{14}H_{18}ClN_3S$	Chlorothen, 2153
295.091669	$C_{11}H_{13}N_5O_5$	Azidoamphenicol, 920
295.095691	$C_{16}H_{13}N_3O_3$	Mebendazole, 5593 Nimetazepam, 6370
295.105587	$C_{14}H_{17}NO_6$	Indican (Plant Indican), 4826 Mandelonitrile Glucoside, 5548
295.120844	$C_{18}H_{17}NO_3$	Pukateine, 7724
295.139472	$C_{19}H_{21}NS$	Dothiepin, A5 Pizotyline, 7298
295.168462	$C_{18}H_{21}N_3O$	Dibenzepin, 2972
295.178358	$C_{16}H_{25}NO_4$	Floredil, 4016
295.193615	$C_{20}H_{25}NO$	Diphemethoxidine, 3310 α,α-Diphenyl-2-piperidinepropanol, 3344 Normethadone, 6517 Pridinol, 7539
295.204848	$C_{19}H_{25}N_3$	Picoperine, 7207
295.230000	$C_{21}H_{29}N$	Diisopromine, 3179
295.287515	$C_{19}H_{37}NO$	Lactobacillic Acid Amide, 5189

Use in conjunction with The Merck Index, Ninth Edition

Molecular Weight	Empirical Formula	Compound Name, Monograph Number
295.741908	BaBr$_2$	Barium Bromide, 971
295.748087	Cl$_5$Sb	Antimony Pentachloride, 733
295.768313	C$_2$H$_3$Br$_3$O$_2$	Bromal Hydrate, 1382
295.843073	C$_4$CdK$_2$N$_4$	Cadmium Potassium Cyanide, 1611
295.944814	C$_6$H$_9$AsN$_2$O$_5$S	Sulfarside, 8735
295.963111	Cl$_2$Ra	Radium Chloride, 7912
295.971421	C$_{14}$H$_4$N$_2$O$_2$S$_2$	Dithianone, 3386
295.975422	C$_8$H$_{10}$AsN$_2$O$_4$Na	Tryparsamide, 9454
296.003784	C$_{10}$H$_8$N$_4$O$_3$S$_2$	Nitrodan, 6419
296.032088	C$_{16}$H$_8$O$_6$	Medicagol, 5614
296.036911	C$_{14}$H$_{14}$ClO$_3$P	Dibenzyl Chlorophosphonate, 2977
296.044249	C$_{10}$H$_{11}$F$_3$N$_2$O$_3$S	Fluoridamid, 4045
296.050936	C$_{10}$H$_{20}$N$_2$S$_4$	Disulfiram, 3382
296.068474	C$_{17}$H$_{12}$O$_5$	Coumestrol, Dimethyl ether, 2549
296.075173	C$_8$H$_{24}$O$_4$Si$_4$	Octamethylcyclotetrasiloxane, 6557
296.076555	C$_{12}$H$_{16}$N$_4$OS$_2$	Thiothiamine, 9104
296.079707	C$_{16}$H$_{12}$N$_2$O$_4$	Isatide, 4951
296.086451	C$_{10}$H$_{20}$N$_2$O$_4$S$_2$	Penicillamine Disulfide, 6882
296.104859	C$_{18}$H$_{16}$O$_4$	Benzyl Fumarate, 1151 Isatropic Acid, 4954 Truxillic Acid, 9451
296.105546	C$_{11}$H$_{16}$N$_6$O$_2$S	Supazine, 8794
296.119464	C$_{14}$H$_{20}$N$_2$O$_3$S	Tolcyclamide, 9210
296.125989	C$_{15}$H$_{20}$O$_6$	Shellolic Acid, 8221
296.131349	C$_{21}$H$_{16}$N$_2$	Paranyline, 6835
296.134721	C$_{18}$H$_{20}$N$_2$S	Methdilazine, 5823 Pyrathiazine, 7739
296.138559	C$_{15}$H$_{16}$N$_6$O	Amicarbalide, 397
296.163711	C$_{17}$H$_{20}$N$_4$O	Propizepine, 7623
296.177630	C$_{20}$H$_{24}$O$_2$	Diethylstilbestrol, Dimethyl ether, 3113 Dimestrol, 3197 Ethynylestradiol, 3817

Molecular Weight	Empirical Formula	Compound Name, Monograph Number
296.188864	$C_{19}H_{24}N_2O$	Aminopentamide, 469 Cinchonamine, 2274 Hydrocinchonidine, 4669 Hydrocinchonine, 4670 Imipramine N-Oxide, 4814
296.235145	$C_{18}H_{32}O_3$	Vernolic Acid, 9620
296.271531	$C_{19}H_{36}O_2$	Lactobacillic Acid, 5189 Oleic Acid, Methyl ester, 6674 Petroselinic Acid, Methyl ester, 6974 Vaccenic Acid, Methyl ester, 9562
296.307916	$C_{20}H_{40}O$	Isophytol, 5059 Phytol, 7196
296.319149	$[C_{19}H_{40}N_2]^{2+}$	Mebezonium, 5595
296.819164	$C_6H_2Br_2ClNO$	2,6-Dibromoquinone-4-chlorimide, 2994
296.962760	$C_8H_9ClNO_5PS$	Chlorthion®, 2184 Dicapthon, 3013
296.964477	$C_7H_8ClN_3O_4S_2$	Hydrochlorothiazide, 4668
297.049465	$C_{13}H_{14}BCl_2NO_2$	8-Quinolineboronic Acid, Chloroethyl diester, 7890
297.060570	$C_{12}H_{15}N_3O_2S_2$	Glybuzole, 4313
297.107319	$C_{11}H_{15}N_5O_5$	Psicofuranine, 7709
297.118552	$C_{10}H_{15}N_7O_4$	Saxitoxin, Neutral, 8141
297.118736	$C_{18}H_{19}NOS$	Tolindate, 9213
297.136494	$C_{18}H_{19}NO_3$	Morphothebaine, 6119 Oripavine, 6707 Tsuduranine, 9460
297.151750	$C_{22}H_{19}N$	Bis(1-naphthylmethyl)amine, 1319
297.172879	$C_{19}H_{23}NO_2$	4-[2-(Benzhydryloxy)ethyl]morpholine, 1081
297.209265	$C_{20}H_{27}NO$	Cyclorphan, 2756
297.233074	$[C_{40}H_{58}N_4]^{2+}$	Bisdequalinium (half mass), 1267
297.339551	$C_{20}H_{43}N$	Dymanthine, 3452
297.818867	BaO_6S_2	Barium Dithionate, 977
297.869500	Cl_3Ir	Iridium Trichloride, 4937
297.922362	HgO_4S	Mercuric Sulfate, 5727
297.941373	F_6W	Tungsten Hexafluoride, 9469
297.983724	$C_{11}H_{11}CuN_2O_2S$	Allocupreide, 249
297.996967	$C_{12}H_{10}O_5S_2$	Benzenesulfonic Anhydride, 1074

Use in conjunction with The Merck Index, Ninth Edition

Molecular Weight	Empirical Formula	Compound Name, Monograph Number
298.047738	$C_{16}H_{10}O_6$	Rhein, Methyl ester, 7966
298.055819	$C_{11}H_{14}N_4O_2S_2$	Glyprothiazol, 4351
298.080101	$C_{12}H_{14}N_2O_7$	Nicotinamide Ascorbate, 6341
298.084124	$C_{17}H_{14}O_5$	Anthragallol, Trimethyl ether, 719 Coumafuryl, 2541 Pterocarpin, 7717
298.095357	$C_{16}H_{14}N_2O_4$	Nybomycin, 6543
298.099380	$C_{21}H_{14}O_2$	Anthranol, Benzoate, 721
298.105253	$C_{14}H_{18}O_7$	Picein, 7201
298.120509	$C_{18}H_{18}O_4$	Benzyl Succinate, 1169 Diphenic Acid, Diethyl ester, 3317 Metochalcone, 6017
298.124065	$C_{13}H_{26}O_2Si_3$	1,1,1,3,5,5,5-Heptamethyl-3-phenyltrisiloxane, 4517
298.141639	$C_{15}H_{22}O_6$	Sesamex, 8214
298.142976	$C_{16}H_{18}N_4O_2$	D-Erythrose, Phenylosazone, 3612 L-Erythrose, Phenylosazone, 3613 L-Erythrulose, Phenylosazone, 3616 Nialamide, 6307 D-Threose, Phenylosazone, 9125 L-Threose, Phenylosazone, 9126
298.146999	$C_{21}H_{18}N_2$	Amarine, 379
298.150371	$C_{18}H_{22}N_2S$	Diethazine, 3079 Trimeprazine, 9371
298.156895	$C_{19}H_{22}O_3$	Doisynoestrol, 3415 Ostruthin, 6731
298.168128	$C_{18}H_{22}N_2O_2$	Phenacaine, 6985
298.175756	$[C_{40}H_{44}N_4O]^{2+}$	C-Curarine I (half mass), 2679
298.193280	$C_{20}H_{26}O_2$	Benzestrol, 1077 Norethindrone, 6506 Norethynodrel, 6507 Promethestrol, 7582
298.204514	$C_{19}H_{26}N_2O$	Geissoschizoline, 4215 Naphthipramide, 6206
298.217090	$[C_{20}H_{28}NO]^+$	Methyl 2-(Benzhydryloxy)triethylammonium, 1082
298.250795	$C_{18}H_{34}O_3$	Ricinoleic Acid, 8005
298.287181	$C_{19}H_{38}O_2$	Methyl Stearate, 5992
298.934134	$C_9H_8Cl_3NO_2S$	Captan, 1770
299.052090	$C_{13}H_{18}BrNO_2$	α-Bromoisovaleryl-p-phenetidine, 1422

Molecular Weight	Empirical Formula	Compound Name, Monograph Number
299.068826	$C_{10}H_{13}N_5O_4S$	Thioguanosine, 9072
299.068939	$C_{10}H_{19}ClNO_5P$	Phosphamidon, 7148
299.072849	$C_{15}H_{13}N_3O_2S$	Fenbendazole, 3891
299.082540	$C_{16}H_{14}ClN_3O$	Chlordiazepoxide, 2054
299.129672	$[C_{19}H_{15}N_4]^+$	Triphenyltetrazolium, 9411
299.134202	$[C_{10}H_{17}N_7O_4]^{2+}$	Saxitoxin, 8141
299.140055	$C_{14}H_{22}ClN_3O_2$	Metoclopramide, 6018
299.152144	$C_{18}H_{21}NO_3$	Codeine, 2420-2424, 2426-2428 2-Dimethylaminoethyl Benzilate, 3221 Hydrocodone, 4672 Metopon, 6023 Morphine, 6-Methyl ether, 6108 Neopine, 6285 Pseudocodeine, 7698 Thebainone, 8989
299.158196	$[C_{18}H_{23}N_2S]^+$	Thiazinamium, 9038
299.163377	$C_{17}H_{21}N_3O_2$	MEMPP, 5648
299.188529	$C_{19}H_{25}NO_2$	Cyclopyrazate, 2755 Nylidrin, 6544
299.282430	$C_{18}H_{37}NO_2$	Palmidrol, 6804 Sphingosine, 8519
299.978591	$C_{15}H_9BrO_2$	Bromindione, 1391
300.052841	$C_{10}H_{12}N_4O_5S$	Carbazochrome Sulfonic Acid, 1792
300.063388	$C_{16}H_{12}O_6$	Acetylsalicylsalicylic Acid, 98 Diosmetin, 3296 Hematein, 4493 Pratensein, 7507 Tectorigenin, 8860
300.066556	$C_{16}H_{13}ClN_2O_2$	Clobazam, 2327
300.068098	$C_{14}H_{12}N_4O_2S$	Sulfaquinoxaline, 8734
300.074622	$C_{15}H_{12}N_2O_5$	Cinnabarine, O-Methyl ether, 2285
300.132137	$C_{13}H_{20}N_2O_6$	Actinobolin, 133
300.139327	$C_{18}H_{21}ClN_2$	Chlorcyclizine, 2050
300.150073	$[C_{20}H_{18}N_3]^+$	Dimidium, 3257
300.158626	$C_{16}H_{20}N_4O_2$	Apazone, 759
300.159969	$[C_{18}H_{22}NO_3]^+$	Methylmorphinium, 6111
300.172545	$C_{19}H_{24}O_3$	Adrenosterone, 162 Testolactone, 8889

Use in conjunction with The Merck Index, Ninth Edition

Molecular Weight	Empirical Formula	Compound Name, Monograph Number
300.196355	$[C_{38}H_{52}N_2O_4]^{2+}$	Diethylbelladonnine (half mass), 1032
300.204908	$C_{15}H_{28}N_2O_4$	Elaiomycin Acetate, 3486
300.208930	$C_{20}H_{28}O_2$	Ethynodiol, 3813 Methandrostenolone, 5808 Norgesterone, 6509 Retinoic Acid, 7961 17α-Vinyl-19-nortestosterone, 9648
300.230060	$C_{17}H_{32}O_4$	Roccellic Acid, 8018
300.245316	$C_{21}H_{32}O$	Allylestrenol, 284
300.873786	$C_9H_5Br_2NO$	Broxyquinoline, 1452
300.946185	F_6Re	Rhenium Hexafluoride, 7969
301.071213	$[C_{16}H_{13}O_6]^+$	Peonidin, 6938
301.082447	$[C_{15}H_{13}N_2O_5]^+$	Gallocyanine, Cation, 4195
301.131408	$C_{17}H_{19}NO_4$	Buphanamine, 1478 Morphine N-Oxide, 6114 Oxymorphone, 6782
301.142642	$C_{16}H_{19}N_3O_3$	Febrifugine, 3882
301.146664	$C_{21}H_{19}NO$	Cyprolidol, 2772
301.157247	$C_{12}H_{23}N_5O_2S$	Lambast®, 5199
301.157898	$C_{20}H_{19}N_3$	Rosaniline, 8024
301.167794	$C_{18}H_{23}NO_3$	Dihydrocodeine, 3148 Dihydroisocodeine, 3153 Dobutamine, 3407 Isoxsuprine, 5100
301.215413	$C_{18}H_{27}N_3O$	Isopentaquine, 5054 Pentaquine, 6920
301.240565	$C_{20}H_{31}NO$	Trihexyphenidyl, 9361
301.697088	P_2Se_3	Phosphorus Hemitriselenide, 7157
301.764555	Mn_2O_8Zn	Zinc Permanganate, 9814
301.770220	$O_7P_2Zn_2$	Zinc Pyrophosphate, 9821
301.801744	$CH_4Ni_3O_7$	Nickel Carbonate Hydroxide, 6317
301.866751	$C_4H_5O_4S_2Sb$	Antimony Thioglycollic Acid, 741
301.873118	$AuCl_3$	Gold Trichloride, 4370
301.940924	Na_2O_3Pb	Sodium Plumbate(IV), 8428
301.957121	$C_{12}H_8Cl_2O_3S$	Genite®, 4223 Ovex, 6737

Molecular Weight	Empirical Formula	Compound Name, Monograph Number
301.961861	$C_6H_{11}N_2O_4PS_3$	Methidathion, 5840
301.998222	$C_4H_6N_4O_{12}$	Erythrityl Tetranitrate, 3598
302.006267	$C_{14}H_6O_8$	Ellagic Acid, 3497
302.011264	$C_{13}H_{12}Cl_2O_4$	Ethacrynic Acid, 3643
302.026604	$C_{11}H_{15}BrN_2O_3$	Butallylonal, 1494 Narcobarbital, 6249
302.030444	O_4U	Uranium Peroxide, 9514
302.042653	$C_{15}H_{10}O_7$	Morin, 6101 Quercetin, 7821
302.065120	$C_{13}H_{10}N_4O_5$	4,4'-Dinitrocarbanilide, 3274, 6309
302.079038	$C_{16}H_{14}O_6$	Hematoxylin, 4496 Hesperetin, 4534 Homoeriodictyol, 4617 Protocotoin, 7680
302.083748	$C_{14}H_{14}N_4O_2S$	Cambendazole, 1729
302.094294	$C_{20}H_{14}O_3$	Florantyrone, 4015
302.097666	$C_{10}H_{11}O_3SC_6H_4CH_3$	Tetralol, p-Tosylate, 8937
302.111401	$C_{12}H_{18}N_2O_7$	Bicyclomycin, 1228
302.139233	$[C_{17}H_{20}NO_4]^+$	Methyllycorinium, 5444
302.141913	$C_{20}H_{18}N_2O$	N-(1,2-Diphenylethyl)nicotinamide, 3332
302.151809	$C_{18}H_{22}O_4$	Equilin Glycol, 3553 Nigellone, 6364 Nordihydroguaiaretic Acid, 6503
302.172939	$C_{15}H_{26}O_6$	Tributyrin, 9299
302.174276	$C_{16}H_{22}N_4O_2$	Aminopropylon, 483
302.188195	$C_{19}H_{26}O_3$	Allethrin I, 242
302.224580	$C_{20}H_{30}O_2$	Abietic Acid, 1 Hepaxanthin, 4511 Isopimaric Acid, 5060 Levopimaric Acid, 5314 Methenolone, 5836 17-Methyltestosterone, 5998 Norethandrolone, 6505 Oxenin, 6757 Palustric Acid, 6808 Pimaric Acid, 7229 Stenbolone, 8586
302.248390	$[C_{20}H_{32}NO]^+$	Methylprocyclidinium, 7561
302.284775	$[C_{21}H_{36}N]^+$	Dichlorobenzalkonium, 3027

Use in conjunction with The Merck Index, Ninth Edition

Molecular Weight	Empirical Formula	Compound Name, Monograph Number
302.987131	$C_{10}H_9NO_6S_2$	Amido-G-Acid, 403 Amido-R-Acid, 405 1-Naphthylamine-2,7-disulfonic Acid, 6227 1-Naphthylamine-4,6(or 7)-disulfonic Acid, 6228, 6229
303.048289	$C_{15}H_{14}NO_2PS$	Cyanophenphos, 2703
303.050478	$[C_{15}H_{11}O_7]^+$	Delphinidin, 2850
303.079284	$C_{14}H_{19}Cl_2NO_2$	Chlorambucil, 2036
303.087176	$C_{12}H_{18}NO_6P$	Diisopropyl p-Nitrophenyl Phosphate, 3184
303.121906	$C_{15}H_{17}N_3O_4$	Pidylon, 7221
303.133140	$C_{14}H_{17}N_5O_3$	Pipemidic Acid, A13
303.137162	$C_{19}H_{17}N_3O$	Evodiamine, 3856
303.138992	$C_{18}H_{22}ClNO$	Chlorphenoxamine, 2171 Phenoxybenzamine, 7055
303.147058	$C_{17}H_{21}NO_4$	Atroscine, 895 Cocaine, 2408-2411 Fenoterol, 3908 Pseudococaine, 7697 Scopolamine, 8158, 8159
303.183444	$C_{18}H_{25}NO_3$	Levomepate, 5311
303.186124	$[C_{21}H_{23}N_2]^+$	Quinaldine Red, Cation, 7835
303.219829	$C_{19}H_{29}NO_2$	Samandarone, 8105
303.796802	AlO_8RbS_2	Aluminum Rubidium Sulfate, 361
303.812429	$BaCl_2O_6$	Barium Chlorate, 973
303.893639	$BiBrO$	Bismuth Bromide Oxide, 1282
303.912462	$C_6FeN_6Na_4$	Sodium Ferrocyanide, 8366
303.914606	$C_6H_6Cl_2N_2O_4S_2$	Dichlorphenamide, 3056
303.917481	$C_6H_{12}N_2S_4Zn$	Ziram, 9838
303.928371	O_4PbS	Lead Sulfate, 5278
303.933810	BiO_4P	Bismuth Phosphate, 1300
303.971146	$C_{10}H_8O_7S_2$	1-Naphthol-4,8-disulfonic Acid, 6213 2-Naphthol-3,6-disulfonic Acid, 6214 2-Naphthol-6,8-disulfonic Acid, 6215
303.972404	BiF_5	Bismuth Pentafluoride, 1299
304.021917	$C_{14}H_8O_8$	Rufigallol, 8052
304.047756	$C_9H_{12}N_4O_6S$	Theophyllineethyl Sulfuric Acid, 7759

Molecular Weight	Empirical Formula	Compound Name, Monograph Number
304.074533	$C_{13}H_{18}Cl_2N_2O_2$	Medphalan, 5616 Melphalan, 5646 Merphalan, 5747
304.084792	$C_{18}H_{12}N_2O_3$	Xanthocillin Y_1, 9724
304.088164	$C_{15}H_{16}N_2O_3S$	N-Sulfanilyl-3,4-xylamide, 8726
304.095663	$C_{10}H_{22}Cl_2N_2O_4$	Mannomustine, Free base, 5577
304.101053	$C_{12}H_{21}N_2O_3PS$	Dimpylate, 3262
304.101899	$C_{10}H_{16}N_4O_7$	Vicine, 9632
304.115818	$C_{13}H_{20}O_8$	Pentaerythritol Tetraacetate, 6907 Thevetose, β-L-Triacetate, 9016 Thevetose, β-D-Triacetate, 9016
304.131074	$C_{17}H_{20}O_5$	Acetylhelenalin, 4481 Matricarin, 5588
304.191269	$[C_{18}H_{26}NO_3]^+$	Methylatropine, 892 N-Methylhyoscyaminium, 4780
304.224974	$C_{16}H_{32}O_5$	Aleuritic Acid, 223
304.240230	$C_{20}H_{32}O_2$	Arachidonic Acid, 793 Etiocholanic Acid, 3823 Mestanolone, 5761 Mesterolone, 5762 Methandriol, 5807
304.300425	$[C_{21}H_{38}N]^+$	Cetylpyridinium, 1987
304.910444	C_9H_5ClINO	Iodochlorhydroxyquin, 4894
304.917829	$C_7H_7Cl_3NO_4P$	Fospirate, 4111
304.954895	$C_9H_8INO_3$	o-Iodohippuric Acid, 4899
304.964807	$C_8H_{10}NO_4Sb$	Stibenyl, Free acid, 8591
305.010642	$C_{12}H_7N_3O_5S$	3,7-Dinitro-5-oxophenothiazine, 3276
305.047028	$C_{13}H_{11}N_3O_4S$	Sulfoniazide, 8755
305.081871	$C_{16}H_{16}ClNO_3$	Clofenpyride, 2341
305.085915	$C_{10}H_{15}N_3O_8$	Convicine, 2486
305.089937	$C_{15}H_{15}NO_6$	Ascorbigen, 856
305.096302	$C_{11}H_{20}N_3O_3PS$	Pyrimithate, 7770
305.105193	$C_{19}H_{15}NO_3$	N-(4-Hydroxyphenyl)-3-phenylsalicylamide, 4754
305.126323	$C_{16}H_{19}NO_5$	Peyonine, 6977
305.152144	$[C_{37}H_{42}N_2O_6]^{2+}$	Tubocurarine (half mass), 9466

Use in conjunction with The Merck Index, Ninth Edition

Molecular Weight	Empirical Formula	Compound Name, Monograph Number
305.162708	$C_{17}H_{23}NO_4$	Atropine N-Oxide, 893 Convolvamine, 2487 Lunacridine, 5419
305.189198	$C_{20}H_{23}N_3$	Ormosinine, 6709
305.199094	$C_{18}H_{27}NO_3$	Capsaicin, 1766
305.235479	$C_{19}H_{31}NO_2$	1-Dodecanol, Phenylurethan, 3410 Samandarine, 8105
305.809650	$Cr_2FeH_4NO_8$	Ammonium Ferric Chromate, 537
305.844831	$Na_3O_9P_3$	Sodium Trimetaphosphate, 8469
305.910231	$C_6H_{12}Br_2O_4$	Mitobronitol, 6059
305.938911	$C_9H_7IO_4$	Acetyl-5-iodosalicylic Acid, 90
305.951907	F_6Os	Osmium Hexafluoride, 6727
306.037567	$C_{14}H_{10}O_8$	Oosporein, 6689
306.067429	$C_{14}H_{14}N_2O_4S$	Acediasulfone, 16
306.073953	$C_{15}H_{14}O_7$	Fusarubin, 4169 Leucocyanidin, 5304
306.078662	$C_{13}H_{14}N_4O_3S$	N^2-Formylsulfisomidine, 4108 Sulfaperine, N^4-Acetyl deriv, 8728
306.094206	$C_{18}H_{20}Cl_2$	1,1-Dichloro-2,2-bis(p-ethylphenyl)ethane, 3035
306.125595	$C_{20}H_{18}O_3$	Phenolphthalol, 7043
306.128783	$C_9H_{18}N_6O_6$	Hexamethylolmelamine, 4561
306.146724	$C_{17}H_{22}O_5$	Globicin, 4275 Pyrethrosin, 7748
306.157957	$C_{16}H_{22}N_2O_4$	Inproquone, 4857
306.170533	$[C_{17}H_{24}NO_4]^+$	Methylconvolvinium, 2488
306.183110	$C_{18}H_{26}O_4$	Isoamyl Phthalate, 4972
306.205576	$C_{16}H_{26}N_4O_2$	Ruscopine, 8056
306.219495	$C_{19}H_{30}O_3$	Androstane-$3\beta,11\beta$-diol-17-one, 672 Oxandrolone, 6749
306.222175	$[C_{22}H_{28}N]^+$	Prifinium, 7540
306.230728	$C_{18}H_{30}N_2O_2$	Butacaine, 1488
306.243304	$[C_{19}H_{32}NO_2]^+$	Valethamate, 9569
306.953362	F_6Ir	Iridium Hexafluoride, 4935
306.970545	$C_9H_{10}INO_3$	3-Iodotyrosine, 4912

Molecular Weight	Empirical Formula	Compound Name, Monograph Number
306.984404	$C_{13}H_{10}BrNO_3$	Resorantel, 7950
307.062678	$C_{13}H_{13}N_3O_4S$	Proflavine Sulfate, 7568
307.083807	$C_{10}H_{17}N_3O_6S$	Glutathione, 4307
307.103086	$C_{19}H_{17}NOS$	Tolnaftate, 9216
307.110297	$C_{13}H_{17}N_5O_2S$	Sulfasymazine, 8738
307.120844	$C_{19}H_{17}NO_3$	Cusparine, 2688
307.157229	$C_{20}H_{21}NO_2$	Eupaverin, 3849
307.159258	$[C_{15}H_{23}N_4OS]^+$	Butylthiamine, Cation, 1581
307.160601	$C_{17}H_{25}NO_2S$	2-Diethylaminoethyl α-(2-Cyclopenten-1-yl)-2-thiophene Acetate, 3085
307.168462	$C_{19}H_{21}N_3O$	Talastine, 8816
307.181039	$[C_{20}H_{23}N_2O]^+$	C-Curarine III, 2680
307.181039	$[C_{40}H_{46}N_4O_2]^{2+}$	C-Toxiferine I (half mass), 9253
307.193615	$C_{21}H_{25}NO$	Benztropine, 1135 Hepzidine, 4530
307.214744	$C_{18}H_{29}NO_3$	Amprotropine, Free base, 626 Butamirate, 1496 p-Butoxybenzoic Acid 3-Diethylaminopropyl Ester, 1515
307.225977	$C_{17}H_{29}N_3O_2$	Amoxecaine, 610
307.238553	$[C_{18}H_{31}N_2O_2]^+$	Methylleucinocaine, 5302
307.804698	$C_4H_7Br_3O$	Tribromo-tert-butyl Alcohol, 9289
307.941137	$C_{10}H_{14}Br_2O$	α,α'-Dibromo-d-camphor, 2987
307.962121	$C_8H_{11}Cl_3O_6$	α-Chloralose, 2034
307.992550	$C_{12}H_8N_2O_4S_2$	Nitrophenide, 6440
308.025560	$C_5H_{11}ClHg$	n-Amylmercuric Chloride, 650
308.030591	$C_{11}H_{17}O_4PS_2$	Fensulfothion, 3917
308.035460	$C_{14}H_{12}O_6S$	Sulisobenzone, 8779
308.043322	$C_{16}H_8N_2O_5$	Catalin, 1896
308.046693	$C_{13}H_{12}N_2O_5S$	4-Sulfanilamidosalicylic Acid, 8719
308.094312	$C_{13}H_{16}N_4O_3S$	Cycothiamin(e), 2760
308.104859	$C_{19}H_{16}O_4$	Warfarin, 9700
308.127326	$C_{17}H_{16}N_4O_2$	Nifenazone, 6354
308.141245	$C_{20}H_{20}O_3$	Equilenin, Acetate, 3551

Use in conjunction with The Merck Index, Ninth Edition

Molecular Weight	Empirical Formula	Compound Name, Monograph Number
308.152478	$C_{19}H_{20}N_2O_2$	Phenylbutazone, 7078
308.162374	$C_{17}H_{24}O_5$	Dihydroglobicin, 4275 Tetrahydromatricarin, 5588
308.188864	$[C_{40}H_{48}N_4O_2]^{2+}$	C-Calebassine (half mass), 1719
308.209993	$C_{17}H_{28}N_2O_3$	Ambucaine, 389 Benoxinate, 1049 Bethoxycaine, 1217 p-Butylaminosalicylic Acid 2-Diethylaminoethyl Ester, 1530 Metabutoxycaine, 5768
308.235145	$C_{19}H_{32}O_3$	Helenynolic Acid, Methyl ester, 4483
308.271531	$C_{20}H_{36}O_2$	Ethyl Linoleate, 3760
308.319149	$C_{20}H_{40}N_2$	Glyodin, Free base, 4345
308.747000	I_2Mn	Manganese Iodide, 5560
308.949810	$C_8H_8INO_4$	3,4-Dihydro-3-iodo-6-methyl-2,4-dioxo-1(2H)-pyridine-acetic Acid, 7923
309.032334	$C_{15}H_{13}Cl_2NO_2$	2-Nitro-1,1-bis(p-chlorophenyl)propane, 6417
309.063723	$C_{17}H_{11}NO_5$	3'-Carboxy-4'-hydroxycinchophen, 1831
309.078328	$C_{13}H_{15}N_3O_4S$	Acetyl Sulfisoxazole, 104 Glymidine, 4344
309.084852	$C_{14}H_{15}NO_7$	Laetrile®, 5197
309.089561	$C_{12}H_{15}N_5O_3S$	Sulfaguanol, 8701
309.105981	$C_{11}H_{19}NO_9$	N-Acetylneuraminic Acid, 8225
309.136494	$C_{19}H_{19}NO_3$	7-[2-(Dimethylamino)ethoxy]flavone, 6951 Laureline, 5227
309.147727	$C_{18}H_{19}N_3O_2$	Diacetazotol, 2917
309.155122	$C_{20}H_{23}NS$	Methixene, 5851
309.172879	$C_{20}H_{23}NO_2$	Amolanone, 607 2-Diethylaminoethyl 9-Fluorenecarboxylate, 3086
309.184113	$C_{19}H_{23}N_3O$	Benzydamine, 1136
309.185942	$C_{18}H_{28}ClNO$	Chlorphencyclan, 2166
309.196689	$[C_{20}H_{25}N_2O]^+$	Macusine B, 5468
309.209265	$C_{21}H_{27}NO$	Benproperine, 1051 Diphenidol, 3319 Isomethadone, 5039 Methadone, 5799
309.266780	$C_{19}H_{35}NO_2$	Dicyclomine, 3074

309.

Molecular Weight	Empirical Formula	Compound Name, Monograph Number
309.703550	As_2S_5	Arsenic Pentasulfide, 826
309.743893	FeI_2	Ferrous Iodide, 3974
309.761920	Ag_2O_4S	Silver Sulfate, 8275
309.794616	$Ca_3O_8P_2$	Calcium Phosphate, Tribasic, 1695
309.844417	$C_8H_4Cl_6$	1,4-Bis(trichloromethyl)benzene, 1326
309.985488	$C_5H_{12}O_{11}P_2$	D-Ribulose, Diphosphate, 8001
310.016350	$C_{15}H_{12}Cl_2O_3$	Sesin, 8217
310.032062	$C_{14}H_9ClF_2N_2O_2$	Difluron, 3127
310.059842	$C_{17}H_{14}N_2S_2$	Fezatione, 3989
310.068868	$C_{14}H_{14}O_8$	Pyromellitic Acid, Tetramethyl ester, 7790
310.073577	$C_{12}H_{14}N_4O_4S$	Sulfadimethoxine, 8696 Sulfadoxine, 8697
310.084124	$C_{18}H_{14}O_5$	Formononetin, Acetate, 4102
310.105253	$C_{15}H_{18}O_7$	Hyenanchin, 4772 Picrotin, 7217, 7218
310.108625	$C_{12}H_{22}O_7S$	Diisobutyl Sulfosuccinic Acid, 3178
310.121847	$C_{20}H_{14}N_4$	Porphine, 7372
310.127209	$C_{10}H_{30}O_3Si_4$	Decamethyltetrasiloxane, 2833
310.144806	$C_{16}H_{23}ClN_2O_2$	Alloclamide, 247
310.150371	$C_{19}H_{22}N_2S$	Mepazine, 5673
310.165448	$[C_{16}H_{24}NO_5]^+$	Sinapine, Cation, 8288
310.168128	$C_{19}H_{22}N_2O_2$	Apoquinine, 778 Cupreine, 2623 Sarpagine, 8132 Wieland-Gumlich Aldehyde, 9709
310.178024	$C_{17}H_{26}O_5$	Tetrahydroglobicin, 4275
310.179361	$C_{18}H_{22}N_4O$	Kethoxal, Phenylosazone, 5149
310.190595	$[C_{17}H_{22}N_6]^{2+}$	Quinapyramine, Cation, 7839
310.193280	$C_{21}H_{26}O_2$	Cannabinol, 1748 Mestranol, 5764
310.204514	$C_{20}H_{26}N_2O$	Astrocasine, 878 Ibogaine, 4792 Tabernanthine, 8808
310.217090	$[C_{21}H_{28}NO]^+$	N-Methylnormethadone, Cation, 6517
310.250795	$C_{19}H_{34}O_3$	Methoprene, 5856

Use in conjunction with The Merck Index, Ninth Edition

Molecular Weight	Empirical Formula	Compound Name, Monograph Number
310.287181	$C_{20}H_{38}O_2$	Ethyl Oleate, 3774 Lactobacillic Acid, Methyl ester, 5189
310.860211	CuO_4W	Cupric Tungstate(VI), 2661
310.944480	$C_7H_{10}AsNO_6S$	Phenarsone Sulfoxylic Acid, 7000
311.093978	$C_{13}H_{17}N_3O_4S$	Dipyrone, Anhydrous free acid, 3369
311.100502	$C_{14}H_{17}NO_7$	Dhurrin, 2915
311.115758	$C_{18}H_{17}NO_4$	Actinodaphnine, 134 Laurepukine, 5229
311.130364	$C_{14}H_{21}N_3O_3S$	Metahexamide, 5773 Tolazamide, 9206
311.144078	$C_{20}H_{22}ClN$	Pyrrobutamine, 7799
311.148121	$C_{14}H_{21}N_3O_5$	Leonurine, 5294
311.152144	$C_{19}H_{21}NO_3$	Isothebaine, 5091 Nalorphine, 6180 Thebaine, 8988
311.159354	$C_{13}H_{21}N_5O_4$	Xanthinol, 9722
311.188529	$C_{20}H_{25}NO_2$	Adiphenine, 150 Benzobutamine, 1093 Cuspareine, 2687 Propanocaine, 7596
311.212339	$[C_{20}H_{27}N_2O]^+$	Ambutonium, 393
311.224915	$C_{21}H_{29}NO$	Biperiden, 1246 Dimepheptanol, 3195 Isomethadol, 5038
311.676604	$AsBr_3$	Arsenic Tribromide, 828
311.744301	I_2Ni	Nickel Iodide, 6325
311.850819	$C_3H_6I_2O$	Iothion, 4923
312.063388	$C_{17}H_{12}O_6$	Aflatoxin B$_1$, 168 Irisolone, 4941
312.071470	$C_{12}H_{16}N_4O_2S_2$	Glybuthiazol(e), 4312
312.085855	$C_{15}H_{12}N_4O_4$	Cinnamaldehyde, 2,4-Dinitrophenylhydrazone, 2286
312.097088	$C_{14}H_{12}N_6O_3$	Pteroic Acid, 7718
312.098744	$C_{12}H_{17}N_4O_4P$	Thiamine Phosphoric Acid Ester, 9026, 9027
312.099774	$C_{18}H_{16}O_5$	Aloe-Emodin, Trimethyl ether, 301 Emodin, Trimethyl ether, 3512
312.105647	$C_{11}H_{20}O_{10}$	Primeverose, 7544 Vicianose, 9631

Molecular Weight	Empirical Formula	Compound Name, Monograph Number
312.112344	$C_{18}H_{12}N_6$	2,4,6-Tripyridyl-s-triazine, 9417
312.114379	$C_{14}H_{20}N_2O_4S$	2-Pentenylpenicillin, 6924
312.120903	$C_{15}H_{20}O_7$	Dihydroisohyenanchin, 3154
312.129635	$C_{18}H_{20}N_2OS$	Diphazine, 3308
312.137497	$C_{20}H_{16}N_4$	Nitron, 6435
312.158626	$C_{17}H_{20}N_4O_2$	Propamidine, 7590
312.159969	$[C_{19}H_{22}NO_3]^+$	N,N-Dimethylsparsiflorinium, 8509 Methylmorphothebainium, 6119 Methyloripavinium, 6707
312.159969	$[C_{38}H_{44}N_2O_6]^{2+}$	d-Dimethylchondrocurine (half mass), 2210
312.166021	$C_{19}H_{24}N_2S$	Ethopropazine, 3673
312.172545	$C_{20}H_{24}O_3$	5,6,8-Estratrien-3β-ol-17-one, Acetate, 3634 Estrone, Acetate, 3636 Trenbolone, Acetate, 9271
312.183778	$C_{19}H_{24}N_2O_2$	Conquinamine, 2480 Quinamine, 7837 Salverine, 8102
312.208930	$C_{21}H_{28}O_2$	Demegestone, 2858 Dydrogesterone, 3451 Ethisterone, 3666 Norgestrel, 6510
312.302831	$C_{20}H_{40}O_2$	Arachidic Acid, 792 Ethyl Stearate, 3797
312.363026	$[C_{21}H_{46}N]^+$	Octadecyltrimethylammonium, 6554
312.742152	CoI_2	Cobaltous Iodide, 2395
312.985994	$C_{10}H_{14}Cl_2NO_2PS$	Zytron®, 9856
313.009090	$C_4Li_2N_4Pt$	Lithium Tetracyanoplatinate(II), 5385
313.086270	$C_{16}H_{12}FN_3O_3$	Flunitrazepam, 4028
313.095023	$C_{17}H_{15}NO_5$	Benorylate, 1048 Osalmid, Diacetate, 6724
313.131408	$C_{18}H_{19}NO_4$	Hydroxycodeinone, 4723
313.137460	$[C_{18}H_{21}N_2OS]^+$	Secergan®, Cation, 8171
313.167794	$C_{19}H_{23}NO_3$	Armepavine, 813 Biphenamine, 1247 Dihydrothebaine, 3161 Ethylmorphine, 3768 Mevalonic Acid, Benzhydrylamide, 6034 Oxyfedrine, 6776
313.180857	$C_{17}H_{28}ClNO_2$	Propivane, 7622

Use in conjunction with The Merck Index, Ninth Edition

Molecular Weight	Empirical Formula	Compound Name, Monograph Number
313.204179	$C_{20}H_{27}NO_2$	N,N-Dimethyl-α-(3-phenylpropyl)veratrylamine, 3243 Fenalcomine, 3889 Tropentane, 9442
313.227989	$[C_{20}H_{29}N_2O]^+$	Methylgeissoschizolinium, 4215
313.719778	O_4Rb_2Se	Rubidium Selenate, 8046
313.886946	$BiCl_3$	Bismuth Chloride, 1284
313.970010	$C_{10}H_{13}Cl_2O_3PS$	Dichlofenthion, 3015
313.978046	$C_{13}H_9Cl_3N_2O$	Triclocarban, 9333
313.978610	C_6H_5ClHg	Phenylmercuric Chloride, 7106
313.984487	$C_{10}H_6N_2O_8S$	Flavianic Acid, 4005
313.984754	$C_{13}H_9Cl_2FN_2S$	Fluonilid®, 4034
314.042653	$C_{16}H_{10}O_7$	Porphyrillic Acid, 7374
314.044399	F_4U	Uranium Tetrafluoride, 9516
314.065120	$C_{14}H_{10}N_4O_5$	Dantrolene, 2803
314.079038	$C_{17}H_{14}O_6$	Aflatoxin B_2, 168 Pectolinarigenin, 6862
314.083748	$C_{15}H_{14}N_4O_2S$	Sulfaphenazole, 8729
314.091142	$C_{17}H_{18}N_2S_2$	Dibenzthione, 2975
314.091927	$C_{14}H_{19}O_6P$	Crotoxyphos, 2594
314.100168	$C_{14}H_{18}O_8$	Glucovanillin, 4297
314.116761	$C_{19}H_{14}N_4O$	1,3-Di-6-quinolylurea, 3371
314.130029	$C_{14}H_{22}N_2O_4S$	Amylpenicillin, 654
314.145285	$C_{18}H_{22}N_2OS$	Methoxypromazine, 5875
314.149124	$C_{15}H_{18}N_6O_2$	Pimefylline, A12
314.151809	$C_{19}H_{22}O_4$	Menadiol Dibutyrate, 5650
314.154977	$C_{19}H_{23}ClN_2$	Clomipramine, 2350 Homochlorcyclizine, 4613
314.165723	$[C_{21}H_{20}N_3]^+$	Homidium, 4608, 4609
314.175619	$[C_{19}H_{24}NO_3]^+$	Methylthebainonium, 8989
314.188195	$C_{20}H_{26}O_3$	Estradiol, 3-Acetate, 3626 Estradiol, 17-Acetate, 3626 Taxodione, 8855
314.212004	$[C_{20}H_{28}NO_2]^+$	Methylcyclopyrazatium, 2755

314.

Molecular Weight	Empirical Formula	Compound Name, Monograph Number
314.224580	$C_{21}H_{30}O_2$	Cannabidiol, 1747 Progesterone, 7569 Tetrahydrocannabinols, 8927 Urushiol IV, 9553 Urushiol V, 9553
314.282095	$C_{19}H_{38}O_3$	Cetyl Lactate, 1985
314.962062	$C_{13}H_8Cl_3NO_2$	3′,4′,5-Trichlorosalicylanilide, 9326
315.000723	$C_{14}H_{10}BrN_3O$	Bromazepam, 1384
315.007001	$C_{10}H_{14}Cl_3N_2O_3$	Fenuron, Trichloroacetate, 3923
315.012750	$C_{13}H_5N_3O_7$	2,4,7-Trinitrofluorenone, 9395
315.041069	$C_{15}H_{10}ClN_3O_3$	Clonazepam, 2352
315.084849	$C_{18}H_{18}ClNS$	Chlorprothixene, 2178
315.092915	$C_{17}H_{17}NO_3S$	Protizinic Acid, 7676
315.106650	$C_{12}H_{17}N_3O_7$	Hamamelose, p-Nitrophenylhydrazone, 4460
315.113840	$C_{17}H_{18}ClN_3O$	Clobenzepam, 2329
315.121906	$C_{16}H_{17}N_3O_4$	Anthramycin, 720
315.147058	$C_{18}H_{21}NO_4$	N-Benzoylmescaline, 5752 Codeine N-Oxide, 2425 Oxycodone, 6774
315.183444	$C_{19}H_{25}NO_3$	Naltropine, 6183
315.198700	$C_{23}H_{25}N$	Fendiline, 3899
315.231063	$C_{19}H_{29}N_3O$	Pamaquine, 6810
315.253535	$C_{18}H_{35}O_4$	9,10-Dihydroxystearic Acid, 3169
316.012988	$C_{12}H_9N_2NaO_5S$	Tropaeoline O, 9438
316.013872	$C_5H_8N_4O_{12}$	Pentaerythritol Tetranitrate, 6908, 2054
316.042254	$C_{12}H_{17}BrN_2O_3$	5-(2-Bromoallyl)-5-(1-methylbutyl)barbituric Acid, 1402
316.058303	$C_{16}H_{12}O_7$	Rhamnetin, 7962
316.080770	$C_{14}H_{12}N_4O_5$	p-Anisaldehyde, 2,4-Dinitrophenylhydrazone, 696
316.090666	$C_{12}H_{16}N_2O_8$	Orotidine, N^3-Methylorotidine methyl ester, 6712
316.103791	$C_{15}H_{25}BrO_2$	d-Bornyl α-Bromoisovalerate, 1352
316.115818	$C_{14}H_{20}O_8$	Taxicatin, 8852
316.121178	$C_{20}H_{16}N_2O_2$	Anilinephthalein, 694 Xanthocillin, Dimethyl ether, 9724
316.124550	$C_{17}H_{20}N_2O_2S$	Tinoridine, 9173

Use in conjunction with The Merck Index, Ninth Edition

Molecular Weight	Empirical Formula	Compound Name, Monograph Number
316.137126	$[C_{18}H_{22}NO_2S]^+$	Heteronium, 4538
316.156913	$C_{13}H_{24}N_4O_3S$	Timolol, 9170
316.178693	$C_{18}H_{24}N_2O_3$	Julocrotine, 5120
316.191269	$[C_{19}H_{26}NO_3]^+$	Methyldihydroisocodeinium, 3153
316.203845	$C_{20}H_{28}O_3$	Cafestol, 1621 Cinerin I, 2283 Mytatrienediol, 6167
316.215078	$C_{19}H_{28}N_2O_2$	4-Androstene-3,17-dione, Dioxime, 675
316.226312	$C_{18}H_{28}N_4O$	Butalamine, 1492
316.240230	$C_{21}H_{32}O_2$	3,20-Allopregnanedione, 257 Bolasterone, 1337 Calusterone, 1726 Cyclopregnol, 2752 Levopimaric Acid, Methyl ester, 5314 Methyl Abietate, 5882 Norbolethone, 6495 Palustric Acid, Methyl ester, 6808 Pimaric Acid, Methyl ester, 7229 3,20-Pregnanedione, 7524 Pregnenolone, 7533 Urushiol III, 9553
316.297745	$C_{19}H_{40}O_3$	Chimyl Alcohol, 2016
316.994540	$C_{11}H_{12}NO_4PS_2$	Phosmet, 7147
317.005773	$C_{10}H_{12}N_3O_3PS_2$	Azinphos-Methyl, 922
317.066128	$[C_{16}H_{13}O_7]^+$	Petunidin, 6975
317.126323	$C_{17}H_{19}NO_5$	Actiphenol, Acetate, 139
317.154642	$C_{19}H_{24}ClNO$	Mecloxamine, 5607
317.162708	$C_{18}H_{23}NO_4$	Cocaethylene, 2407
317.199094	$C_{19}H_{27}NO_3$	Tetrabenazine, 8899
317.235479	$C_{20}H_{31}NO_2$	Drofenine, 3435 Metcaraphen, 5788
317.259289	$[C_{20}H_{33}N_2O]^+$	Hexocyclium, Cation, 4579
317.271865	$C_{21}H_{35}NO$	Funtumine, 4144
317.669713	Br_2Se_2	Selenium Bromide, 8180
317.738099	I_2Zn	Zinc Iodide, 9801
317.761420	Na_3S_4Sb	Sodium Thioantimonate(V), 8463
317.865634	$C_8H_2Cl_4O_5$	Mucochloric Anhydride, 6125
317.920924	$C_2HgN_2S_2$	Mercuric Thiocyanate, 5730

317.

Molecular Weight	Empirical Formula	Compound Name, Monograph Number
317.953661	$C_{14}H_{10}Cl_4$	1,1-Dichloro-2,2-bis(p-chlorophenyl)ethane, 3034 Mitotane, 6061
318.021518	$C_{11}H_{15}BrN_2O_4$	Xylulose, D-Isomer p-bromophenylhydrazone, 9750
318.037567	$C_{15}H_{10}O_8$	Myricetin, 6158 Quercetagetin, 7820
318.073160	$C_{16}H_{19}BrN_2$	Brompheniramine, 1446 Dexbrompheniramine, 2900
318.089209	$C_{20}H_{14}O_4$	Phenolphthalein, 7040 Phenyl Phthalate, 7112
318.095748	$C_{17}H_{19}ClN_2S$	Chlorpromazine, 2175
318.132805	$C_{15}H_{18}N_4O_4$	Phorone, 2,4-Dinitrophenylhydrazone, 7144
318.138658	$C_{19}H_{23}ClO_2$	16α-Chloroestrone 3-Methyl Ether, 2113
318.146724	$C_{18}H_{22}O_5$	Zearalenone, 9775
318.152776	$[C_{18}H_{24}NO_2S]^+$	Tiemonium, 9159
318.170533	$[C_{18}H_{24}NO_4]^+$	Methscopolamine, 5879, 8158
318.206919	$[C_{19}H_{28}NO_3]^+$	Glycopyrrolate, Cation, 4337
318.219495	$C_{20}H_{30}O_3$	Oxymesterone, 6778 Steviol, 8588
318.230728	$C_{19}H_{30}N_2O_2$	Bietamiverine, 1229
318.240624	$C_{17}H_{34}O_5$	Aleuritic Acid, Methyl ester, 223
318.251858	$C_{16}H_{34}N_2O_4$	Aleuritic Acid, Hydrazide, 223
318.251858	$[C_{16}H_{34}N_2O_4]^{2+}$	Suxethonium, 8798
318.255881	$C_{21}H_{34}O_2$	Allopregnan-3α-ol-20-one, 267 Allopregnan-3β-ol-20-one, 268 Allopregnan-20α-ol-3-one, 269 Allopregnan-20β-ol-3-one, 270 Arachidonic Acid, Methyl ester, 793 Etiocholanic Acid, Methyl ester, 3823 Pregnan-3α-ol-20-one, 7525 Pregnan-3β-ol-20-one, 7526 Pregnan-20α-ol-3-one, 7527 Pregnan-20β-ol-3-one, 7528 Urushiol II, 9553
318.279690	$[C_{21}H_{36}NO]^+$	Tridihexethyl, 9339
318.316075	$[C_{22}H_{40}N]^+$	Halimide®, Cation, 4448
318.982045	$C_{10}H_9NO_7S_2$	1-Naphthol-8-amino-3,6-disulfonic Acid, 6212
319.060273	$C_{12}H_{14}N_3NaO_4S$	Sulfamipyrine, 8715
319.101565	$C_{11}H_{17}N_3O_8$	Tetrodotoxin, 8961

Use in conjunction with The Merck Index, Ninth Edition

Molecular Weight	Empirical Formula	Compound Name, Monograph Number
319.141973	$C_{17}H_{21}NO_5$	Buphanitine, 1479 Scopolamine N-Oxide, 8160
319.167794	$[C_{39}H_{46}N_2O_6]^{2+}$	Tubocurarine, Dimethyl ether (half mass), 9466
319.181526	$C_{18}H_{26}ClN_3$	Chloroquine, 2146
319.214744	$C_{19}H_{29}NO_3$	Cyclodrine, 2724
319.287515	$C_{21}H_{37}NO$	Funtumidine, 4144
319.817531	F_7K_2NbO	Niobium Potassium Oxypentafluoride, 6379
319.821972	$HfCl_4$	Hafnium Tetrachloride, 4442
319.899738	$C_8H_8Cl_3O_3PS$	Ronnel, 8022
319.966061	$C_{10}H_8O_8S_2$	Chromotropic Acid, 2236
319.970770	$C_8H_8N_4O_4S_3$	Benzolamide, 1104
319.973790	$C_{14}H_9Cl_3F_2$	DFDT, 2914
319.999075	$C_{14}H_8O_7S$	Alizarinesulfonic Acid, 8312
320.011933	$C_{15}H_{10}Cl_2N_2O_2$	Lorazepam, 5407
320.053218	$C_{15}H_{12}O_8$	Ampelopsin, 612
320.079707	$C_{18}H_{12}N_2O_4$	Xanthocillin Y_2, 9724
320.092283	$[C_{19}H_{14}NO_4]^+$	Coptisine, 2504
320.104859	$C_{20}H_{16}O_4$	Dehydrootobain, 6733 Phenolphthalin, 7042
320.116093	$C_{19}H_{16}N_2O_3$	Fantan, 3875
320.125989	$C_{17}H_{20}O_6$	Mycophenolic Acid, 6152
320.140594	$C_{13}H_{24}N_2O_5S$	S-Acetylpantetheine, 6816
320.152478	$C_{20}H_{20}N_2O_2$	Prenazone, 7535
320.162374	$C_{18}H_{24}O_5$	Curvularin, O,O-Dimethyl ether, 2685
320.186184	$[C_{18}H_{26}NO_4]^+$	Methylconvolvaminium, 2487
320.198760	$C_{19}H_{28}O_4$	Atractyligenin, 885
320.235145	$C_{20}H_{32}O_3$	Tetrahydrocafestol, 1621 Grindelic Acid, 4390
320.246378	$C_{19}H_{32}N_2O_2$	Camylofine, 1739

320.

Molecular Weight	Empirical Formula	Compound Name, Monograph Number
320.271531	$C_{21}H_{36}O_2$	Allopregnane-3α,20α-diol, 251 Allopregnane-3α,20β-diol, 252 Allopregnane-3β,20α-diol, 253 Allopregnane-3β,20β-diol, 254 3-Pentadecylcatechol, 6903 5β-Pregnane-3α,20α-diol, 7523 Uranediol, 9510 Urushiol I, 9553
321.041942	$C_{13}H_{11}N_3O_5S$	Salazosulfamide, 8081
321.100108	$C_{19}H_{15}NO_4$	3-Methyl-6,7-methylenedioxy-1-piperonylisoquinoline, 5963
321.122108	$C_{17}H_{23}NOS_2$	Thihexinol, 9044
321.132471	$C_{15}H_{19}N_3O_5$	Carboquone, 1827
321.136494	$C_{20}H_{19}NO_3$	Oxazidione, 6752
321.160790	$C_{17}H_{24}ClN_3O$	Clamoxyquin, 2316
321.172879	$C_{21}H_{23}NO_2$	Norlobelanine, 6515
321.194009	$C_{18}H_{27}NO_4$	Carbetidine, 1798
321.209265	$C_{22}H_{27}NO$	Ethybenztropine, 3683 Phenazocine, 7003
321.266780	$C_{20}H_{35}NO_2$	Dihexyverine, 3145
321.782221	O_5Sb_2	Antimony Pentoxide, 736
321.865546	$C_6H_8Cl_6O_2$	Chloralacetonechloroform, 2026
322.012327	$C_{11}H_{12}Cl_2N_2O_5$	Chloramphenicol, 2040
322.017150	$C_9H_{18}Cl_3N_2O_2P$	Trofosfamide, 9431
322.032482	$C_{14}H_{10}O_9$	Digallic Acid, 3128
322.047738	$C_{18}H_{10}O_6$	Hydrindantin, 4662
322.073577	$C_{13}H_{14}N_4O_4S$	Acetyl Sulfamethoxypyrazine, 99 Sulfamethoxypyridazine, N^1-Acetyl deriv, 8710
322.116487	$C_{15}H_{18}N_2O_6$	Binapacryl, 1238
322.120509	$C_{20}H_{18}O_4$	Cyclocumarol, 2723 Phaseolin, 6981
322.131743	$C_{19}H_{18}N_2O_3$	Kebuzone, 5140
322.135115	$C_{16}H_{22}N_2O_3S$	Glyhexamide, 4343
322.168128	$C_{20}H_{22}N_2O_2$	Akuammicine, 190 Diocaine®, Free base, 3286 Gelsemine, 4218 Quininone, 7884
322.169958	$C_{19}H_{27}ClO_2$	Clostebol, 2368

Use in conjunction with The Merck Index, Ninth Edition

Molecular Weight	Empirical Formula	Compound Name, Monograph Number
322.178024	$C_{18}H_{26}O_5$	Zeranol, 9781
322.185235	$C_{12}H_{26}N_4O_6$	Neamine, 6261
322.204514	$C_{21}H_{26}N_2O$	Fenpipramide, 3913
322.240899	$C_{22}H_{30}N_2$	Aprindine, 782
322.359952	$C_{23}H_{46}$	Muscalure, 6134
323.038118	$C_{14}H_{14}NO_4PS$	EPN, 3549
323.047984	$C_{16}H_{15}Cl_2NO_2$	Bulan®, 1469
323.051853	$C_9H_{14}N_3O_8P$	2'-Cytidylic Acid, 2784 3'-Cytidylic Acid, 2785
323.115758	$C_{19}H_{17}NO_4$	Stylopine, 8654
323.130364	$C_{15}H_{21}N_3O_3S$	Gliclazide, 4271
323.145620	$C_{19}H_{21}N_3S$	Cyamemazine, 2692
323.152144	$C_{20}H_{21}NO_3$	Dimefline, 3190 Galipine, 4185
323.173273	$C_{17}H_{25}NO_5$	Cycloheximide, Acetate, 2733 3-Dibutylaminomethyl-4,5,6-trihydroxy-1-isobenzo- furanone, 3007
323.188529	$C_{21}H_{25}NO_2$	Piperidolate, 7264
323.199763	$C_{20}H_{25}N_3O$	Lysergide, 5453 Prodigiosin, 7563
323.209659	$C_{18}H_{29}NO_4$	Bufetolol, 1462 Lycofawcine, 5436
323.212339	$[C_{21}H_{27}N_2O]^+$	Pyronine® B, Free base, Cation, 7791
323.797327	F_6SnZr	Stannous Hexafluorozirconate(IV), 8564
323.871037	AuI	Gold Monoiodide, 4359
323.896809	CrO_4Pb	Lead Chromate(VI), 5255
323.926933	$C_2N_2PbS_2$	Lead Thiocyanate, 5284
323.957036	$C_8H_{11}Cl_3O_7$	Urochloralic Acid, 9547
324.032000	$C_{16}H_{14}Cl_2O_3$	Chlorobenzilate®, 2098
324.035869	$C_9H_{13}N_2O_9P$	5'-Uridylic Acid, 9544
324.084518	$C_{15}H_{16}O_8$	Leucodrin, 5305 Skimmin, 8299
324.099774	$C_{19}H_{16}O_5$	Efloxate, 3479
324.102941	$C_{19}H_{17}ClN_2O$	Prazepam, 7508

324.

Molecular Weight	Empirical Formula	Compound Name, Monograph Number
324.105647	$C_{12}H_{20}O_{10}$	Thrombo-Holzinger, Free acid, 9129
324.106984	$C_{13}H_{16}N_4O_6$	Furaltadone, 4145
324.114379	$C_{15}H_{20}N_2O_4S$	Acetohexamide, 48
324.129635	$C_{19}H_{20}N_2OS$	Propyromazine, Free base, 7661
324.133142	$C_8H_{20}Pb$	Tetraethyllead, 8921
324.133474	$C_{16}H_{16}N_6O_2$	Benzil, Disemicarbazone, 1084
324.136159	$C_{20}H_{20}O_4$	Otobain, 6733
324.142033	$C_{13}H_{24}O_9$	Strophanthobiose, 8647
324.147393	$C_{19}H_{20}N_2O_3$	Amphotalide, 622 Ditazol, 3385 Oxyphenbutazone, 6785
324.157289	$C_{17}H_{24}O_6$	Shellolic Acid, Dimethyl ester, 8221
324.168522	$C_{16}H_{24}N_2O_5$	Tricetamide, 9302
324.175917	$C_{18}H_{28}O_3S$	Sulfoxide, 8762
324.183778	$C_{20}H_{24}N_2O_2$	Epiquinidine, 3546 Epiquinine, 3547 Quinidine, 7850 Quinine, 7853 Viquidil, 9656
324.204908	$C_{17}H_{28}N_2O_4$	Embelin, Dioxime, 3502
324.220164	$C_{21}H_{28}N_2O$	Diampromide, 2948 N-(3-Diethylaminopropyl)-2,2-diphenylacetamide, 3090
324.232740	$[C_{22}H_{30}NO]^+$	Triethyl(2-P-styrylphenoxyethyl)ammonium, 9351
324.245316	$C_{23}H_{32}O$	MON-0585, 6075
324.290255	$[C_{20}H_{38}NO_2]^+$	Diponium, 3357
324.363026	$[C_{22}H_{46}N]^+$	Ethyldimethyl-9-octadecenylammonium, 3727
324.869809	LaN_3O_9	Lanthanum Nitrate, 5209
324.950517	$C_{13}H_9BrClNO_2$	5-Bromo-4'-chlorosalicylanilide, 1416
324.995777	$C_9H_{12}ClN_3O_4S_2$	Ethiazide, 3660
325.020755	$C_{10}H_{16}NO_5PS_2$	Famophos, 3874
325.030333	$C_{11}H_{11}N_5O_3S_2$	Urothion, 9550
325.084476	$C_{12}H_{15}N_5O_4S$	Sigumid, 8228

Use in conjunction with The Merck Index, Ninth Edition

Molecular Weight	Empirical Formula	Compound Name, Monograph Number
325.131408	$C_{19}H_{19}NO_4$	Methylactinodaphnine, 134 Actinodaphnine, Methyl ether, 134 Bulbocapnine, 1470 Domesticine, 3418 Nandinine, 6184
325.134575	$C_{19}H_{20}ClN_3$	Clemizole, 2320
325.152538	$C_{16}H_{23}NO_6$	Capobenic Acid, 1754 Monocrotaline, 6086
325.179027	$C_{19}H_{23}N_3O_2$	Ergometrinine, 3570 Ergonovine, 3571, 3572
325.204179	$C_{21}H_{27}NO_2$	Aprophen, 785 2,3-Diphenylpropionic Acid 2-Diethylaminoethyl Ester, 3345 Etafenone, 3638 Ifenprodil, 4804
325.227989	$[C_{21}H_{29}N_2O]^+$	Methylastrocasinium, 878 Denatonium, 2860
325.795610	CCs_2O_3	Cesium Carbonate, 1966
325.797454	La_2O_3	Lanthanum Oxide, 5209
325.842546	$C_7H_4Br_2O_5$	Dibromogallic Acid, 2989
325.858027	K_2O_4W	Potassium Tungstate(VI), 7491
325.868896	CeN_3O_9	Cerous Nitrate, 1956
325.946268	HgN_2O_6	Mercuric Nitrate, 5717
325.964445	$C_8H_{14}Cl_3O_5P$	Butonate, 1511
325.986112	$C_{13}H_8Cl_2N_2O_4$	Niclosamide, 6332
326.010859	$C_9H_{11}AsN_6O_3$	N-(4,6-Diamino-s-triazin-2-yl)arsanilic Acid, 2945
326.039501	$C_{13}H_{14}N_2O_4S_2$	Diphenylmethane-4,4'-disulfonamide, 3340 Gliotoxin, 4272
326.070798	$C_{18}H_{15}O_4P$	Triphenyl Phosphate, 9409
326.079038	$C_{18}H_{14}O_6$	Irisolone, Methyl ether, 4941 Rhein, Methyl ester dimethyl ether, 7966
326.100168	$C_{15}H_{18}O_8$	Melilotoside, 5641
326.111401	$C_{14}H_{18}N_2O_7$	Dinobuton, 3283
326.121297	$C_{12}H_{22}O_{10}$	Rutinose, 8063 Scillabiose, 8150
326.129824	$C_{18}H_{19}ClN_4$	Clozapine, 2374
326.145285	$C_{19}H_{22}N_2OS$	Acepromazine, 23 Thiazesim, 9037

326.

Molecular Weight	Empirical Formula	Compound Name, Monograph Number
326.163043	$C_{19}H_{22}N_2O_3$	Bumadizon, 1471
326.175619	$[C_{40}H_{48}N_2O_6]^{2+}$	Metocurine (half mass), 6019
326.181671	$C_{20}H_{26}N_2S$	Etymemazine, 3836
326.188195	$C_{21}H_{26}O_3$	Estrone, Propionate, 3636 Moxestrol, 6121 Octabenzone, 6551 Thymol Carbonate, 9143
326.191567	$C_{18}H_{30}O_3S$	Dodecylbenzenesulfonic Acid, 8361
326.199428	$C_{20}H_{26}N_2O_2$	Ajmaline, 185 Hydroquinidine, 4703 Hydroquinine, 4704
326.212004	$[C_{21}H_{28}NO_2]^+$	Methyladipheninium, 150 Methylcuspareinium, 2687
326.235814	$C_{21}H_{30}N_2O$	Bunaftine, 1473
326.318481	$C_{21}H_{42}O_2$	Arachidic Acid, Methyl ester, 792
326.789382	$C_6H_4Br_3N$	2,4,6-Tribromoaniline, 9288
327.034750	$C_{12}H_{13}N_3O_4S_2$	N^4-Sulfanilylsulfanilamide, 8724
327.065358	$C_{14}H_{14}N_3NaO_3S$	Methyl Orange, 5970
327.071135	$C_{13}H_{17}N_3O_3S_2$	Isobuzole, 5011
327.113840	$C_{18}H_{18}ClN_3O$	Loxapine, 5409
327.125929	$C_{22}H_{17}NO_2$	Normolaxol®, Base, 6519
327.147058	$C_{19}H_{21}NO_4$	Boldine, 1339 Corytuberine, 2531 Laurotetanine, 5233 Naloxone, 6182 Salutaridine, 8100
327.165686	$C_{20}H_{25}NOS$	Thiphenamil, 9110
327.174691	$C_{19}H_{22}FN_3O$	Azaperone, 912
327.176920	$C_{19}H_{25}N_3S$	Aminopromazine, 478
327.183444	$C_{20}H_{25}NO_3$	Benactyzine, 1034 Difemerine, 3120 Dimenoxadol, 3194
327.183444	$[C_{40}H_{50}N_2O_6]^{2+}$	Dimethyldauricinium (half mass), 2816
327.280024	$[C_{22}H_{35}N_2]^+$	Laurolinium, 5232
327.773398	$C_6H_3Br_3O$	2,4,6-Tribromophenol, 9292
327.871185	N_3NdO_9	Neodymium Nitrate, 6273
327.994260	C_7H_7ClHg	p-Tolylmercuric Chloride, 9242

Use in conjunction with The Merck Index, Ninth Edition

Molecular Weight	Empirical Formula	Compound Name, Monograph Number
328.037799	$C_{12}H_{14}N_2O_5P_2$	sym-Diphenylpyrophosphorodiamidic Acid, 3347
328.051779	$C_{16}H_{12}N_2O_4S$	Orange I, Free acid, 6699 Orange II, Free acid, 6700
328.053434	$C_{14}H_{17}O_5PS$	Hymecromone O,O-Diethyl Phosphorothioate, 4778
328.058303	$C_{17}H_{12}O_7$	Aflatoxin G_1, 169 Aflatoxin M_1, 170
328.064842	$C_{14}H_{17}ClN_2O_3S$	Clorexolone, 2362
328.079432	$C_{14}H_{16}O_9$	Bergenin, 1181
328.097856	$C_{18}H_{17}ClN_2O_2$	Oxazolam, 6753
328.099398	$C_{16}H_{16}N_4O_2S$	Sulfazamet, 8743
328.109945	$C_{22}H_{16}O_3$	Ethyl Biscoumacetate, 3701
328.115085	$C_6H_{26}CdN_6O_2$	Tris(ethylenediamine)cadmium Dihydroxide, 9418
328.115818	$C_{15}H_{20}O_8$	Androsin, 670
328.129731	$[C_{17}H_{18}N_3O_4]^+$	Celestine Blue, Cation, 1915
328.132412	$[C_{40}H_{32}N_8O_2]^{2+}$	Blue Tetrazolium, Cation (half mass), 1335
328.152203	$C_{16}H_{24}O_7$	Rhododendrin, 7981
328.153541	$C_{17}H_{20}N_4O_3$	Arabinose, Phenylosazone, 789 D-Ribose, Phenylosazone, 7997 Xylose, Phenylosazone, 9749
328.160935	$C_{19}H_{24}N_2OS$	Fencarbamide, 3895 Methotrimeprazine, 5859
328.167459	$C_{20}H_{24}O_4$	Crocetin, 2583 Guaiaretic Acid, 4405
328.191269	$[C_{20}H_{26}NO_3]^+$	Ethyl-N-methylmorphinium, 3768 Lachesine, Cation, 5180
328.203845	$C_{21}H_{28}O_3$	Boldenone, Acetate, 1338 Cyclethrin, 2711 Estradiol, 17-Propionate, 3626 11-Ketoprogesterone, 5155 Pyrethrin I, 7747
328.240230	$C_{22}H_{32}O_2$	Retinoic Acid, Ethyl ester, 7961 Synhexyl, 8804 Vitamin A, Acetate, 9666
328.251464	$C_{21}H_{32}N_2O$	Stanozolol, 8574
328.764074	$CoK_2O_8S_2$	Cobaltous Potassium Sulfate, 2400
328.844507	$C_6FeK_3N_6$	Potassium Ferricyanide, 7409
328.954390	$C_{11}H_{11}Cl_4NO_2$	Chlorbetamide, 2049

Molecular Weight	Empirical Formula	Compound Name, Monograph Number
328.975184	$C_8H_6F_3N_3O_4S_2$	Flumethiazide, 4026
329.052522	$C_{10}H_{12}N_5O_6P$	Cyclic AMP, 2714
329.101171	$C_{16}H_{15}N_3O_5$	Saluzid, 8101
329.137556	$C_{17}H_{19}N_3O_4$	Anthramycin, Methyl ether, 720
329.141579	$C_{22}H_{19}NO_2$	Indoxole, 4847
329.154642	$C_{20}H_{24}ClNO$	Cloperastine, 2358
329.162708	$C_{19}H_{23}NO_4$	Cinnamoylcocaine, 2292 Reticuline, 7959 Sinomenine, 8290
329.168760	$[C_{19}H_{25}N_2OS]^+$	N-Hydroxyethylpromethazine, 4726
329.177964	$C_{23}H_{23}NO$	WL 8008, 9713
329.201774	$[C_{23}H_{25}N_2]^+$	Malachite Green, Cation, 5527
329.214350	$C_{24}H_{27}N$	Prenylamine, 7536
329.235479	$C_{21}H_{31}NO_2$	Androisoxazole, 669
329.246713	$C_{20}H_{31}N_3O$	Vitamin A Aldehyde, Semicarbazone, 9668
329.304228	$C_{18}H_{39}N_3O_2$	Dodicin, 3413
329.730358	Ag_2CrO_4	Silver Chromate(VI), 8252
329.766761	$K_4O_7P_2$	Potassium Pyrophosphate, 7451
329.800058	O_3Pr_2	Praseodymium Oxide, 7506
329.819703	$C_6H_4Br_2O_4S$	3,5-Dibromo-4-hydroxybenzenesulfonic Acid, 2990
329.902020	$C_{10}H_6Cl_4O_4$	DCPA, 2820
329.973525	C_6H_5ClHgO	o-Hydroxyphenylmercuric Chloride, 4753
330.007721	$C_{12}H_{11}ClN_2O_5S$	Furosemide, 4161
330.036071	$C_{10}H_{19}O_6PS_2$	Malathion, 5528
330.070801	$C_{13}H_{18}N_2O_4S_2$	Penicillin O, 6888
330.073953	$C_{17}H_{14}O_7$	Aflatoxin G_2, 169 Aflatoxin M_2, 170
330.102272	$C_{19}H_{19}ClO_3$	Caprochlorone, 1758
330.132805	$C_{16}H_{18}N_4O_4$	Carvone, 2,4-Dinitrophenylhydrazone, 1867 Safranal, 2,4-Dinitrophenylhydrazone, 8075
330.138658	$C_{20}H_{23}ClO_2$	Ethynerone, 3812
330.140200	$C_{18}H_{22}N_2O_2S$	Oxomemazine, 6763
330.167853	$C_{16}H_{26}O_7$	Picrocrocin, 7211

Use in conjunction with The Merck Index, Ninth Edition

Molecular Weight	Empirical Formula	Compound Name, Monograph Number
330.206919	$[C_{20}H_{28}NO_3]^+$	Endobenzyline, 3517
330.219495	$C_{21}H_{30}O_3$	Deoxycorticosterone, 2863 17α-Hydroxyprogesterone, 4756 Jasmolin I, 5109 Nandrolone Propionate, 6189 Testosterone, Acetate, 8890
330.230728	$C_{20}H_{30}N_2O_2$	Dipiproverine, 3355 Furazabol, 4149
330.255881	$C_{22}H_{34}O_2$	Anagestone, 662 Pregnenolone Methyl Ether, 7534
330.328652	$C_{24}H_{42}$	Cholane, 2186
330.990834	$C_8H_8F_3N_3O_4S_2$	Hydroflumethiazide, 4681
331.075741	$C_{13}H_{18}ClN_3O_3S$	Glypinamide, 4350
331.081778	$[C_{17}H_{15}O_7]^+$	Malvidin, 5543
331.126717	$C_{14}H_{21}NO_8$	1,3,4-Tri-O-acetyl-N-acetyl-6-desoxy-β-D-glucosamine, 2890 Tetraacetylmycosamine, 6153
331.133907	$C_{19}H_{22}ClNO_2$	Difencloxazine, 3122
331.141973	$C_{18}H_{21}NO_5$	Protokylol, 7683 Tazettine, 8856
331.157229	$C_{22}H_{21}NO_2$	Benzylimidobis(p-methoxyphenyl)methane, 1156
331.170292	$C_{20}H_{26}ClNO$	Clofenetamine, 2340
331.189592	$C_{18}H_{25}N_3O_3$	1-(2-Diethylaminoethyl)phenobarbital, 3087
331.214744	$C_{20}H_{29}NO_3$	N-Ethyl-3-piperidyl Phenylcyclopentylglycolate, 3789 Promandeline 263, 7577 Propenzolate, 7604
331.298748	$C_{21}H_{37}N_3$	Funtumine, Hydrazone, 4144
331.800206	Nd_2O_3	Neodymium Oxide, 6273
331.808903	CeO_8S_2	Ceric Sulfate, 1949
331.836898	Cl_4Os	Osmium Tetrachloride, 6728
331.842766	$C_6CoK_3N_6$	Potassium Hexacyanocobaltate(III), 7418
331.878887	ITl	Thallium Iodide, 8977
331.952277	N_2O_6Pb	Lead Nitrate, 5268
331.957036	$C_{14}H_8N_2S_4$	2,2′-Dithiobis[benzothiazole], 3388
332.028065	$C_{14}H_8N_2O_8$	5,5′-Dinitrosalicil, 3282
332.056385	$C_{16}H_{13}ClN_2O_4$	Clorazepate, 2361

Molecular Weight	Empirical Formula	Compound Name, Monograph Number
332.068474	$C_{20}H_{12}O_5$	Fluorescein, 4040, 4042
332.074347	$C_{13}H_{16}O_{10}$	α-Glucogallin, 4284 β-Glucogallin, 4285
332.083079	$C_{16}H_{16}N_2O_4S$	Acedapsone, 15
332.091241	$C_{20}H_{12}MgN_4$	Porphine, Magnesium salt, 7372
332.092283	$[C_{20}H_{14}NO_4]^+$	Sanguinarine, 8112
332.158351	$C_{14}H_{24}N_2O_7$	Spectinomycin, 8513
332.162374	$C_{19}H_{24}O_5$	Illudin S, Diacetate, 4806 Trichothecin, 9331
332.198760	$C_{20}H_{28}O_4$	Marrubiin, 5582
332.235145	$C_{21}H_{32}O_3$	Alfaxalone, 226 Androstenediol, 3-Acetate, 673 Androstenediol, 17-Acetate, 673 Androsterone, Acetate, 676 Epiandrosterone, Acetate, 3535 Oxymetholone, 6780
332.256275	$C_{18}H_{36}O_5$	Aleuritic Acid, Ethyl ester, 223 Phloionolic Acid, 7135
332.271531	$C_{22}H_{36}O_2$	Bisnorcholanic Acid, 1321
332.961310	$C_{10}H_7NO_8S_2$	Nitroso-R Salt, Free acid, 6463
333.054801	$C_{15}H_{13}Cl_2N_5$	Robenidine, 8016
333.072825	$C_{17}H_{20}BrNO$	Bromodiphenhydramine, 1417
333.084059	$C_{16}H_{20}BrN_3$	p-Bromtripelennamine, 1449
333.194009	$C_{19}H_{27}NO_4$	Drotebanol, 3440
333.196689	$[C_{44}H_{50}N_4O_2]^{2+}$	Alcuronium (half mass), 214
333.197176	$C_{19}H_{28}ClN_3$	7-Chloro-4-(4-diethylamino-1-methylbutylamino)-3-methylquinoline, 2110
333.209265	$C_{23}H_{27}NO$	Deptropine, 2872
333.230394	$C_{20}H_{31}NO_3$	Carbetapentane, 1797
333.760022	$K_2O_8S_2Zn$	Potassium Zinc Sulfate, 7496
333.783527	I_2O_5	Iodine Pentoxide, 4880
333.868561	K_2O_4Os	Potassium Osmate(VI), 7436
333.899272	$C_{14}H_8Br_2$	9,10-Dibromoanthracene, 2985
333.981711	$C_{11}H_{10}O_8S_2$	Menadiol Disulfate, 5652
334.000744	$C_{11}H_{12}O_8P_2$	Menadiol Diphosphate, 5651

Use in conjunction with The Merck Index, Ninth Edition

Molecular Weight	Empirical Formula	Compound Name, Monograph Number
334.056604	$C_{11}H_{15}N_2O_8P$	NMN, 6480
334.083473	$C_{12}H_{18}N_2O_7S$	N^4-β-D-Glucosylsulfanilamide, 4296
334.084124	$C_{20}H_{14}O_5$	Fluorescin, 4043
334.090663	$C_{17}H_{19}ClN_2OS$	Opromazine, 6697
334.098729	$C_{16}H_{18}N_2O_4S$	Benzylpenicillinic Acid, 1161, 1043, 1066, 1159, 1160, 1162, 1163, 4640, 7122, 7557, 7750 Sulfaproxyline, 8730
334.100559	$C_{15}H_{23}ClO_4S$	Aramite®, 797
334.127720	$C_{15}H_{18}N_4O_5$	Mitomycin C, 6060
334.164105	$C_{16}H_{22}N_4O_4$	Rhodinal, 2,4-Dinitrophenylhydrazone, 7976
334.168128	$C_{21}H_{22}N_2O_2$	Strychnine, 8649, 8650, 8652, 8653
334.179361	$C_{20}H_{22}N_4O$	Difenamizole, 3121
334.189258	$C_{18}H_{26}N_2O_4$	Proglumide, 7570
334.214410	$C_{20}H_{30}O_4$	Atractyligenin, Methyl ester, 885
334.225643	$C_{19}H_{30}N_2O_3$	Fenalamide, 3888
334.250795	$C_{21}H_{34}O_3$	Allopregnane-3β,17α-diol-20-one, 256 Grindelic Acid, Methyl ester, 4390
334.310990	$[C_{22}H_{40}NO]^+$	Domiphen, 3419
335.057593	$C_{14}H_{13}N_3O_5S$	Sulfanitran, 8727
335.068826	$C_{13}H_{13}N_5O_4S$	3,5-Diamino-2-[p-(sulfamoylphenyl)azo]benzoic Acid, 2944
335.109291	$C_{13}H_{16}F_3N_3O_4$	Trifluralin, 9356
335.173273	$C_{18}H_{25}NO_5$	Senecionine, 8197
335.176440	$C_{18}H_{26}ClN_3O$	Hydroxychloroquine, 4718
335.188529	$C_{22}H_{25}NO_2$	Lobelanine, 5395 Tropacine, 9436
335.224915	$C_{23}H_{29}NO$	Norpipanone, 6523
335.246044	$C_{20}H_{33}NO_3$	Ganglerone, 4200 Oxeladin, 6756
335.796676	$C_4H_2I_2S$	Thiophene Diiodide, 9092
335.802259	$BaCl_2O_8$	Barium Perchlorate, 992
335.877905	O_3PbSe	Lead Selenite, 5273
335.928565	$C_4H_{10}CrN_7S_4$	Reinecke Salt, 7926

336.

Molecular Weight	Empirical Formula	Compound Name, Monograph Number
336.063388	$C_{19}H_{12}O_6$	Dicumarol, 3066 Pachyrrhizin, 6795
336.099774	$C_{20}H_{16}O_5$	Psoralidin, 7714
336.112837	$C_{18}H_{21}ClO_4$	Barthrin, 1010
336.123583	$[C_{20}H_{18}NO_4]^+$	Berberine, 1177
336.145583	$[C_{18}H_{26}NOS_2]^+$	Methylthihexinolium, 9044
336.162649	$C_{24}H_{20}N_2$	N,N'-Diphenylbenzidine, 3327
336.204908	$C_{18}H_{28}N_2O_4$	Acebutolol, 14
336.210073	$C_{20}H_{29}FO_3$	Fluoxymesterone, 4070
336.220164	$C_{22}H_{28}N_2O$	2,2-Diphenyl-4-hexamethyleneiminobutyramide, 3334 Fentanyl, 3918
336.241293	$C_{19}H_{32}N_2O_3$	2-Diethylaminoethyl 4-Butylamino-2-ethoxybenzoate, 3084
336.266445	$C_{21}H_{36}O_3$	Allopregnane-3β,17α,20α-triol, 262 Allopregnane-3β,17α,20β-triol, 263
336.894915	$C_9H_9Br_2NO_3$	3,5-Dibromotyrosine, 2998
337.137282	$C_{13}H_{23}NO_9$	Streptobiosamine, 8607
337.188923	$C_{18}H_{27}NO_5$	Platyphylline, 7310 Propanidid, 7594
337.204179	$C_{22}H_{27}NO_2$	Danazol, 2799 Lobeline, 5397 Pifenate, 7222
337.227989	$[C_{22}H_{29}N_2O]^+$	Fenpiverinium, 3915
337.723863	S_3Sb_2	Antimony Trisulfide, 751
337.812582	$C_6Cu_2FeN_6$	Cupric Ferrocyanide, 2638
337.849796	$AuCl_4H$	Gold Trichloride, Acid, 4371
337.855980	$N_4O_{12}Zr$	Zirconium Nitrate, 9845
337.882870	$PbTe$	Lead Telluride, 5280
337.934573	$C_2H_5AuO_3S_2$	Auromercaptoethane Sulfonic Acid, 8324
337.935127	$H_4O_4P_2Pb$	Lead Hypophosphite, 5263
337.986676	$C_7H_4HgO_3$	Mercuric Salicylate, 5722
338.012807	$C_{14}H_{11}ClN_2O_4S$	Chlorthalidone, 2182
338.023062	$C_8H_8HgO_2$	Phenylmercuric Acetate, 7105
338.062752	$C_9H_{15}N_4O_8P$	AICAR, 180

Molecular Weight	Empirical Formula	Compound Name, Monograph Number
338.079038	$C_{19}H_{14}O_6$	Diacetoxychrysin, 2254 Rubiadin, Diacetate, 8039
338.139233	$[C_{20}H_{20}NO_4]^+$	Columbamine, 2459 Jatrorrhizine, 5111
338.139925	$C_{14}H_{26}O_7S$	Diamyl Sulfosuccinic Acid, 2950
338.147787	$C_{16}H_{22}N_2O_6$	2,5-Bis(1-aziridinyl)-3,6-bis(2-methoxyethoxy)-p-benzo-quinone, 1259
338.163043	$C_{20}H_{22}N_2O_3$	Perivine, 6963
338.175619	$[C_{21}H_{24}NO_3]^+$	Methylgalipinium, 4185
338.188195	$C_{22}H_{26}O_3$	Bioresmethrin, 1242 Ethynylestradiol, 3-Acetate, 3817
338.209324	$C_{19}H_{30}O_5$	Piperonyl Butoxide, 7270
338.210662	$C_{20}H_{26}N_4O$	Lysuride, 5461
338.233134	$[C_{19}H_{32}NO_4]^+$	Methyllycofawcinium, 5436
338.293329	$[C_{20}H_{38}N_2O_2]^{2+}$	Dipropamine, Cation, 3360
338.318481	$C_{22}H_{42}O_2$	Brassidic Acid, 1372
338.884697	EuN_3O_9	Europic Nitrate, 3853
339.040481	$C_{14}H_{18}BrN_3S$	2-[(5-Bromo-2-thenyl)(2-dimethylaminoethyl)amino]-pyridine, 1442
339.052507	$C_{13}H_{13}N_3O_6S$	Cephacetrile, 1930
339.125278	$C_{15}H_{21}N_3O_4S$	1-Butyl-3-(N-crotonoylsulfanilyl)urea, 1552
339.147058	$C_{20}H_{21}NO_4$	Bulbocapnine, Methyl ether, 1470 Canadine, 1740 Dicentrine, 3014 Papaverine, 6823-6825
339.153110	$[C_{20}H_{23}N_2OS]^+$	Propyromazine, Cation, 7661
339.176920	$C_{20}H_{25}N_3S$	Perazine, 6946
339.183444	$C_{21}H_{25}NO_3$	Piperilate, 7265
339.194677	$C_{20}H_{25}N_3O_2$	Methylergonovine, 5938
339.207253	$[C_{21}H_{27}N_2O_2]^+$	Methylquinidinium, 7850
339.219829	$C_{22}H_{29}NO_2$	Levopropoxyphene, 5315 Lobelanidine, 5394 Noracymethadol, 6493 Propoxyphene, 7627
339.231063	$C_{21}H_{29}N_3O$	Disopyramide, 3378
339.361349	$C_{21}H_{45}N_3$	Hexetidine, 4575

339.

Molecular Weight	Empirical Formula	Compound Name, Monograph Number
339.811674	$C_6FeN_6Zn_2$	Zinc Ferrocyanide, 9794
339.978321	Cl_2O_2U	Uranyl Chloride, 9520
339.996053	$C_6H_{14}O_{12}P_2$	Fructose-1,6-diphosphate, 4123
340.019456	$C_6H_7BHgO_3$	Phenylmercury Borate, 7108
340.027714	$C_9H_{20}Cl_3N_2O_3P$	Defosfamide, 2838
340.079432	$C_{15}H_{16}O_9$	Cichoriin, 2264 Daphnin, 2806 Esculin, 3621
340.092003	$C_{15}H_{12}N_6O_4$	Rhizopterin, 7972
340.105922	$C_{18}H_{16}N_2O_5$	Nybomycin, Acetate, 6543
340.109945	$C_{23}H_{16}O_3$	Diphenadione, 3311
340.136947	$C_{13}H_{24}O_{10}$	Gaultherioside, 4210
340.138285	$C_{14}H_{20}N_4O_6$	Inhasan, 4850
340.160935	$C_{20}H_{24}N_2OS$	Lucanthone, 5410 Propiomazine, 7612 Propionylpromazine, 7619
340.163437	$C_{16}H_{24}N_2O_6$	Pyrisuccideanol, 7772
340.167459	$C_{21}H_{24}O_4$	Ostruthin, Acetate, 6731
340.176013	$[C_{17}H_{26}NO_6]^+$	Methylmonocrotalinium, 6086
340.178693	$C_{20}H_{24}N_2O_3$	Lochneridine, 5399
340.189926	$C_{19}H_{24}N_4O_2$	Pentamidine, 6912
340.191269	$[C_{21}H_{26}NO_3]^+$	Mepenzolate, 5674 Methantheline, 5815 Poldine, 7335
340.194641	$[C_{18}H_{30}NO_3S]^+$	Penthienate, 6926
340.203845	$C_{22}H_{28}O_3$	Canrenone, 1750 17-Hydroxy-16-methylene-Δ^6-progesterone, 4737 Norethindrone, Acetate, 6506 Santalyl Salicylate, 8115
340.215078	$C_{21}H_{28}N_2O_2$	N-Formyl-N-deacetylaspidospermine, 873 Ethylhydrocupreine, 3747
340.240230	$C_{23}H_{32}O_2$	Dimethisterone, 3205 Medrogestone, 5617
340.334131	$C_{22}H_{44}O_2$	Arachidic Acid, Ethyl ester, 792 Behenic Acid, 1029 Butyl Stearate, 1579 Isobutyl Stearate, 5005
340.964802	$C_3H_{13}N_3O_{10}Zr$	Ammonium Zirconyl Carbonate, 604

Use in conjunction with The Merck Index, Ninth Edition

Molecular Weight	Empirical Formula	Compound Name, Monograph Number
341.050400	$C_{13}H_{15}N_3O_4S_2$	4'-(Methylsulfamoyl)sulfanilanilide, 5994
341.053552	$C_{17}H_{11}NO_7$	Aristolochic Acid, 811
341.140928	$C_{15}H_{23}N_3O_4S$	Cyclacillin, 2706 Sulpiride, 8786
341.154642	$C_{21}H_{24}ClNO$	Clobenztropine, 2331
341.162708	$C_{20}H_{23}NO_4$	Corydine, 2527 Corypalmine, 2530 Dihydrocodeinone Enol Acetate, 3149 Isocorydine, 5017 Isocorypalmine, 5018
341.185175	$C_{18}H_{23}N_5O_2$	Fenethylline, 3901
341.199094	$C_{21}H_{27}NO_3$	Benapryzine, 1036
341.688994	$C_2H_2Br_4$	sym-Tetrabromoethane, 8902
341.711454	Ag_2O_3Se	Silver Selenite, 8273
341.711785	Br_2O_6Sr	Strontium Bromate, 8626
341.714579	I_2Sr	Strontium Iodide, 8635
341.789048	$C_7H_5Br_3O$	2,4,6-Tribromo-m-cresol, 9290
341.818274	$Al_2O_{12}S_3$	Aluminum Sulfate, 368
341.825713	$C_8Co_2O_8$	Dicobalt Octacarbonyl, 3061
341.879711	$AsHHgO_4$	Mercuric Arsenate, 5707
341.973861	$C_{11}H_{16}ClO_2PS_3$	Carbophenothion, 1826
341.985926	$C_{12}H_6O_{12}$	Mellitic Acid, 5645
342.045649	$C_{12}H_{14}N_4O_4S_2$	Thiophanate-Methyl, 9089
342.065887	$C_{19}H_{15}ClO_4$	Coumachlor, 2540
342.073953	$C_{18}H_{14}O_7$	Gentisin, Diacetate, 4231
342.110339	$C_{19}H_{18}O_6$	Tetramethylbaptigenin, 959 Tetramethoxyfisetin, 4003
342.116212	$C_{12}H_{22}O_{11}$	Cellobiose, 1916 Gentiobiose, 4228 Lactose, 5192 β-Lactose, 5192 Lactulose, 5196 Maltose, 5541 Melibiose, 5639 Sophorose, 8494 Sucrose, 8681 Trehalose, 9267 Turanose, 9472
342.131468	$C_{16}H_{22}O_8$	Coniferin, 2473

342.

Molecular Weight	Empirical Formula	Compound Name, Monograph Number
342.155576	$C_{24}H_{20}BNa$	Sodium Tetraphenylborate, 8461
342.161329	$C_{16}H_{26}N_2O_4S$	Heptylpenicillin, 4529
342.167853	$C_{17}H_{26}O_7$	Rhododendrin, Methyl ether, 7981
342.169191	$C_{18}H_{22}N_4O_3$	D-Fucose, Phenylosazone, 4129 L-Fucose, Phenylosazone, 4130
342.170533	$[C_{20}H_{24}NO_4]^+$	Magnoflorine, 5524
342.206919	$[C_{21}H_{28}NO_3]^+$	Methylbenactyzinium, 1034
342.219495	$C_{22}H_{30}O_3$	Siccanin, 8226
342.957344	Cl_3U	Uranium Trichloride, 9517
343.084458	$C_{21}H_{13}NO_4$	1-Aminoanthraquinone-2-carboxylic Acid, Phenyl ester, 420
343.090997	$C_{18}H_{18}ClN_3S$	Clothiapine, 2369
343.095691	$C_{20}H_{13}N_3O_3$	Violacein, 9649
343.115478	$C_{15}H_{15}N_6O_4$	Etofylline Nicotinate, 3829
343.170292	$C_{21}H_{26}ClNO$	Clemastine, 2319
343.171834	$C_{19}H_{25}N_3OS$	Thiambutosine, 9021
343.178358	$C_{20}H_{25}NO_4$	Codamine, 2419 Laudanidine, 5223 Laudanine, 5224
343.225977	$C_{20}H_{29}N_3O_2$	4-Androstene-3,17-dione, 3-Semicarbazone, 675 Dibucaine, 3000
343.251130	$C_{22}H_{33}NO_2$	Atisine, 883 Garryine, 4207 Veatchine, 9598
343.650273	Br_4Si	Silicon Tetrabromide, 8240
343.887565	GdN_3O_9	Gadolinium Nitrate, 4174
343.960292	$C_{12}H_6Cl_2N_2O_6$	Menichlopholan, 5661
344.013763	$C_{16}H_{15}Cl_3O_2$	Methoxychlor, 5865
344.089603	$C_{18}H_{16}O_7$	Eupatorin, 3847 Usnic Acid, 9556
344.108026	$C_{22}H_{17}ClN_2$	Clotrimazole, 2370
344.138093	$C_{19}H_{24}N_2S_2$	Methiomeprazine, 5843
344.154308	$C_{21}H_{25}ClO_2$	Trengestone, 9272
344.163711	$[C_{21}H_{20}N_4O]^{2+}$	Quinuronium, 3371
344.178117	$[C_{21}H_{27}ClNO]^+$	Methylcloperastinium, 2358

Use in conjunction with The Merck Index, Ninth Edition

Molecular Weight	Empirical Formula	Compound Name, Monograph Number
344.198760	$C_{21}H_{28}O_4$	11-Dehydrocorticosterone, 2844 Formyldienolone, 4105
344.209993	$C_{20}H_{28}N_2O_3$	Oxyphencyclimine, 6786
344.235145	$C_{22}H_{32}O_3$	Anacardic Acid, 660 Medroxyprogesterone, 5618 Medrysone, 5620 Methenolone, 17-Acetate, 5836 Stenbolone, Acetate, 8586 Testosterone Propionate, 8896
344.246378	$C_{21}H_{32}N_2O_2$	Progesterone, Dioxime, 7569
344.258955	$[C_{22}H_{34}NO_2]^+$	Cyclonium, 2742
344.329046	$C_{21}H_{44}O_3$	Batyl Alcohol, 1018
344.851153	$C_6H_5I_2N$	2,4-Diiodoaniline, 3171
344.888804	N_3O_9Tb	Terbium Nitrate, 8877
345.047436	$C_{10}H_{12}N_5O_7P$	Cyclic GMP, 2715
345.091391	$C_{14}H_{20}ClN_3O_3S$	Clopamide, 2356
345.149557	$C_{20}H_{24}ClNO_2$	Diaphen, 2954 Metofoline, 6020
345.157623	$C_{19}H_{23}NO_5$	Metaphanine, 5781 Trimetoquinol, 9389
345.303165	$C_{23}H_{39}NO$	N,N-Dimethylfuntumine, 4144
345.957416	$C_4H_5AuO_4S$	Gold Thiomalic Acid, 4365
345.968439	$C_6H_5ClHgO_2$	Hydroxymercurichlorophenols, p-Form (n = 1), 4734
346.007633	$C_{11}H_{17}Cl_3N_2O_2S$	Chlordantoin, 2052
346.037305	$C_{14}H_{16}ClO_6P$	Coralox®, 2505
346.102101	$C_{14}H_{22}N_2O_4S_2$	Penicillin BT, 6886
346.105253	$C_{18}H_{18}O_7$	Frenolicin, 4118
346.116487	$C_{17}H_{18}N_2O_6$	Nifedipine, 6352
346.120509	$C_{22}H_{18}O_4$	Benzyl Phthalate, 1167 o-Cresolphthalein, 2573
346.126383	$C_{15}H_{22}O_9$	Aucubin, 896
346.127048	$C_{19}H_{23}ClN_2S$	Chlorproethazine, 2173
346.141639	$C_{19}H_{22}O_6$	Gibberellic Acid, 4250
346.143601	$C_{12}H_{24}N_6O_2P_2$	Dipin, 3353
346.165443	$C_{18}H_{18}N_8$	Bismark Brown Y, Free base, 1276

346.

Molecular Weight	Empirical Formula	Compound Name, Monograph Number
346.165448	$[C_{19}H_{24}NO_5]^+$	Methyltazettinium, 8856
346.178024	$C_{20}H_{26}O_5$	Allethrin II, 242
346.213067	$[C_{19}H_{28}N_3O_3]^+$	Defenal, Cation, 2836
346.214410	$C_{21}H_{30}O_4$	Algestone, 227 Corticosterone, 2513 11-Desoxy-17-hydroxycorticosterone, 2891 4-Pregnene-20,21-diol-3,11-dione, 7529
346.215747	$C_{22}H_{26}N_4$	Calycanthine, 1728 Chimonanthine, 2015
346.225643	$C_{20}H_{30}N_2O_3$	Morpheridine, 6106
346.235539	$C_{18}H_{34}O_6$	Phloionic Acid, 7134
346.250795	$C_{22}H_{34}O_3$	Androsterone, Propionate, 676 Mesterolone, Acetate, 5762
346.262029	$C_{21}H_{34}N_2O_2$	3,20-Pregnanedione, Dioxime, 7524
346.271925	$C_{19}H_{38}O_5$	Phloionolic Acid, Methyl ester, 7135
346.287181	$C_{23}H_{38}O_2$	Norcholanic Acid, 6499
346.298414	$C_{22}H_{38}N_2O$	Funtessine, 4143
346.334800	$C_{23}H_{42}N_2$	Chonemorphine, 2214
346.773595	CHI_2O_3S	Dimethiodal, 3203
346.830418	$C_5H_3I_2NO$	Iopydone, 4921
346.910811	$C_{10}H_9Cl_4NO_2S$	Captafol, 1769
346.963108	$C_{11}H_{14}AsNO_3S_2$	Arsthinol, 841
347.063086	$C_{10}H_{14}N_5O_7P$	3'-Adenylic Acid, 146 5'-Adenylic Acid, 147
347.088476	$C_{18}H_{22}BrNO$	Embramine, 3504
347.093978	$C_{16}H_{17}N_3O_4S$	Cephalexin, 1932
347.173273	$C_{19}H_{25}NO_5$	Dihydrometaphanine, 5781
347.199763	$C_{22}H_{25}N_3O$	Benzpiperylon, 1131
347.224915	$C_{24}H_{29}NO$	Phenomorphan, 7047
347.846151	$H_8MgN_2O_8Se_2$	Ammonium Magnesium Selenate, 557
348.004403	$C_{14}H_9Cl_2F_3N_2O$	Cloflucarban, 2344
348.043233	$C_{17}H_{14}Cl_2N_2O_2$	Cloxazolam, 2373
348.047102	$C_{10}H_{13}N_4O_8P$	Inosinic Acid, 4852
348.077994	$C_{16}H_{16}N_2O_5S$	Succisulfone, 8680

Molecular Weight	Empirical Formula	Compound Name, Monograph Number
348.111007	$C_{20}H_{16}N_2O_4$	Camptothecin, 1738
348.123583	$[C_{21}H_{18}NO_4]^+$	Chelerythrine, 2007
348.132137	$C_{17}H_{20}N_2O_6$	α-Terpineol, Dinitrobenzoate, 8886
348.136159	$C_{22}H_{20}O_4$	Phenolphthalol, Monoacetate, 7043
348.143370	$C_{16}H_{20}N_4O_5$	Porfiromycin, 7370
348.147393	$C_{21}H_{20}N_2O_3$	Alstonine, 310 Serpentine (Alkaloid), 8211
348.175917	$C_{20}H_{28}O_3S$	Ethyl Dibunate, 3723
348.193674	$C_{20}H_{28}O_5$	Pododacric Acid, 7322
348.220164	$C_{23}H_{28}N_2O$	Leiopyrrole, 5290
348.230060	$C_{21}H_{32}O_4$	Allopregnane-3β,21-diol-11,20-dione, 255 Androstane-3β,11β-diol-17-one, 3-Acetate, 672 4-Pregnene-17α,20β,21-triol-3-one, 7532
348.253869	$[C_{21}H_{34}NO_3]^+$	Oxyphenonium, 6788
348.857833	Cl_4Po	Polonium Tetrachloride, 7340
348.923089	$C_{13}H_7Cl_4NO_2$	3,3',4',5-Tetrachlorosalicylanilide, 8909
348.926287	$C_9H_{11}Cl_3NO_3PS$	Chlorpyrifos, 2179
349.071526	$C_{16}H_{16}NO_6P$	Naftalofos, 6177
349.109628	$C_{16}H_{19}N_3O_4S$	Ampicillin, 624 Cephradine, 1943
349.127386	$C_{16}H_{19}N_3O_6$	Mitomycin A, 6060 Mitomycin B, 6060
349.131408	$C_{21}H_{19}NO_4$	Cinmetacin, 2284
349.207551	$C_{20}H_{31}NO_2S$	Cetiedil, 1975
349.240565	$C_{24}H_{31}NO$	Dipipanone, 3354 Pipidone, 7274
349.843277	$CeH_4NO_8S_2$	Ammonium Cerous Sulfate, 527
349.892637	DyN_3O_9	Dysprosium Nitrate, 3457
349.943989	F_6PbSi	Lead Hexafluorosilicate, 5260
350.043523	$C_{20}H_{14}O_2S_2$	DDD (Analytical), 2821
350.079038	$C_{20}H_{14}O_6$	3,3'-Ethylidenebis(4-hydroxycoumarin), 3749
350.093644	$C_{16}H_{18}N_2O_5S$	p-Hydroxybenzylpenicillin, 4713 Penicillin V, 6890, 1067, 4641, 6894
350.118796	$C_{18}H_{22}O_5S$	Estrone, Sulfate, 3636

Molecular Weight	Empirical Formula	Compound Name, Monograph Number
350.151809	$C_{22}H_{22}O_4$	Dienestrol, Diacetate, 3076
350.163043	$C_{21}H_{22}N_2O_3$	Strychnine N^6-Oxide, 8651
350.167065	$C_{26}H_{22}O$	Benzohydrol, Diphenylmethyl ether, 1099
350.209324	$C_{20}H_{30}O_5$	Andrographolide, 668 Taxicin-II, 8853
350.245710	$C_{21}H_{34}O_4$	Allopregnane-3β,11β,21-triol-20-one, 265 Allopregnane-3β,17α,21-triol-20-one, 266
350.329714	$C_{22}H_{42}N_2O$	2-(Heptadecenyl)-2-imidazoline-1-ethanol, 4515 2-(8-Heptadecenyl)-2-imidazoline-1-ethanol, 4516
350.906232	$C_9H_6INO_4S$	8-Hydroxy-7-iodo-5-quinolinesulfonic Acid, 4728
351.110673	$C_{20}H_{17}NO_5$	Worenine, 9716
351.125278	$C_{16}H_{21}N_3O_4S$	Epicillin, 3539
351.147058	$C_{21}H_{21}NO_4$	Diacetylapomorphine, 775
351.158292	$C_{20}H_{21}N_3O_3$	Moquizone, 6098
351.183444	$C_{22}H_{25}NO_3$	Pipoxolan, 7278 Tropine Benzylate, 9447
351.219829	$C_{23}H_{29}NO_2$	Phenadoxone, 6991
351.243639	$[C_{23}H_{31}N_2O]^+$	Metazepium, 3334
351.361349	$C_{22}H_{45}N_3$	Hexedine, 4569
351.665472	Se_2S_6	Selenium Sulfides, 8187
351.742536	Cr_2O_8Sn	Stannic Chromate(VI), 8553
351.790435	$AlCsO_8S_2$	Aluminum Cesium Sulfate, 332
351.824226	O_3Sm_2	Samarium Oxide, 8106
351.872820	O_4PbSe	Lead Selenate, 5272
351.879780	$BiIO$	Bismuth Iodide Oxide, 1292
351.893759	ErN_3O_9	Erbium Nitrate, 3556
351.914689	$C_{14}H_9Cl_5$	DDT, 2822
351.955890	$C_{10}H_8O_{10}S_2$	4-Methylesculetindisulfonic Acid, 5939
352.041205	F_6U	Uranium Hexafluoride, 9513
352.043047	$C_{15}H_{12}O_{10}$	Methylenedigallic Acid, 5933
352.056776	$C_{17}H_{18}Cl_2N_2S$	Dichlorpromazine, 3057
352.058303	$C_{19}H_{12}O_7$	Coumestrol, Diacetate, 2549
352.071366	$C_{17}H_{17}ClO_6$	Griseofulvin, 4392

Use in conjunction with The Merck Index, Ninth Edition

Molecular Weight	Empirical Formula	Compound Name, Monograph Number
352.084142	$C_{14}H_{16}N_4O_5S$	Sulfadimethoxine, N^4Acetyl deriv, 8696
352.094688	$C_{20}H_{16}O_6$	Dithranol, Triacetate, 3394 Elliptone, 3499 Viridin, 9660
352.122105	$C_{18}H_{19}F_3N_2S$	Triflupromazine, 9355
352.132411	$C_{22}H_{16}N_4O$	Sudan III, 8683
352.154883	$[C_{21}H_{22}NO_4]^+$	Palmatine, Cation, 6803
352.162703	$C_{16}H_{33}I$	Cetyl Iodide, 1984
352.167459	$C_{22}H_{24}O_4$	Diethylstilbestrol, Diacetate, 3113
352.178693	$C_{21}H_{24}N_2O_3$	Ajmalicine, 184 Diaboline, 2916 Lochnericine, 5398
352.191269	$[C_{22}H_{26}NO_3]^+$	Clidinium, 2323
352.215078	$C_{22}H_{28}N_2O_2$	Anileridine, 691
352.224974	$C_{20}H_{32}O_5$	Grayanotoxin II, 4388 Prostaglandin E_2, 7665
352.240230	$C_{24}H_{32}O_2$	Quinbolone, 7841
352.261360	$C_{21}H_{36}O_4$	Allopregnane-3β,17α,20β,21-tetrol, 259
352.751487	$Cu_3H_4O_8S$	Copper Sulfate Dibasic, 2658
352.997575	$C_7H_5HgNO_3$	Nitromersol, 6432
353.027078	$C_{11}H_{16}ClN_3O_4S_2$	Buthiazide, 1508
353.083413	$C_{18}H_{15}N_3O_3S$	Metanil Yellow, Free acid, 5780
353.089937	$C_{19}H_{15}NO_6$	Acenocoumarin, 20
353.126323	$C_{20}H_{19}NO_5$	Chelidonine, 2009 Papaveraldine, 6822 Protopine, 7684
353.140928	$C_{16}H_{23}N_3O_4S$	Dibupyrone, Free acid, 3002
353.183838	$C_{18}H_{27}NO_6$	Hygrophylline, 4776
353.199094	$C_{22}H_{27}NO_3$	Dioxaphetyl Butyrate, 3301
353.201774	$[C_{25}H_{25}N_2]^+$	Quinaldine Blue, Cation, 7834
353.210327	$C_{21}H_{27}N_3O_2$	Methysergid(e), 6011
353.235479	$C_{23}H_{31}NO_2$	Acetylmethadol, 91
353.259289	$[C_{23}H_{33}N_2O]^+$	Isopropamide, Cation, 5063
353.827230	Eu_2O_3	Europium Sesquioxide, 3853

353.

Molecular Weight	Empirical Formula	Compound Name, Monograph Number
353.884262	$C_{12}H_6Cl_4O_2S$	Bithionol, 1328 Tetradifon, 8916
353.957850	$C_{10}H_{16}Br_2N_2O_2$	Pipobroman, 7275
353.971274	$C_9H_{11}IN_2O_5$	Idoxuridine, 4802
354.017976	$C_8H_8HgO_3$	Afridol, Free acid, 171
354.044107	$C_{15}H_{15}ClN_2O_4S$	Xipamide, 9741
354.056196	$C_{19}H_{14}O_5S$	Phenolsulfonphthalein, 7045
354.095082	$C_{16}H_{18}O_9$	Chlorogenic Acid, 2121 4-Methylesculin, 3621 Scopolin, 8162
354.110339	$C_{20}H_{18}O_6$	Asarinin, 846 Sesamin, 8215
354.146724	$C_{21}H_{22}O_5$	Pleurotin(e), 7315
354.194343	$C_{21}H_{26}N_2O_3$	Corynanthine, 2529 Pseudoyohimbine, 7708 Vincamine, 9638 Yohimbine, 9769 allo-Yohimbine, 9770 α-Yohimbine, 9771
354.206919	$[C_{22}H_{28}NO_3]^+$	Benzilonium, 1087 Bevonium, 1221 Pipenzolate, 7252
354.219495	$C_{23}H_{30}O_3$	Melengestrol, 5637
354.222867	$C_{20}H_{34}O_3S$	Oxychlorosene, (less HOCl), 6772
354.230728	$C_{22}H_{30}N_2O_2$	Aspidospermine, 873 Eprozinol, 3550
354.240624	$C_{20}H_{34}O_5$	Prostaglandin E$_1$, 7664 Prostaglandin F$_{2\alpha}$, 7666
354.313395	$C_{22}H_{42}O_3$	Butyl Ricinoleate, 1577 Erucic Acid, 3594
354.349781	$C_{23}H_{46}O_2$	Behenic Acid, Methyl ester, 1029
355.004800	$C_{12}H_{15}Cl_2NO_5S$	Thiamphenicol, 9034
355.029664	$C_{13}H_{13}N_3O_5S_2$	Succinylsulfathiazole, 8679
355.066050	$C_{14}H_{17}N_3O_4S_2$	4'-(Dimethylsulfamoyl)sulfanilanilide, 3245
355.069202	$C_{18}H_{13}NO_7$	Aristolochic Acid, Methyl ester, 811
355.075741	$C_{15}H_{18}ClN_3O_3S$	Tiaramide, 9157
355.111461	$C_{12}H_{21}NO_{11}$	Chondrosine, 2213 Hyalobiuronic Acid, 4633

Use in conjunction with The Merck Index, Ninth Edition

Molecular Weight	Empirical Formula	Compound Name, Monograph Number
355.145140	$C_{20}H_{22}ClN_3O$	Amodiaquin, 606
355.178358	$C_{21}H_{25}NO_4$	Corybulbine, 2521 Glaucine, 4267 Isocorybulbine, 5016 Tetrahydropalmatine, 8932
355.184410	$[C_{21}H_{27}N_2OS]^+$	Methylpropiomazinium, 7612 Methylpropionylpromazinium, 7619
355.194758	$C_{22}H_{26}FNO_2$	Moperone, 6097
355.196987	$C_{22}H_{29}NOS$	Diprophen, 3361
355.238553	$[C_{22}H_{31}N_2O_2]^+$	Benzomethamine, 1105
355.792278	Cl_5Ta	Tantalum Pentachloride, 8832
355.994259	$C_{12}H_{13}BrN_4O_2S$	Sulfabromomethazine, 8690
356.000049	$C_{11}H_{14}Cl_2N_2O_5S$	Tevenel®, 8964
356.038060	$[C_{14}H_{20}Cl_4N_2]^{2+}$	Chlorisondamine, 2072
356.088245	$C_{20}H_{17}FO_3S$	Sulindac, 8778
356.110733	$C_{16}H_{20}O_9$	Gentiopicrin, 4229
356.125989	$C_{20}H_{20}O_6$	Cubebin, 2608
356.134070	$C_{15}H_{24}N_4O_2S_2$	Prosultiamine, 7667
356.155850	$C_{20}H_{24}N_2O_2S$	Hycanthone, 4636
356.178117	$[C_{22}H_{27}ClNO]^+$	Methylclobenztropinium, 2331
356.186184	$[C_{21}H_{26}NO_4]^+$	Methyldihydrothebainium, 3161
356.190006	$C_{21}H_{25}FN_2O_2$	Fluanisone, 4019
356.198760	$C_{22}H_{28}O_4$	Crocetin, Dimethyl ester, 2583 Estradiol, Diacetate, 3626 α-Estradiol, Diacetate, 3627 Guaiaretic Acid, Dimethyl ether, 4405 Hexahydroequilenin, 4555
356.235145	$C_{23}H_{32}O_3$	Estradiol, 17-Valerate, 3626 Quinestradiol, 7843
356.319149	$C_{24}H_{40}N_2$	Conessine, 2465
356.992490	$C_6H_5HgNO_4$	Mercurophen, Free acid, 5732
357.076786	$C_{19}H_{16}ClNO_4$	Clometacin, 2347 Indomethacin, 4845
357.142367	$C_{16}H_{23}NO_8$	Bakankosin, 948
357.157623	$C_{20}H_{23}NO_5$	Cyproquinate, 2773
357.194009	$C_{21}H_{27}NO_4$	Laudanosine, 5225

Molecular Weight	Empirical Formula	Compound Name, Monograph Number
357.230394	$C_{22}H_{31}NO_3$	Amicibone, 399 Oxybutynin, 6771 Songorine, 8491
357.658832	Br_3Sb	Antimony Tribromide, 744
357.706369	Ag_2O_4Se	Silver Selenate, 8271
357.711662	$Fe_3O_8P_2$	Ferrous Phosphate, 3978
357.765944	$C_3O_9Y_2$	Yttrium Carbonate, 9773
357.948376	$C_{10}H_4N_2Na_2O_8S$	Naphthol Yellow S, 6220
357.969531	$C_{12}H_{14}Cl_3O_4P$	Chlorfenvinphos, 2057
358.060801	$C_{19}H_{15}ClO_5$	Clobenfurol, Hemisuccinate, 2328
358.062344	$C_{17}H_{14}N_2O_5S$	Calmagite, 1722
358.068868	$C_{18}H_{14}O_8$	Succinylsalicylic Acid, 8678
358.105253	$C_{19}H_{18}O_7$	Benfurodil Hemisuccinate, 1045 Sal Ethyl® Carbonate, 8083
358.111126	$C_{12}H_{22}O_{12}$	Lactobionic Acid, 5190
358.135115	$C_{19}H_{22}N_2O_3S$	Dimethoxanate, 3209
358.141639	$C_{20}H_{22}O_6$	Columbin, 2460 Isocolumbin, 2460
358.164105	$C_{18}H_{22}N_4O_4$	D-Allose, Phenylosazone, 272 D-Altrose, Phenylosazone, 313 DL-Fructose, Phenylosazone, 4122 D-Glucose Phenylosazone, 4292 D-Gulose, Phenylosazone, 4429 Idose, Phenylosazone, 4801 D-Psicose, Phenylosazone, 7710 D-Tagatose, Phenylosazone, 8814
358.178024	$C_{21}H_{26}O_5$	Prednisone, 7518
358.193280	$C_{25}H_{26}O_2$	Diethylstilbestrol, Monobenzyl ether, 3113
358.214410	$C_{22}H_{30}O_4$	Cafestol, Acetate, 1621
358.250795	$C_{23}H_{34}O_3$	Pregnenolone, Acetate, 7533 Testosterone, Isobutyrate, 8890
358.334800	$C_{24}H_{42}N_2$	Phenetamine, 7012
358.956805	$C_9H_{11}Cl_2N_3O_4S_2$	Methyclothiazide, 5881
359.034475	$C_{10}H_{17}NO_9S_2$	Sinigrin, Free acid, 8289
359.115107	$C_{14}H_{21}N_3O_6S$	Penicillin N, 6887
359.174144	$C_{21}H_{29}NS_2$	Captodiamine, 1771

Molecular Weight	Empirical Formula	Compound Name, Monograph Number
359.246044	$C_{22}H_{33}NO_3$	Ajaconine, 183 Cyclomethycaine, 2740 Napelline, 6191
359.282430	$C_{23}H_{37}NO_2$	N-Acetylfuntumine, 4144
359.807304	Br_2Hg	Mercuric Bromide, 5709
359.831741	$AuCl_4Na$	Sodium Tetrachloroaurate(III), 8459
359.837250	$C_9H_4Cl_3IO$	Haloprogin, 4452
359.902327	N_3O_9Yb	Ytterbium Nitrate, 9772
360.081832	$C_{14}H_{12}N_6O_6$	Nitrovin, 6474
360.084518	$C_{18}H_{16}O_8$	Irigenin, 4939
360.106313	$C_{19}H_{21}ClN_2OS$	Chloracizine, 2025 Halethazole, 4446
360.157289	$C_{20}H_{24}O_6$	Gibberellic Acid, Methyl ester, 4250
360.193674	$C_{21}H_{28}O_5$	Aldosterone, 219 Cinerin II, 2283 Cortisone, 2514 Prednisolone, 7510
360.214804	$C_{18}H_{32}O_7$	Butyl Citrate, 1551
360.217484	$[C_{21}H_{30}NO_4]^+$	N-Butylscopolammonium, 1578
360.230060	$C_{22}H_{32}O_4$	Algestone, 16α-Methyl ether, 227
360.266445	$C_{23}H_{36}O_3$	Allopregnan-3α-ol-20-one, Acetate, 267 Allopregnan-3β-ol-20-one, Acetate, 268 Allopregnan-20α-ol-3-one, Acetate, 269 Allopregnan-20β-ol-3-one, Acetate, 270 Dromostanolone Propionate, 3436 Pregnan-3α-ol-20-one, Acetate, 7525 Pregnan-3β-ol-20-one, Acetate, 7526 Pregnan-20α-ol-3-one, Acetate, 7527
360.302831	$C_{24}H_{40}O_2$	Cholanic Acid, 2187 Norcholanic Acid, Methyl ester, 6499
360.363026	$[C_{25}H_{46}N]^+$	Cetalkonium, 1973
361.109628	$C_{17}H_{19}N_3O_4S$	Metampicillin, 5776
361.131408	$C_{22}H_{19}NO_4$	Bisacodyl, 1256
361.188923	$C_{20}H_{27}NO_5$	Buquinolate, 1482 Chromonar, 2235
361.204179	$C_{24}H_{27}NO_2$	Levophenacylmorphan, 5313
361.215413	$C_{23}H_{27}N_3O$	Libexin, Free base, 5320
361.225309	$C_{21}H_{31}NO_4$	Furethidine, 4154

Molecular Weight	Empirical Formula	Compound Name, Monograph Number
361.272928	$C_{21}H_{35}N_3O_2$	Mestanolone, Semicarbazone, 5761
361.321889	$[C_{23}H_{41}N_2O]^+$	Dodecarbonium, 3411
361.334465	$C_{24}H_{43}NO$	Linoleic Acid, Cyclohexylamide, 5343
361.762596	Cs_2O_4S	Cesium Sulfate, 1972
361.833972	CdO_4W	Cadmium Tungstate(VI), 1619
361.875717	$C_{12}H_8Cl_6$	Aldrin, 220 Isodrin, 5022
362.009455	$C_8H_6N_6O_{11}$	Pentryl®, 6933
362.014462	$C_{14}H_{16}ClO_5PS$	Coumaphos, 2543
362.057258	$C_{16}H_{14}N_2O_6S$	Phthalylsulfacetamide, 7183
362.067012	$C_{22}H_{19}Br$	Broparoestrol, 1451
362.100168	$C_{18}H_{18}O_8$	Oosporein, Tetramethyl ether, 6689
362.126657	$C_{21}H_{18}N_2O_4$	α-Aminoorcein, 6704
362.136553	$C_{19}H_{22}O_7$	Samaderin B, 8104
362.170253	$C_{16}H_{22}N_6O_4$	TRH, 9274
362.199428	$C_{23}H_{26}N_2O_2$	Benzetimide, 1079 Dexetimide, 2902
362.209324	$C_{21}H_{30}O_5$	Humulon, 4631 Hydrocortisone, 4674 4-Pregnene-17α,20β,21-triol-3,11-dione, 7531
362.212492	$C_{21}H_{31}ClN_2O$	Viminol, 9635
362.245710	$C_{22}H_{34}O_4$	Estrane-3α,17α-diol, Diacetate, 3633
362.248390	$[C_{25}H_{32}NO]^+$	Methylphenomorphanium, 7047
362.282095	$C_{23}H_{38}O_3$	5β-Pregnane-3α,20α-diol, 3-Acetate, 7523 5β-Pregnane-3α,20α-diol, 20-Acetate, 7523
362.999221	C_8H_8AuNOS	Aurothioglycanide, 903
363.023983	$C_{16}H_5N_5O_6$	9-Dicyanomethylene-2,4,7-trinitrofluorene, 3070
363.058001	$C_{10}H_{14}N_5O_8P$	3'-Guanylic Acid, 4421 5'-Guanylic Acid, 4422
363.110673	$C_{21}H_{17}NO_5$	α-Hydroxyorcein, 6704
363.231063	$C_{23}H_{29}N_3O$	Opipramol, 6692
363.252192	$C_{20}H_{33}N_3O_3$	Androstane-3β,11β-diol-17-one, Semicarbazone, 672
363.621291	Br_4Ti	Titanium Tetrabromide, 9187
363.644615	P_4Se_3	Phosphorus Triselenide, 7171

Use in conjunction with The Merck Index, Ninth Edition

Molecular Weight	Empirical Formula	Compound Name, Monograph Number
363.712885	$Ni_3O_8P_2$	Nickel Phosphate, 6329
363.832966	Gd_2O_3	Gadolinium Oxide, 4174
363.899258	$C_{10}H_9Cl_4O_4P$	Tetrachlorvinphos, 8910
363.899433	$C_{11}H_9Cl_5O_3$	Erbon, 3558
364.027889	$C_{12}H_{20}Cl_4N_2O_2$	Fertilysin, 3985
364.058303	$C_{20}H_{12}O_7$	Gallein, 4189
364.060132	$C_{10}H_{13}FN_6O_6S$	Nucleocidin, 6538
364.071366	$C_{18}H_{17}ClO_6$	Monorden, 6089
364.094688	$C_{21}H_{16}O_6$	Justicidin B, 5123
364.109294	$C_{17}H_{20}N_2O_5S$	Bumetanide, 1472 Phenethicillin, 7014, 5685
364.138285	$C_{16}H_{20}N_4O_6$	Queen Substance, 2,4-Dinitrophenylhydrazone, 7819
364.152203	$C_{19}H_{24}O_7$	Samaderin C, 8104
364.163437	$C_{18}H_{24}N_2O_6$	Dinocap, 3284
364.167459	$C_{23}H_{24}O_4$	Cyclofenil, 2725
364.178693	$C_{22}H_{24}N_2O_3$	α-Colubrine, 2457 β-Colubrine, 2458
364.180523	$C_{21}H_{29}ClO_3$	Clostebol, Acetate, 2368
364.188589	$C_{20}H_{28}O_6$	Amarolide, 381 Phorbol, 7143
364.224974	$C_{21}H_{32}O_5$	Allopregnane-3β,17α,21-triol-11,20-dione, 264 Allotetrahydrocortisone, 273 Hydrallostane, 4646 4-Pregnene-11β,17α,20β,21-tetrol-3-one, 7530 Tetrahydrocortisone, 8928
364.240230	$C_{25}H_{32}O_2$	Quinestrol, 7844
364.978485	$C_7H_6AuNO_2S$	4-Amino-2-aurothiosalicylic Acid, 8316
365.060091	$C_{16}H_{16}ClN_3O_3S$	Indapamide, 4822 Metolazone, 6021
365.104543	$C_{16}H_{19}N_3O_5S$	Amoxicillin, 611
365.108565	$C_{21}H_{19}NO_3S$	Alkofanone, 238
365.156184	$C_{21}H_{23}N_3OS$	Pericyazine, 6955
365.162708	$C_{22}H_{23}NO_4$	Nequinate, 6296
365.168760	$[C_{22}H_{25}N_2OS]^+$	Trimethaphan, 9374
365.210327	$C_{22}H_{27}N_3O_2$	Spadon, 8505

365.

Molecular Weight	Empirical Formula	Compound Name, Monograph Number
365.683235	K_3S_4Sb	Potassium Thioantimonate(V), 7483
365.813313	Br_2Pb	Lead Bromide, 5250
365.835444	O_3Tb_2	Terbium, Oxide, 8877
365.933069	$C_{12}H_{12}As_2N_2O_2$	Arsphenamine, Free base, 840
365.992345	O_6SU	Uranyl Sulfate, 9523
366.001862	HO_6PU	Uranyl Phosphate, 9522
366.136177	$C_{16}H_{22}N_4O_4S$	Acetiamine, 42
366.146724	$C_{22}H_{22}O_5$	Cyclovalone, 2759
366.161329	$C_{18}H_{26}N_2O_4S$	Glibornuride, 4270
366.161980	$C_{26}H_{22}O_2$	Benzopinacol, 1110
366.171225	$C_{16}H_{30}O_7S$	Bis(1-methylamyl) Sulfosuccinate, 1277
366.194343	$C_{22}H_{26}N_2O_3$	Corynantheine, 2528
366.204239	$C_{20}H_{30}O_6$	Taxicin-I, 8853
366.230728	$C_{23}H_{30}N_2O_2$	Piminodine, 7234
366.240624	$C_{21}H_{34}O_5$	Allopregnane-3α,11β,17α,21-tetrol-20-one, 260 Allopregnane-3β,11β,17α,21-tetrol-20-one, 261 Cortolone, 2519
366.277010	$C_{22}H_{38}O_4$	Citrullol, 2313
366.706437	$Co_3O_8P_2$	Cobaltous Phosphate, 2399
366.986868	$C_{12}H_{15}ClNO_4PS_2$	Phosalone, 7145
367.105587	$C_{20}H_{17}NO_6$	Adlumidine, 152 Bicuculline, 1227
367.141973	$C_{21}H_{21}NO_5$	Corycavamine, 2522
367.178358	$C_{22}H_{25}NO_4$	Dimoxyline, 3261
367.202168	$[C_{22}H_{27}N_2O_3]^+$	Macusine A, 5467 Macusine C, 5469 Melinonine A, 5642
367.214744	$C_{23}H_{29}NO_3$	Fenbutrazate, 3892 Phenoperidine, 7048
367.712315	CdI_2	Cadmium Iodide, 1608
367.808215	$C_6FeK_4N_6$	Potassium Ferrocyanide, 7410
367.813510	$C_9H_2Cl_6O_3$	Chlorendic Anhydride, 2056
367.819285	$Na_5O_{10}P_3$	Sodium Tripolyphosphate, 8470
367.909604	$C_{14}H_9Cl_5O$	4,4'-Dichloro-α-(trichloromethyl)benzhydrol, 3054

Molecular Weight	Empirical Formula	Compound Name, Monograph Number
367.945137	$C_3H_7AuO_4S_2$	3-Aurothio-2-propanol-1-sulfonic Acid, 1641
368.021537	$C_5H_{11}ClHgN_2O_2$	Chlormerodrin, 2074
368.022314	$C_{18}H_{16}OSn$	Triphenyltin Hydroxide, 9412
368.050065	$C_{15}H_{16}N_2O_5S_2$	Gliotoxin, Monoacetate, 4272
368.084734	$C_{17}H_{21}O_5PS$	Coumithoate, 2552
368.086451	$C_{16}H_{20}N_2O_4S_2$	Pyrithioxin, 7776
368.089603	$C_{20}H_{16}O_7$	Acacetin, Diacetate, 9 Diacetyloroxylin, 6713 Prunetin, Diacetate, 7693
368.110733	$C_{17}H_{20}O_9$	Chlorogenic Acid, 3'-Methyl ether, 2121
368.116093	$C_{23}H_{16}N_2O_3$	α-Phenylcinchoninylanthranilic Acid, 7084
368.117748	$C_{21}H_{21}O_4P$	Tritolyl Phosphate, 9426, 9427
368.125989	$C_{21}H_{20}O_6$	Curcumin, 2681
368.147118	$C_{18}H_{24}O_8$	Heptolide, 4526
368.149798	$[C_{21}H_{22}NO_5]^+$	Methylprotopinium, 7684
368.173607	$C_{21}H_{24}N_2O_4$	Cyclarbamate, 2709
368.209993	$C_{22}H_{28}N_2O_3$	Dihydrocorynantheine, 2528 Voacangine, 9696
368.222569	$[C_{23}H_{30}NO_3]^+$	N-Ethylpipethanate, 7265 Propantheline, 7597
368.225249	$C_{26}H_{28}N_2$	Cinnarizine, 2295
368.256275	$C_{21}H_{36}O_5$	Allopregnane-3β,11β,17α,20β,21-pentol, 258 Cortol, 2518
368.365431	$C_{24}H_{48}O_2$	Behenic Acid, Ethyl ester, 1029 Lignoceric Acid, 5327
368.389240	$[C_{24}H_{50}NO]^+$	Cethexonium, 1974
369.045314	$C_{14}H_{15}N_3O_5S_2$	Acetosulfone, 61
369.121238	$C_{20}H_{19}NO_6$	Ochratoxin B, 6549
369.127111	$C_{13}H_{23}NO_{11}$	Chondrosine, Methyl ester, 2213
369.157623	$C_{21}H_{23}NO_5$	Allocryptopine, 248 Cryptopine, 2602 Diacetylmorphine, 2925, 2926 Homochelidonine, 4612
369.194009	$C_{22}H_{27}NO_4$	Corydaline, 2525
369.254204	$[C_{23}H_{33}N_2O_2]^+$	Prajmaline, 7501

369.

Molecular Weight	Empirical Formula	Compound Name, Monograph Number
369.821094	$C_{10}H_5Cl_7$	Heptachlor, 4514
369.861705	MoO_4Pb	Lead Molybdate(VI), 5266
370.029330	$C_{14}H_{14}N_2O_6S_2$	Amsonic Acid, 628
370.076949	$C_{14}H_{18}N_4O_4S_2$	Thiophanate, 9089
370.089997	$C_{16}H_{18}O_{10}$	Fraxin, 4117
370.093966	$C_{10}H_{30}O_5Si_5$	Decamethylcyclopentasiloxane, 2831
370.105253	$C_{20}H_{18}O_7$	Sesamolin, 8216
370.153743	$C_{21}H_{26}N_2S_2$	Thioridazine, 9098
370.156895	$C_{25}H_{22}O_3$	Equilenin, Benzoate, 3551
370.178024	$C_{22}H_{26}O_5$	Colpormon, 2455 Equilin Glycol, Diacetate, 3553
370.182733	$C_{20}H_{26}N_4OS$	Oxypendyl, 6783
370.201834	$[C_{22}H_{28}NO_4]^+$	Methylisocorybulbinium, 5016
370.217090	$[C_{26}H_{28}NO]^+$	Phenacridane, 6988
370.235539	$C_{20}H_{34}O_6$	Grayanotoxin III, 4388
370.287181	$C_{25}H_{38}O_2$	17α-Methyltestosterone 3-Cyclopentyl Enol Ether, 5999
370.908944	$Cl_4H_8N_2Pt$	Ammonium Platinous Chloride, 579
371.035796	$C_{20}H_{12}N_4Cu$	Porphine, Copper salt, 7372
371.096479	$C_{14}H_{17}N_3O_9$	6-Azauridine, 2′,3′,5′-Triacetate, 917
371.151288	$C_{19}H_{22}ClN_5O$	Trazodone, 9266
371.173273	$C_{21}H_{25}NO_5$	Demecolcine, 2857 Diacetyldihydromorphine, 2924
371.209659	$C_{22}H_{29}NO_4$	Tetrahydroaureothamine, 899
371.224915	$C_{26}H_{29}NO$	Tamoxifen, 8823
371.282430	$C_{24}H_{37}NO_2$	2-(1-Pyrrolidinyl)methyltestosterone, 7803
371.850022	HfO_8S_2	Hafnium Sulfate, 4442
371.913465	Cl_4Th	Thorium Chloride, 9117
371.993087	$C_9H_{11}N_6O_3Sb$	Melaminylphenylstibonic Acid, 5630
372.084518	$C_{19}H_{16}O_8$	5,5′-Methylenedisalicylic Acid, Diacetyl deriv, 5934
372.087685	$C_{19}H_{17}ClN_2O_4$	Glaphenine, 4264
372.120903	$C_{20}H_{20}O_7$	Herqueinone, 4533
372.142033	$C_{17}H_{24}O_9$	Syringin, 8806

Use in conjunction with The Merck Index, Ninth Edition

Molecular Weight	Empirical Formula	Compound Name, Monograph Number
372.168522	$C_{20}H_{24}N_2O_5$	Codoxime, 2430 Diamfenetide, 2932
372.169859	$C_{21}H_{20}N_6O$	N,N'-Bis(4-amino-2-methyl-6-quinolyl)urea, 1258
372.172545	$C_{25}H_{24}O_3$	Equilin, Benzoate, 3552
372.183127	$C_{16}H_{28}N_4O_4S$	Biocytin, 1239
372.193674	$C_{22}H_{28}O_5$	Estradiol, Hemisuccinate, 3626 Meprednisone, 5689 Prednylidene, 7520 Pyrethrin II, 7747
372.222481	$[C_{21}H_{36}Cl_2N]^+$	Aralkonium, 796
372.230060	$C_{23}H_{32}O_4$	Deoxycorticosterone Acetate, 2864 17α-Hydroxyprogesterone, Acetate, 4756
372.243973	$[C_{25}H_{30}N_3]^+$	Gentian Violet, Cation, 4227
372.266445	$C_{24}H_{36}O_3$	Anagestone, Acetate, 662
372.314064	$C_{24}H_{40}N_2O$	Holarrhenine, 4601
372.350450	$C_{25}H_{44}N_2$	Kurchessine, 5174
372.375602	$C_{27}H_{48}$	Cholestane, 2190 Coprostane, 2500
372.966929	$C_{12}H_{12}Cl_2CoN_4O_2$	Coamid, 2378
373.094372	$C_{14}H_{19}N_3O_7S$	N-2,4-Dinitrophenylfelinine, 3883
373.137947	$C_{20}H_{24}ClN_3S$	Prochlorperazine, 7560
373.188923	$C_{21}H_{27}NO_5$	Hasubanonine, 4473
373.272928	$C_{22}H_{35}N_3O_2$	3,20-Pregnanedione, Monosemicarbazone, 7524
373.711153	I_2Sn	Stannous Iodide, 8565
373.883834	Cl_2O_6Pb	Lead Chlorate, 5253
373.926550	$C_{12}H_{12}Br_2N_2O_2$	5-(α,β-Dibromophenethyl)-5-methylhydantoin, 2991
373.999321	$C_{14}H_{20}Br_2N_2$	Bromhexine, 1389
374.070972	$C_{23}H_{15}ClO_3$	Chlorophacinone, 2131
374.100168	$C_{19}H_{18}O_8$	Atranorin, 886 Diacetyljavanicin, 5112
374.121297	$C_{16}H_{22}O_{10}$	D-Fucose, Pentaacetate, 4129 Swertiamarin, 8800
374.164873	$C_{22}H_{27}ClO_3$	Cyproterone, 2774
374.174276	$C_{22}H_{22}N_4O_2$	Forbisen, 4092
374.176106	$C_{21}H_{27}ClN_2O_2$	Hydroxyzine, 4771

374.

Molecular Weight	Empirical Formula	Compound Name, Monograph Number
374.188195	$C_{25}H_{26}O_3$	5,6,8-Estratrien-3β-ol-17-one, Benzoate, 3634 8-Isoestrone, dl-Benzoate, 5025
374.209324	$C_{22}H_{30}O_5$	Jasmolin II, 5109 6α-Methylprednisolone, 5980
374.245710	$C_{23}H_{34}O_4$	21-Acetoxypregnenolone, 67 Androstenediol, Diacetate, 673 5-Androstene-3β,16α-diol, Diacetate, 674 Digitoxigenin, 3138 Oxymetholone, Enol acetate, 6780 Uzarigenin, 9561
374.266839	$C_{20}H_{38}O_6$	Phloionic Acid, Dimethyl ester, 7134
374.282095	$C_{24}H_{38}O_3$	Stanolone, 17-Valerate, 8573
374.318481	$C_{25}H_{42}O_2$	Cholanic Acid, Methyl ester, 2187 Norcholanic Acid, Ethyl ester, 6499
375.065358	$C_{18}H_{14}N_3NaO_3S$	Tropaeolin OO, 9439
375.100126	$C_{16}H_{17}N_5O_4S$	Azidocillin, 921
375.140135	$C_{21}H_{23}ClFNO_2$	Haloperidol, 4450
375.147058	$C_{23}H_{21}NO_4$	Xenalamine, 9736
375.183444	$C_{24}H_{25}NO_3$	Benzylmorphine, 1158
375.232206	$C_{21}H_{30}FN_3O_2$	Pipamperone, 7248
375.288578	$C_{22}H_{37}N_3O_2$	Pregnan-3α-ol-20-one, Semicarbazone, 7525 Pregnan-20α-ol-3-one, Semicarbazone, 7527 Pregnan-20β-ol-3-one, Semicarbazone, 7528
375.678378	N_4Se_4	Nitrogen Selenide, 6428
375.704728	O_6Se_2Sn	Stannic Selenite, 8558
375.740646	$BaMn_2O_8$	Barium Permanganate, 993
375.805679	$AuCl_4K$	Potassium Tetrachloroaurate(III), 7472
375.843110	Dy_2O_3	Dysprosium Oxide, 3457
376.094688	$C_{22}H_{16}O_6$	Resistomycin, 7948
376.138285	$C_{17}H_{20}N_4O_6$	D-Araboflavin, 791 Isoriboflavine, 5085 Lyxoflavine, 5462 Riboflavine, 7993, 7994
376.152203	$C_{20}H_{24}O_7$	Olivil, 6684
376.203845	$C_{25}H_{28}O_3$	α-Estradiol, 3-Benzoate, 3627 Estradiol Benzoate, 3628 Isoestradiol, 3-Benzoate, 5024

Use in conjunction with The Merck Index, Ninth Edition

Molecular Weight	Empirical Formula	Compound Name, Monograph Number
376.204988	$C_{22}H_{29}FO_4$	Desoximetasone, 2888 Fluocortolone, 4032 Fluorometholone, 4061
376.248784	$[C_{22}H_{34}NO_4]^+$	Methylfurethidinium, 4154
376.261360	$C_{23}H_{36}O_4$	Allopregnane-3β,17α-diol-20-one, 3-Acetate, 256
376.295060	$C_{20}H_{36}N_6O$	Lauroguadine, 5231
376.297745	$C_{24}H_{40}O_3$	Lithocholic Acid, 5388
376.304956	$[C_{18}H_{40}N_4O_4]^{2+}$	Hexacarbacholine, 4543
376.660450	Br_3Ce	Cerous Bromide, 1951
376.904497	$C_4K_2N_4Pt$	Potassium Tetracyanoplatinate(II), 7476
376.937287	$C_{11}H_{12}AsNO_5S_2$	Arsenamide, 819
376.993865	$C_5H_7N_5O_{15}$	Arabitol, Pentanitrate, 790
377.156184	$C_{22}H_{23}N_3OS$	Thiocarbanidin, 9055
377.210327	$C_{23}H_{27}N_3O_2$	Morazone, 6100
377.767018	$O_{12}S_3Sc_2$	Scandium Sulfate, 8145
377.782575	$C_4H_7Br_2Cl_2O_4P$	Naled, 6178
377.845408	Ho_2O_3	Holmium Oxide, 4602
377.870631	$C_{12}H_8Cl_6O$	Dieldrin, 3075 Endrin, 3522
377.914791	$Cl_4H_8HgN_2$	Ammonium Mercuric Chloride, 560
377.926197	Cl_4U	Uranium Tetrachloride, 9515
378.042648	$C_{13}H_{19}BrN_2O_6$	D-manno-Heptulose, p-Bromophenylhydrazone, 4528
378.047479	$C_{14}H_{19}ClN_2O_4S_2$	Penicillin S, 6889
378.088558	$C_{17}H_{18}N_2O_6S$	Carbenicillin, 1795
378.124944	$C_{18}H_{22}N_2O_5S$	Propicillin, Free acid, 7607
378.167853	$C_{20}H_{26}O_7$	Cnicin, 2376
378.184252	$C_{21}H_{27}FO_5$	Descinolone, 2879 Fluprednisolone, 4075
378.207611	$C_{18}H_{34}O_6S$	Ricinolsulfuric Acid, 8005
378.219495	$C_{25}H_{30}O_3$	Nandrolone, Benzoate, 6185
378.230728	$C_{24}H_{30}N_2O_2$	Doxapram, 3426
378.240624	$C_{22}H_{34}O_5$	Pleuromutilin, 7314
378.695642	$Cu_3O_8P_2$	Cupric Phosphate, 2651

378.

Molecular Weight	Empirical Formula	Compound Name, Monograph Number
378.902183	$C_8H_8Cl_3N_3O_4S_2$	Trichlormethiazide, 9304
379.015986	$C_{14}H_{12}F_3NO_4S_2$	Perfluidone, 6952
379.042728	$C_{13}H_{18}ClN_3O_4S_2$	Cyclopenthiazide, 2749
379.169605	$C_{22}H_{22}FN_3O_2$	Droperidol, 3437
379.181730	$C_{20}H_{29}NO_4S$	Camphamedrine, 1731
379.189592	$C_{22}H_{25}N_3O_3$	Solypertine, 8488
379.202168	$[C_{23}H_{27}N_2O_3]^+$	β-Methylcolubrine, 2458
379.225977	$C_{23}H_{29}N_3O_2$	Oxypertine, 6784
379.845354	Er_2O_3	Erbium Oxide, 3556
380.024996	$C_{18}H_{15}Cl_3N_2O$	Econazole, 3472
380.089603	$C_{21}H_{16}O_7$	Coumetarol, 2550
380.104208	$C_{17}H_{20}N_2O_6S$	Methicillin, 5839
380.117922	$C_{23}H_{21}ClO_3$	Chlorotrianisene, 2162
380.163711	$C_{24}H_{20}N_4O$	Scarlet Red, 8146
380.173607	$C_{22}H_{24}N_2O_4$	Alstonidine, 309 Kopsine, 5167 Vomicine, 9698
380.198760	$C_{24}H_{28}O_4$	Diethylstilbestrol Dipropionate, 3115
380.199903	$C_{21}H_{29}FO_5$	Fludrocortisone, 4022
380.307916	$C_{27}H_{40}O$	Neoergosterol, 6275
381.185255	$C_{22}H_{24}FN_3O_2$	Benperidol, 1050
381.194009	$C_{23}H_{27}NO_4$	Tropenzile, 9443
381.233074	$[C_{27}H_{29}N_2]^+$	Dicyanine, Cation, 3067
381.694280	$O_8P_2Zn_3$	Zinc Phosphate, 9817
382.042393	$C_{13}H_{19}ClN_2O_5S_2$	Mefruside, 5624
382.065850	$C_8H_{14}O_4Pb$	Lead Butyrate, 5251
382.087496	$C_{21}H_{18}O_5S$	Cresol Red, 2574
382.127720	$C_{19}H_{18}N_4O_5$	Tremetone, 2,4-Dinitrophenylhydrazone, 9268
382.156244	$C_{18}H_{26}N_2O_5S$	S-Benzoylpantetheine, 6816
382.189258	$C_{22}H_{26}N_2O_4$	Akuammine, 191 Aricine, 810 1,4-Bis(1,4-benzodioxan-2-ylmethyl)piperazine, 1262 Methoxylochnericine, 5398

Use in conjunction with The Merck Index, Ninth Edition

Molecular Weight	Empirical Formula	Compound Name, Monograph Number
382.214410	$C_{24}H_{30}O_4$	17-Hydroxy-16-methylene-Δ^6-progesterone, 17-Acetate, 4737 Promethestrol, Diacetate, 7582
382.228323	$[C_{26}H_{28}N_3]^+$	Pyrvinium, 7810, 7811
382.262029	$C_{24}H_{34}N_2O_2$	Euprocin, 3851
382.287181	$C_{26}H_{38}O_2$	Quingestrone, 7847
382.298414	$C_{25}H_{38}N_2O$	Bunamidine, 1474
382.381081	$C_{25}H_{50}O_2$	Lignoceric Acid, Methyl ester, 5327
382.983499	$C_{11}H_{14}ClN_3O_4S_3$	Althiazide, 312
383.078096	$[C_{24}H_{20}As]^+$	Tetraphenylarsonium, 8953, 8954
383.136888	$C_{21}H_{21}NO_6$	Adlumine, 153 Hydrastine, 4650 Rhoeadine, 7988
383.159354	$C_{19}H_{21}N_5O_4$	Prazosin, 7509
383.173273	$C_{22}H_{25}NO_5$	Corycavidine, 2523
383.246044	$C_{24}H_{33}NO_3$	Nafronyl, 6176
383.751084	$O_{12}S_3Ti_2$	Titanium Sesquisulfate, 9185
383.862078	Br_2Ra	Radium Bromide, 7911
383.910344	$C_4HgK_2N_4$	Potassium Tetracyanomercurate(II), 7474
383.980358	$C_{16}H_{10}Cl_2O_7$	Erdin, 3559
383.987623	$C_9H_{22}O_4P_2S_4$	Ethion, 3663
384.010783	$C_9H_{10}HgO_2S$	Thimerosal, Free acid, 9046
384.019911	$C_{17}H_{15}Cl_3N_2O_2$	Triclodazol, 9334
384.080183	$C_{14}H_{14}Hg$	p-Ditolylmercury, 3397
384.146002	$C_{12}H_{36}O_4Si_5$	Dodecamethylpentasiloxane, 3409
384.168522	$C_{21}H_{24}N_2O_5$	Dipropaesin, 3358
384.181098	$[C_{22}H_{26}NO_5]^+$	Methyldiacetylmorphinium, 2925
384.187150	$C_{22}H_{28}N_2O_2S$	Becanthone, 1023 Perimethazine, 6957
384.204908	$C_{22}H_{28}N_2O_4$	Catharosine, 1900 Rhynchophylline, 7990
384.227374	$C_{20}H_{28}N_6O_2$	Fencamine, 3894

384.

Molecular Weight	Empirical Formula	Compound Name, Monograph Number
384.230060	$C_{24}H_{32}O_4$	Estradiol, Dipropionate, 3626 Ethynodiol, Diacetate, 3813 Guaiaretic Acid, Diethyl ether, 4405 Megestrol Acetate, 5626 Resibufogenin, 7938 Scillarenin, 8152
384.241293	$C_{23}H_{32}N_2O_3$	Zipeprol, 9837
384.266445	$C_{25}H_{36}O_3$	Estradiol, 17-Heptanoate, 3626 Nandrolone, Cyclohexanecarboxylate, 6185
384.302831	$C_{26}H_{40}O_2$	Ethyl Choladienate, 3717
384.339216	$C_{27}H_{44}O$	7-Dehydrocholesterol, 2842 Desmosterol, 2884 Vitamin D_3, 9679
385.152538	$C_{21}H_{23}NO_6$	Colchiceine, 2435
385.211390	$C_{20}H_{27}N_5O_3$	Bamifylline, 957
385.212733	$[C_{22}H_{29}N_2O_4]^+$	Echitamine, 3469
385.251798	$C_{26}H_{31}N_3$	Triethylrosaniline, 4600
385.264374	$[C_{27}H_{33}N_2]^+$	Brilliant Green, Cation, 1378
385.853194	O_3Tm_2	Thulium Oxide, 9135
386.081760	$C_{12}H_{22}N_2O_8S_2$	Piposulfan, 7276
386.136553	$C_{21}H_{22}O_7$	Ostruthol, 6732
386.148657	$C_{21}H_{26}N_2OS_2$	Mesoridazine, 5757
386.150954	$C_{20}H_{23}ClN_4O_2$	Clonitazene, 2354
386.199428	$C_{25}H_{26}N_2O_2$	Medibazine, 5613
386.209324	$C_{23}H_{30}O_5$	11-Dehydrocorticosterone, Acetate, 2844
386.245710	$C_{24}H_{34}O_4$	Algestone, Cyclic acetal with acetone, 227 Medroxyprogesterone, 17-Acetate, 5618
386.282095	$C_{25}H_{38}O_3$	Anagestone, Propionate, 662
386.329714	$C_{25}H_{42}N_2O$	Cyclobuxine, 2722
386.354866	$C_{27}H_{46}O$	Allocholesterol, 246 Cholesterol, 2192 Epicholesterol, 3538
386.391252	$C_{28}H_{50}$	Coproergostane, 2498 Ergostane, 3575
387.011427	$C_{14}H_{14}ClN_3O_4S_2$	Benzylhydrochlorothiazide, 1152
387.098789	$C_{16}H_{21}NO_8S$	Naphthionine N-Glucoside, 6230
387.125278	$C_{19}H_{21}N_3O_4S$	Hypoglycine B, Phenylhydantoin deriv, 4788

Use in conjunction with The Merck Index, Ninth Edition

Molecular Weight	Empirical Formula	Compound Name, Monograph Number
387.136757	$C_{12}H_{24}N_9P_3$	Apholate, 761
387.151368	$C_{21}H_{23}ClFN_3O$	Flurazepam, 4078
387.171355	$C_{21}H_{26}ClN_3O_2$	Acranil, Free base, 117
387.728370	$C_4Cl_2O_4Rh_2$	Rhodium Carbonyl Chloride, 7979
388.094688	$C_{23}H_{16}O_6$	Pamoic Acid, 6811, 6810, 7811
388.115818	$C_{20}H_{20}O_8$	Rufigallol, Hexamethyl ether, 8052
388.118985	$C_{20}H_{21}ClN_2O_4$	Fipexide, 4001
388.136947	$C_{17}H_{24}O_{10}$	Verbenalin, 9614
388.152203	$C_{21}H_{24}O_7$	Visnadin, 9663
388.188589	$C_{22}H_{28}O_6$	Isoquassin, 5082 Quassin, 7815
388.199822	$C_{21}H_{28}N_2O_5$	Trimethobenzamide, 9376
388.212393	$C_{21}H_{24}N_8$	Bismark Brown R, Free base, 1275
388.224974	$C_{23}H_{32}O_5$	Corticosterone, 21-Acetate, 2513 11-Desoxy-17-hydroxycorticosterone, Acetate, 2891 Tanghinigenin, 8826
388.261360	$C_{24}H_{36}O_4$	Bolandiol, Dipropionate, 1336 Methandriol, Diacetate, 5807 Oxymetholone, Enol propionate, 6780
388.308979	$C_{24}H_{40}N_2O_2$	Funtessine, N-Acetyl deriv, 4143
388.334131	$C_{26}H_{44}O_2$	Cholanic Acid, Ethyl ester, 2187 Tocol, 9195
388.345364	$C_{25}H_{44}N_2O$	Azacosterol, 908 N-Acetylchonemorphine, 2214
388.370516	$C_{27}H_{48}O$	Cholestanol, 2191 Coprosterol, 2501 Epicholestanol, 3537
389.027078	$C_{14}H_{16}ClN_3O_4S_2$	Cyclothiazide, 2758
389.140928	$C_{19}H_{23}N_3O_4S$	Hetacillin, 4536
389.592753	As_2Se_3	Arsenic Triselenide, 833
389.704819	$K_2MgO_8Se_2$	Magnesium Potassium Selenate, 5508
389.733076	$K_2O_8S_2Sn$	Potassium Stannosulfate, 7461
389.741033	CaI_2O_6	Calcium Iodate, 1673
389.743116	$O_{12}S_3V_2$	Vanadium Trisulfate, 9582
389.824998	$C_7H_4I_2O_3$	3,5-Diiodosalicylic Acid, 3174

389.

Molecular Weight	Empirical Formula	Compound Name, Monograph Number
389.825239	$C_{10}Mn_2O_{10}$	Manganese Carbonyl, 5555
390.116212	$C_{16}H_{22}O_{11}$	Monotropein, 6092
390.121572	$C_{22}H_{18}N_2O_5$	Camptothecin, Acetate, 1738
390.131468	$C_{20}H_{22}O_8$	Polydatin, 7346 Populin, 7369
390.152597	$C_{17}H_{26}O_{10}$	Loganin, 5402
390.186277	$C_{25}H_{27}ClN_2$	Meclizine, 5602
390.204239	$C_{22}H_{30}O_6$	Neoquassin, 6288
390.216343	$[C_{22}H_{34}N_2S_2]^{2+}$	Brintobal, Cation, 1379
390.240624	$C_{23}H_{34}O_5$	Alfadolone Acetate, 225 Androstane-3β,11β-diol-17-one, Diacetate, 672 Digoxigenin, 3143 Gitoxigenin, 4261 Periplogenin, 6962 Sarmentogenin, 8129
390.277010	$C_{24}H_{38}O_4$	Bis(2-ethylhexyl) Phthalate, 1270
390.313395	$C_{25}H_{42}O_3$	Lithocholic Acid, Methyl ester, 5388
390.820247	$C_6H_3I_2NO_3$	Disophenol, 3377
390.904885	C_4LiN_4PtRb	Lithium Rubidium Tetracyanoplatinate(II), 5379
390.907905	$C_6H_{12}N_3O_3S_3Sb$	Antimony Thioglycollamide, 743
391.074199	$C_{20}H_{19}Cl_2NO_3$	Benzmalecene, 1091
391.130267	$[C_{23}H_{23}N_2S_2]^+$	Dithiazanine, 3387
391.138821	$C_{19}H_{25}N_3O_2S_2$	Fonazine, 4090
391.178358	$C_{24}H_{25}NO_4$	Flavoxate, 4012
391.189592	$C_{23}H_{25}N_3O_3$	Phenodianisyl, 7037
391.711396	$BaBr_2O_6$	Barium Bromate, 970
391.714190	BaI_2	Barium Iodide, 985
391.736210	$Cr_2O_{12}S_3$	Chromic Sulfate, 2228
391.948186	$C_{11}H_{13}AsN_2O_5S_2$	Thiocarbarsone, 9056
391.999280	$C_6H_{11}AuO_5S$	Aurothioglucose, 902
392.110733	$C_{19}H_{20}O_9$	Cervicarcin, 1961
392.139052	$C_{21}H_{25}ClO_5$	Chloroprednisone, 2139
392.194737	$C_{20}H_{28}N_2O_6$	Cinepazet, 2281

Use in conjunction with The Merck Index, Ninth Edition

Molecular Weight	Empirical Formula	Compound Name, Monograph Number
392.199903	$C_{22}H_{29}FO_5$	Betamethasone, 1211 Dexamethasone, 2899 Paramethasone, 6834
392.246378	$C_{25}H_{32}N_2O_2$	Dextromoramide, 2911
392.256275	$C_{23}H_{36}O_5$	Dihydroperiplogenin, 6962
392.292660	$C_{24}H_{40}O_4$	Chenodeoxycholic Acid, 2010 Deoxycholic Acid, 2862 Hyodeoxycholic Acid, 4779 Ursodeoxycholic Acid, 9551
392.993901	$C_{14}H_{21}Br_2NO_2$	Spasmolytol, 8511
393.002518	$C_{14}H_{17}ClNO_4PS_2$	Dialifor, 2928
393.021992	$C_{13}H_{16}ClN_3O_5S_2$	Ambuside, 392
393.061238	$C_6H_{13}HgN_3O_4$	Merbiurelidin, 5696
393.063072	$C_{16}H_{15}N_3O_7S$	Sulphaloxic Acid, 8781
393.138344	$C_{14}H_{23}N_3O_{10}$	Pentetic Acid, 6925
393.194009	$C_{24}H_{27}NO_4$	Tylocrebrine, 9484 Tylophorine, 9485
393.221641	$C_{24}H_{28}FN_3O$	Spirilene, 8526
393.230394	$C_{25}H_{31}NO_3$	Testosterone Nicotinate, 8894
393.721256	CHI_3	Iodoform, 4896
393.797772	$C_9H_6Cl_8$	Alodan (Hoechst)®, 299
393.989001	$C_7H_5BiO_6$	Bismuth Subgallate, 1309
394.016251	N_2O_8U	Uranyl Nitrate, 9521
394.105253	$C_{22}H_{18}O_7$	Justicidin A, 5123
394.141639	$C_{23}H_{22}O_6$	Deguelin, 2839 Rotenone, 8031
394.179167	$C_{21}H_{27}FO_6$	Triamcinolone, 9279
394.186572	$C_{19}H_{22}N_8O_2$	WY-3654, 9718
394.189258	$C_{23}H_{26}N_2O_4$	Brucine, 1453 Cupreine, Diacetate, 2623 Vomicine, Methyl ether, 9698
394.201834	$[C_{24}H_{28}NO_4]^+$	Phenactropinium, 6989
394.214410	$C_{25}H_{30}O_4$	Bixin, 1330
394.250795	$C_{26}H_{34}O_3$	Androstenediol, 17-Benzoate, 673 Androsterone, Benzoate, 676 Epiandrosterone, Benzoate, 3535

394.

Molecular Weight	Empirical Formula	Compound Name, Monograph Number
394.323566	$C_{28}H_{42}O$	Dehydroergosterol, 2847 Neoergosterol, Methyl ether, 6275
394.943842	BiN_3O_9	Bismuth Nitrate, 1295
395.136888	$C_{22}H_{21}NO_6$	O-Acetylchelidonine, 2009
395.145676	$C_{19}H_{20}F_3N_3O_3$	Niflumic Acid, β-Morpholinoethyl ester, 6355
395.197083	$C_{23}H_{27}N_2O_4$	Methylkopsinium, 5167
395.200906	$C_{23}H_{26}FN_3O_2$	Spiperone, 8524
395.209659	$C_{24}H_{29}NO_4$	Ethaverine, 3655
395.589865	Br_4Se	Selenium Tetrabromide, 8188
395.862490	O_3Yb_2	Ytterbium Oxide, 9772
396.044980	$C_{16}H_{16}N_2O_6S_2$	Cephalothin, 1939
396.084518	$C_{21}H_{16}O_8$	Aloe-Emodin, Triacetate, 301 Emodin, Triacetate, 3512 Genistein, Triacetate, 4221 Sulfuretin, Triacetate, 8767
396.106984	$C_{19}H_{16}N_4O_6$	Euparin, 2,4-Dinitrophenylhydrazone, 3846
396.114379	$C_{21}H_{20}N_2O_4S$	Diphenicillin, 3318
396.168522	$C_{22}H_{24}N_2O_5$	Methylhydrastimide, 5953
396.178418	$C_{20}H_{28}O_8$	Bonducin, 1345
396.204908	$C_{23}H_{28}N_2O_4$	Akuammine, Methyl ether, 191 Quinine Ethylcarbonate, 7864
396.216141	$C_{22}H_{28}N_4O_3$	Etonitazene, 3831
396.217484	$[C_{24}H_{30}NO_4]^+$	Tropenzilium, 9443
396.221307	$C_{24}H_{29}FN_2O_2$	Aceperone, 21
396.230060	$C_{25}H_{32}O_4$	Melengestrol, Acetate, 5637 Nandrolone, Furylpropionate, 6185
396.266445	$C_{26}H_{36}O_3$	Estradiol 17β-Cypionate, 3629
396.287575	$C_{23}H_{40}O_5$	Nonoxynol 4, 6487
396.308704	$C_{20}H_{44}O_7$	Cucurbitacin L, 2613
396.326640	$[C_{27}H_{42}NO]^+$	Phenoctide, Cation, 7036

Use in conjunction with The Merck Index, Ninth Edition

Molecular Weight	Empirical Formula	Compound Name, Monograph Number
396.339216	$C_{28}H_{44}O$	Ergosterol, 3580 Fungisterol, 4142 Isopyrocalciferol, 5081 Lumisterol, 5418 Pyrocalciferol, 7778 Suprasterol II, 8795 Tachysterol, 8811 Vitamin D_2, 9678
396.846068	$C_9H_5I_2NO$	Diiodohydroxyquin, 3173
397.122026	$C_{14}H_{23}NO_{12}$	N-Acetylhyalobiuronic acid, 4633
397.152538	$C_{22}H_{23}NO_6$	Aureothin, 899
397.188923	$C_{23}H_{27}NO_5$	Octaverine, 6566
397.200157	$C_{22}H_{27}N_3O_4$	Diperodon, 3307
397.210053	$C_{20}H_{31}NO_7$	Heliosupine, 4485
397.334465	$C_{27}H_{43}NO$	Solanidine, 8478
397.690280	$As_2Ca_3O_8$	Calcium Arsenate, 1638
397.866314	Lu_2O_3	Lutetium Oxide, 5429
397.878931	$C_{14}H_8Br_2O_4$	Dibromsalicil, 2999
397.984369	$C_{12}H_{11}AsN_4O_7$	NPA Acid, 6535
397.996008	$C_{17}H_{12}Cl_2O_7$	Geodin, 4234
397.996473	$C_{12}H_{15}AsN_6OS_2$	Melarsoprol, 5634
398.043627	$C_{17}H_{16}Cl_2N_2O_5$	Chlorphenoxamide, 2170
398.068492	$C_{18}H_{14}N_4O_5S$	Salicylazosulfapyridine, 8091
398.100168	$C_{21}H_{18}O_8$	Daunomycinone, 2815 Triacetylnaringenin, 6251 Pectolinarigenin, Diacetate, 6862
398.144635	$C_{17}H_{26}N_4O_3S_2$	Fursultiamine, 4163
398.220558	$C_{23}H_{30}N_2O_4$	Mitragynine, 6062 Pholcodine, 7140
398.282095	$C_{26}H_{38}O_3$	17α-Hydroxyprogesterone 3-Cyclopentyl Enol Ether, 4758
398.289306	$C_{20}H_{38}N_4O_4$	Dimorpholamine, 3260
398.354866	$C_{28}H_{46}O$	Dihydrotachysterol, 3159 22,23-Dihydroergosterol, 3580 Vitamin D_4, 9680
398.362077	$C_{22}H_{46}N_4O_2$	Pithecolobine, 7288
398.879032	$C_{13}H_6Cl_5NO_3$	Oxyclozanide, 6773
399.119226	$C_{20}H_{19}N_2O_7$	Nybomycin, Succinate, 6543

Molecular Weight	Empirical Formula	Compound Name, Monograph Number
399.131802	$C_{21}H_{21}NO_7$	Narcotoline, 6250
399.145065	$[C_{15}H_{23}N_6O_5S]^+$	Methionine, Active, 5846
399.161664	$C_{21}H_{25}N_3O_3S$	Pipazethate, 7249
399.168188	$C_{22}H_{25}NO_6$	Colchicine, 2436
399.180292	$C_{22}H_{29}N_3S_2$	Thiethylperazine, 9043
399.207741	$C_{23}H_{30}ClN_3O$	Quinacrine, 7830, 7831
399.725070	$Fe_2O_{12}S_3$	Ferric Sulfate, 3953
399.990529	$C_{20}H_{10}Cl_2O_5$	4',5'-Dichlorofluorescein, 3041
400.115818	$C_{21}H_{20}O_8$	α-Peltatin, 6870
400.137613	$C_{22}H_{25}ClN_2OS$	Clopenthixol, 2357
400.188589	$C_{23}H_{28}O_6$	Prednisone, 21-Acetate, 7518
400.199822	$C_{22}H_{28}N_2O_5$	Mycelianamide, 6146 Reserpic Acid, 7935 Venoxidine, 9601
400.203845	$C_{27}H_{28}O_3$	Ethynylestradiol, 3-Benzoate, 3817
400.218450	$C_{23}H_{32}N_2O_2S$	Thiocarlide, 9057
400.261360	$C_{25}H_{36}O_4$	Cephalonic Acid, 1935
400.262503	$C_{22}H_{37}FO_5$	Fluocortolone, 21-Pivalate, 4032
400.297745	$C_{26}H_{40}O_3$	Testosterone Enanthate, 8893
400.334131	$C_{27}H_{44}O_2$	Gefarnate, 4212 1α-Hydroxycholecalciferol, 4719 25-Hydroxycholecalciferol, 4720
400.370516	$C_{28}H_{48}O$	Campesterol, 1730 Cholesterol, Methyl ether, 2192 α-Ergostenol, 3577 β-Ergostenol, 3578 γ-Ergostenol, 3579
401.046456	$C_{10}H_{10}BiNO_3$	Mebiquine, 5597
401.074681	$C_{19}H_{15}NO_9$	4-Hydroxytetracycloxide, 4768
401.104543	$C_{19}H_{19}N_3O_5S$	Oxacillin, 6738
401.110274	$C_{20}H_{24}BrN_3O$	D-2-Bromolysergic Acid Diethylamide, 1423
401.126323	$C_{24}H_{19}NO_5$	Oxyphenisatin Acetate, 6787
401.132862	$C_{21}H_{24}ClN_3OS$	Pipamazine, 7247
401.150620	$C_{21}H_{24}ClN_3O_3$	Fominoben, 4089
401.283098	$[C_{27}H_{35}N_3]^{2+}$	Methyl Green, Cation 5949

Use in conjunction with The Merck Index, Ninth Edition

Molecular Weight	Empirical Formula	Compound Name, Monograph Number
401.668006	S_5Sb_2	Antimony Pentasulfide, 735
401.685340	Br_3Ho	Holmium Bromide, 4602
402.095082	$C_{20}H_{18}O_9$	Cetraric Acid, 1979 Frangulin B, 4114
402.110339	$C_{24}H_{18}O_6$	Pamoic Acid, Methyl ether, 6811
402.143572	$C_{21}H_{26}N_2O_2S_2$	Sulforidazine, 8760
402.159787	$C_{23}H_{27}ClO_4$	Delmadinone Acetate, 2849
402.167853	$C_{22}H_{26}O_7$	Zearalenone, l-Form diacetate, 9775
402.171021	$C_{22}H_{27}ClN_2O_3$	Ajmaline, 17-Chloroacetate, 185
402.173213	$C_{28}H_{22}N_2O$	Amarine, N-Benzoyl deriv, 379
402.186277	$C_{26}H_{27}ClN_2$	Clocinizine, 2336
402.204239	$C_{23}H_{30}O_6$	Aldosterone, 21-Acetate, 219 Cortisone, Monoacetate (C_{21}-acetate), 2514 Prednisolone, 21-Acetate, 7510
402.222867	$C_{24}H_{34}O_3S$	Spiroxasone, 8540
402.240624	$C_{24}H_{34}O_5$	Bufogenin B, 1464 Dehydrocholic Acid, 2843 Gamabufotalin, 4196
402.349781	$C_{27}H_{46}O_2$	Dihydrotestosterone n-Octyl Enol Ether, 3160 24-Hydroxycholesterol, 4721 δ-Tocopherol, 9198 η-Tocopherol, 9202
402.386166	$C_{28}H_{50}O$	Cholestanol, Methyl ether, 2191 Ergostanol, 3576
402.685313	Br_3Er	Erbium Bromide, 3556
403.029664	$C_{17}H_{13}N_3O_5S_2$	Phthalylsulfathiazole, 7184
403.082265	$C_{20}H_{18}ClNO_6$	Ochratoxin A, 6549
403.147846	$C_{17}H_{25}NO_{10}$	N-Methyl-α-L-glucosamine, Pentaacetyl deriv, 5946
403.148512	$C_{21}H_{26}ClN_3OS$	Perphenazine, 6967
403.156578	$C_{20}H_{25}N_3O_4S$	Hetacillin, Methyl ester, 4536
403.214744	$C_{26}H_{29}NO_3$	Amotriphene, 609
403.225977	$C_{25}H_{29}N_3O_2$	Metergoline, 5791
403.816882	$C_9H_6Cl_6O_3S$	Endosulfan, 3519
403.849896	$C_{13}H_6Cl_6O_2$	Hexachlorophene, 4546
403.859927	$CoN_6Na_3O_{12}$	Sodium Cobaltinitrite, 8351

404.

Molecular Weight	Empirical Formula	Compound Name, Monograph Number
404.002201	$C_9H_{14}N_2O_{12}P_2$	Uridine 5'-Diphosphate, 9541
404.017052	$C_{14}H_{16}N_2O_6S_3$	Sulfoxone, 8763
404.057865	$C_{19}H_{18}O_6P_2$	O-Benzylphosphorous O,O-Diphenylphosphoric Anhydride, 1166
404.119464	$C_{23}H_{20}N_2O_3S$	Sulfinpyrazone, 8744
404.162374	$C_{25}H_{24}O_5$	Osajin, 6723
404.175437	$C_{23}H_{29}ClO_4$	Chlormadinone Acetate, 2073
404.219889	$C_{23}H_{32}O_6$	Atractyligenin, Diacetate, 885 Hydrocortisone Acetate, 4675 Strophanthidin, 8645
404.231122	$C_{22}H_{32}N_2O_5$	Benzquinamide, 1133
404.292660	$C_{25}H_{40}O_4$	Allopregnane-3α,20α-diol, Diacetate, 251 Allopregnane-3β,20α-diol, Diacetate, 253 Allopregnane-3β,20β-diol, Diacetate, 254 5β-Pregnane-3α,20α-diol, Diacetate, 7523 Uranediol, Diacetate, 9510
404.320979	$C_{27}H_{45}Cl$	Cholesteryl Chloride, 2194
404.329046	$C_{26}H_{44}O_3$	Lithocholic Acid, Ethyl ester, 5388
404.340279	$C_{25}H_{44}N_2O_2$	Kurcholessine, 5175
404.835897	$C_7H_5I_2NO_3$	3,5-Diiodo-4-pyridone-N-acetic Acid, 4907
405.002006	$C_{14}H_{13}ClFN_3O_4S_2$	Paraflutizide, 6830
405.099457	$C_{18}H_{19}N_3O_6S$	Cephaloglycin, 1934
405.185942	$C_{26}H_{28}ClNO$	Clomiphene, 2349
405.189986	$C_{20}H_{27}N_3O_6$	Febarbamate, 3881
405.287909	$C_{24}H_{39}NO_4$	Cassaine, 1884
405.797772	$C_{10}H_6Cl_8$	Chlordan(e), 2051
405.834054	O_6PbV_2	Lead Vanadate(V), 5286
406.114042	$C_{20}H_{17}F_3N_2O_4$	Floctafenine, A7
406.114356	$C_{21}H_{27}BrO_3$	9α-Bromo-11-oxoprogesterone, 1428
406.119858	$C_{19}H_{22}N_2O_6S$	Penamecillin, 6876
406.148849	$C_{18}H_{22}N_4O_7$	Galactoflavin, 4178
406.156895	$C_{28}H_{22}O_3$	Diphenylacetic Acid, Anhydride, 3324
406.199154	$C_{22}H_{30}O_7$	Amarolide, Monoacetate, 381
406.213759	$C_{18}H_{34}N_2O_6S$	Lincomycin, 5340

Use in conjunction with The Merck Index, Ninth Edition

Molecular Weight	Empirical Formula	Compound Name, Monograph Number
406.215553	$C_{23}H_{31}FO_5$	Flurogestone Acetate, 4079
406.235539	$C_{23}H_{34}O_6$	Diginatigenin, 3130 Hydrallostane, 21-Acetate, 4646 Dihydrostrophanthidin, 8645 Tetrahydrocortisone, 21-Acetate, 8928
406.250795	$C_{27}H_{34}O_3$	Nandrolone Phenpropionate, 6188 Testosterone 17-Phenylacetate, 8895
406.298414	$C_{27}H_{38}N_2O$	Rythmol®, 8066
406.308310	$C_{25}H_{42}O_4$	Chenodeoxycholic Acid, Methyl ester, 2010
406.793551	Cl_6H_2Pt	Platinic Chloride, 7304
407.164304	$C_{21}H_{24}F_3N_3S$	Trifluoperazine, 9353
407.303559	$C_{24}H_{41}NO_4$	Cassaidine, 1883
407.694972	AlI_3	Aluminum Iodide, 344
407.777037	$C_9H_4Cl_8O$	Isobenzan, 4979
408.103330	$C_{21}H_{24}O_3Si_3$	2,4,6-Trimethyl-2,4,6-triphenylcyclotrisiloxane, 9388
408.105647	$C_{19}H_{20}O_{10}$	Khellol Glucoside, 5157
408.142033	$C_{20}H_{24}O_9$	Nodakenin, 6484
408.150366	$C_{22}H_{26}ClFO_4$	Clobetasone, 2333
408.170352	$C_{22}H_{29}ClO_5$	Beclomethasone, 1025
408.178418	$C_{21}H_{28}O_8$	Scabiolide, 8143
408.214804	$C_{22}H_{32}O_7$	Cascarillin, 1878
408.230060	$C_{26}H_{32}O_4$	Bixin, Methyl ester, 1330
408.248504	$C_{19}H_{32}N_6O_4$	Embelin, Disemicarbazone, 3502
408.287575	$C_{24}H_{40}O_5$	Cholic Acid, 2197
408.339216	$C_{29}H_{44}O$	Dehydroergosterol, Methyl ether, 2847
409.166492	$C_{22}H_{23}F_4NO_2$	Trifluperidol, 9354
409.218785	$C_{24}H_{31}N_3OS$	Butaperazine, 1499
409.298080	$C_{27}H_{39}NO_2$	Veratramine, 9606
409.707045	Cs_2O_4Se	Cesium Selenate, 1971
410.021244	$C_{14}H_{19}BrO_9$	Acetobromglucose, 47
410.032879	$C_{20}H_{18}O_2Sn$	Triphenyltin, Acetate, 9412
410.118796	$C_{23}H_{22}O_5S$	Xylenol Blue, 9745
410.125670	$C_{21}H_{28}BrFO_2$	Haloprogesterone, 4451

Molecular Weight	Empirical Formula	Compound Name, Monograph Number
410.136553	$C_{23}H_{22}O_7$	Lactucin, p-Hydroxyphenylacetate, 5195 Sumatrol, 8790 Tephrosin, 8875
410.166016	$C_{22}H_{28}ClFO_4$	Clobetasol, 2332 Clocortolone, 2337
410.168183	$C_{18}H_{35}IO_2$	2-Iodostearic Acid, 1676
410.172939	$C_{24}H_{26}O_6$	Mangostin, 5573
410.190481	$C_{22}H_{28}F_2O_5$	Flumethasone, 4025
410.202800	$C_{24}H_{30}N_2O_2S$	Piperacetazine, 7253
410.245710	$C_{26}H_{34}O_4$	Promethestrol Dipropionate, 7583
410.256943	$C_{25}H_{34}N_2O_3$	Tilorone, 9169
410.318481	$C_{28}H_{42}O_2$	ϵ-Tocopherol, 9199
410.391252	$C_{30}H_{50}$	Squalene, 8547
410.448767	$C_{28}H_{58}O$	Octacosanol, 6552
411.171560	$C_{20}H_{29}NO_6S$	Sultroponium, 8788
411.198049	$C_{23}H_{29}N_3O_2S$	Acetophenazine, 58
411.240959	$C_{25}H_{33}NO_4$	Etorphine, 3832
411.253535	$C_{26}H_{35}O_4$	Trenbolone, Cyclohexylmethylcarbonate, 9271
411.313730	$C_{27}H_{41}NO_2$	Tomatillidine, 9246
411.955679	$C_8H_4AuN_2NaO_2S$	Triphal®, 9405
412.120815	$C_{21}H_{26}Cl_2O_4$	Dichlorisone, 3019
412.156913	$C_{21}H_{24}N_4O_3S$	Vintiamol, 9643
412.199822	$C_{23}H_{28}N_2O_5$	Reserpiline, 7936
412.246104	$C_{22}H_{36}O_7$	Grayanotoxin I, 4388
412.297745	$C_{27}H_{40}O_3$	Hexadecyl 3-Hydroxy-2-naphthoate, 4549 Nandrolone, Cyclohexanepropionate, 6185 Testosterone 17β-Cypionate, 8892
412.321555	$[C_{27}H_{42}NO_2]^+$	Benzethonium, 1078
412.357940	$[C_{28}H_{46}NO]^+$	Methylsolanidinium, 8478
412.370516	$C_{29}H_{48}O$	Chondrillasterol, 2209 7-Dehydrositosterol, 2848 Fucosterol, 4131 α-Spinasterol, 8521 Stigmasterol, 8596
412.767612	Cl_4K_2Pt	Potassium Tetrachloroplatinate(II), 7473

Use in conjunction with The Merck Index, Ninth Edition

Molecular Weight	Empirical Formula	Compound Name, Monograph Number
412.863210	$C_8H_7Cl_4N_3O_4S_2$	Tetrachlormethiazide, 8905
413.147452	$C_{22}H_{23}NO_7$	Gnoscopine, 4352 Noscapine, 6528, 6253
413.183838	$C_{23}H_{27}NO_6$	Colchiceine, Ethyl ether, 2435
413.329380	$C_{27}H_{43}NO_2$	Isorubijervine, 5086 Rubijervine, 8048 Solasodine, 8483 Veralkamine, 9603
413.707587	I_2O_6Zn	Zinc Iodate, 9800
413.716327	$O_7P_2Sn_2$	Stannous Pyrophosphate, 8568
413.959360	$C_{14}H_{14}Cl_3O_6P$	Haloxon, 4456
413.986024	$C_{18}H_{14}Cl_4N_2O$	Miconazole, 6043
414.055545	$C_{16}H_{18}N_2O_7S_2$	Sulbenicillin, 8686
414.095082	$C_{21}H_{18}O_9$	Adriamycinone, 3428
414.106316	$C_{20}H_{18}N_2O_8$	8,6'-Dinitrootobain, 6733
414.116212	$C_{18}H_{22}O_{11}$	Asperuloside, 867
414.124944	$C_{21}H_{22}N_2O_5S$	Nafcillin, 6174
414.131468	$C_{22}H_{22}O_8$	β-Peltatin, 6871 Picropodophyllin, 7215 Podophyllotoxin, 7325
414.142701	$C_{21}H_{22}N_2O_7$	Sancycline, 8108
414.161329	$C_{22}H_{26}N_2O_4S$	Diltiazem, 3187
414.204239	$C_{24}H_{30}O_6$	Estriol, Triacetate, 3635 Meprednisone, 21-Acetate, 5689
414.215472	$C_{23}H_{30}N_2O_5$	Reserpic Acid, Methyl ester, 7935
414.277010	$C_{26}H_{38}O_4$	Gestonorone Caproate, 4246 4-Hydroxy-19-nortestosterone, 17-Cyclopentanepropio- nate, 4748 Lupulon, 5427
414.313395	$C_{27}H_{42}O_3$	Diosgenin, 3295 Methenolone, 17-Enanthate, 5836
414.324629	$C_{26}H_{42}N_2O_2$	O-Acetylholarrhenine, 4601
414.361014	$C_{27}H_{46}N_2O$	N,N'-Dimethylcyclobuxine, 2722
414.386166	$C_{29}H_{50}O$	α-Ergostenol, Methyl ether, 3577 β-Sitosterol, 8294 γ-Sitosterol, 8295
415.066050	$C_{19}H_{17}N_3O_4S_2$	Cephaloridine, 1936

Molecular Weight	Empirical Formula	Compound Name, Monograph Number
415.079821	$C_{19}H_{17}ClF_3NO_4$	Halofenate, 4449
415.104937	$C_{16}H_{21}N_3O_8S$	Cephalosporin C, 1937
415.145345	$C_{22}H_{25}NO_5S$	Thiocolchicine, 9059
415.147846	$C_{18}H_{25}NO_{10}$	Linamarin, Tetraacetate, 5337
415.283492	$C_{24}H_{37}N_3O_3$	Pregnenolone, Acetate semicarbazone, 7533
415.345030	$C_{27}H_{45}NO_2$	Tomatidine, 9245
415.668707	Ag_3O_4P	Silver Phosphate, 8269
415.917443	$C_9H_{18}FeN_3S_6$	Ferbam, 3924
416.029040	$C_{15}H_{14}Cl_2N_4O_6$	Dichlororiboflavin, 3053
416.079056	$C_{18}H_{16}N_4O_6S$	Quinacillin, Free acid, 7829
416.102492	$C_{21}H_{21}O_7P$	Guaiacol Phosphate, 4403
416.110733	$C_{21}H_{20}O_9$	Chrysophanic Acid, Glucoside, 2260 Daidzein, 7-Glucoside, 2796 Frangulin A, 4114 Toringin, 9249
416.121232	$C_{19}H_{29}IO_2$	Iophendylate, 4918
416.163711	$C_{27}H_{20}N_4O$	Pyrinoline, 7771
416.168913	$C_{23}H_{29}ClN_2OS$	Xanthiol, 9723
416.175437	$C_{24}H_{29}ClO_4$	Cyproterone, Acetate, 2774
416.202132	$C_{24}H_{32}O_4S$	Spironolactone, 8537
416.219889	$C_{24}H_{32}O_6$	Desonide, 2886 6α-Methylprednisolone, 21-Acetate, 5980
416.242356	$[C_{22}H_{32}N_4O_4]^{2+}$	Distigmine, 3380
416.256275	$C_{25}H_{36}O_5$	3-Acetyldigitoxigenin, 3138 Pregnenolone, Succinate, 7533 3-O-Acetyluzarigenin, 9561
416.277404	$C_{22}H_{40}O_7$	Agaricic Acid, 174
416.292660	$C_{26}H_{40}O_4$	Methandriol, Dipropionate, 5807
416.329046	$C_{27}H_{44}O_3$	1α,25-Dihydroxycholecalciferol, 3167 Sarsasapogenin, 8134 Smilagenin, 8301 Tigogenin, 9163
416.365431	$C_{28}H_{48}O_2$	β-Tocopherol, 9196 γ-Tocopherol, 9197 ζ_2-Tocopherol, 9201
416.401817	$C_{29}H_{52}O$	Stigmastanol, 8595

Use in conjunction with The Merck Index, Ninth Edition

Molecular Weight	Empirical Formula	Compound Name, Monograph Number
417.121238	$C_{24}H_{19}NO_6$	Bisoxatin Acetate, 1323
417.189986	$C_{21}H_{27}N_3O_6$	Casimiroedine, 1881
417.226371	$C_{22}H_{31}N_3O_5$	Cinepazide, 2282
417.251524	$C_{24}H_{35}NO_5$	Batrachotoxinin A, 1017 Decoquinate, 2834
417.816054	Cr_2HgO_7	Mercuric Dichromate(VI), 5713
417.856298	$C_9H_8I_2O_3$	3,5-Diiodosalicylic Acid, Ethyl ester, 3174
418.059859	$C_4H_{16}HgN_4O_4S$	Sublamine®, 8662
418.120509	$C_{28}H_{18}O_4$	α-Naphtholphthalein, 6216
418.126383	$C_{21}H_{22}O_9$	Aloin, 303 Gardenin, 4203
418.199154	$C_{23}H_{30}O_7$	Sarverogenin, 8136
418.202321	$C_{23}H_{31}ClN_2O_3$	Etodroxizine, 3827
418.215553	$C_{24}H_{31}FO_5$	Descinolone, Acetonide, 2879 Fluocortolone, 21-Acetate, 4032
418.308310	$C_{26}H_{42}O_4$	Acetyllithocholic acid, 5388
419.135754	$F_5H_{12}N_3O_2U$	Ammonium Uranium Fluoride, 600
419.230788	$C_{23}H_{33}NO_6$	Strophanthidin oxime, 8645
419.246044	$C_{27}H_{33}NO_3$	Ethamoxytriphetol, 3647
419.677139	I_3K	Potassium Triiodide, 7488
419.699559	$C_7H_4Br_4O$	3,4,5,6-Tetrabromo-o-cresol, 8901
419.977770	$C_8H_{10}HgO_3S_2$	Thimerfonic Acid, 9045
419.982572	$C_{14}H_4N_4O_{12}$	Chrysamminic Acid, 2246
420.120903	$C_{24}H_{20}O_7$	Psoralidin, Diacetate, 7714
420.138010	$C_{16}H_{24}N_2O_{11}$	Cycasin, Tetraacetate, 2705
420.142033	$C_{21}H_{24}O_9$	Rhapontin, 7965
420.194817	$C_{23}H_{29}FO_6$	Fluprednisolone, 21-Acetate, 4075
420.214804	$C_{23}H_{32}O_7$	16-Hydroxystrophanthidin, 8645
420.226037	$C_{22}H_{32}N_2O_6$	Hexoprenaline, 4581
420.252526	$C_{25}H_{32}N_4O_2$	Pipebuzone, 7250
420.287575	$C_{25}H_{40}O_5$	Allopregnane-3β,17α,20α-triol, 3,20-Diacetate, 262 Allopregnane-3β,17α,20β-triol, 3,20-Diacetate, 263
420.867197	$C_8H_9I_2NO_3$	Iopydol, 4920

421.

Molecular Weight	Empirical Formula	Compound Name, Monograph Number
421.037784	$C_{15}H_{14}F_3N_3O_4S_2$	Bendroflumethiazide, 1042
421.119524	$C_{20}H_{23}NO_7S$	Erysothiovine, 3595
421.282824	$C_{24}H_{39}NO_5$	Erythrophleine, 3609
421.309313	$[C_{27}H_{39}N_3O]^{2+}$	Pentacynium, 6902
421.915317	$C_{17}H_{12}Br_2O_3$	Benzbromarone, 1068
421.928234	$C_{16}H_{10}Cl_4O_5$	Diploicin, 3356
421.987859	$C_{16}H_{10}N_2O_8S_2$	Indigo Carmine, Free acid, 4828
422.028371	$C_{15}H_{16}Cl_2N_2O_8$	Chloramphenicol, Monosuccinate, 2040, 8202
422.174276	$C_{26}H_{22}N_4O_2$	Salicil, Osazone, 8084
422.184172	$C_{24}H_{26}N_2O_5$	Alstonidine, Acetyl deriv, 309
422.202616	$C_{17}H_{26}N_8O_5$	Blasticidin S, 1333
422.210467	$C_{23}H_{31}FO_6$	Fludrocortisone, 21-Acetate, 4022
422.233826	$C_{20}H_{38}O_7S$	Dioctyl Sulfosuccinic Acid, 3287, 3288
422.245710	$C_{27}H_{34}O_4$	Bixin, Ethyl ester, 1330
422.303225	$C_{25}H_{42}O_5$	Cholic Acid, Methyl ester, 2197
422.485152	$C_{30}H_{62}$	Squalane, 8546
423.055879	$C_{17}H_{17}N_3O_6S_2$	Cephapirin, 1941
423.088325	$C_{17}H_{24}Cl_3N_3O_3$	Butylchloralaminopyrine, 1544
423.240959	$C_{26}H_{33}NO_4$	Cyprenorphine, 2769
423.262088	$C_{23}H_{37}NO_6$	Tocamphyl, 9194
423.941515	O_8S_2Th	Thorium Sulfate, 9121
424.082600	$C_{22}H_{17}ClN_2O_5$	Camptothecin, Chloroacetate, 1738
424.179872	$C_{18}H_{33}ClN_2O_5S$	Clindamycin, 2324
424.203194	$C_{21}H_{32}N_2O_5S$	Camphotamide, 1737
424.215078	$C_{28}H_{28}N_2O_2$	Difenoxin, 3123
424.308979	$C_{27}H_{40}N_2O_2$	Isooctylhydrocupreine, 5053
424.334131	$C_{29}H_{44}O_2$	ζ_1-Tocopherol, 9200
424.652972	$CoK_2O_8Se_2$	Potassium Cobaltous Selenate, 7403
424.955234	$C_{10}H_{11}ClF_3N_3O_4S_3$	Epithiazide, 3548
425.044973	$[C_{12}H_{19}N_4O_7P_2S]^+$	Cocarboxylase, Cation, 2412
425.213699	$C_{24}H_{31}N_3O_2S$	Carphenazine, 1858

Use in conjunction with The Merck Index, Ninth Edition

Molecular Weight	Empirical Formula	Compound Name, Monograph Number
425.241353	$C_{22}H_{35}NO_7$	Amoproxan, 608
425.292994	$C_{27}H_{39}NO_3$	Jervine, 5115
425.314124	$C_{24}H_{43}NO_5$	Triacetylsphingosine, 8519
425.565444	Br_5P	Phosphorus Pentabromide, 7160
425.791985	$C_6H_4I_2O_4S$	Sozoiodolic Acid, 8504
425.905192	Au_2S	Gold Monosulfide, 4360
425.943735	OTl_2	Thallium Oxide, 8979
426.095082	$C_{22}H_{18}O_9$	Triacetylpratensein, 7507 Tectorigenin, Triacetate, 8860
426.124944	$C_{22}H_{22}N_2O_5S$	Phenbenicillin, Free acid, 7005
426.157307	$C_{18}H_{26}N_4O_6S$	Cetotiamine, 1977
426.215472	$C_{24}H_{30}N_2O_5$	Demethoxyvindoline, 9641
426.240624	$C_{26}H_{34}O_5$	Scillarenin, 3-Acetate, 8152
426.277010	$C_{27}H_{38}O_4$	Azafrin, 910
426.337205	$[C_{28}H_{44}NO_2]^+$	Methylbenzethonium, 5897
426.349781	$C_{29}H_{46}O_2$	7-Dehydrocholesterol, Acetate, 2842
426.370910	$C_{26}H_{50}O_4$	Bis(2-ethylhexyl) Sebacate, 1271 Glycol Dilaurate, 4334
426.386166	$C_{30}H_{50}O$	α-Amyrin, 656 β-Amyrin, 657 Friedelin, 4120 Lanosterol, 5208 Lupeol, 5424 Shionone, 8223 α_1-Sitosterol, 8293 Taraxasterol, 8837 Taraxerol, 8839
427.029419	$C_{10}H_{15}N_5O_{10}P_2$	Adenosine Diphosphate, 144
427.050794	$C_{16}H_{17}N_3O_7S_2$	Cefoxitin, 1912
427.095041	$C_{19}H_{17}N_5O_5S$	Salazosulfadimidine, 8080
427.137950	$C_{21}H_{21}N_3O_7$	Cacotheline, 1594
427.147846	$C_{19}H_{25}NO_{10}$	Vicianin, 9630
427.229349	$C_{24}H_{33}N_3O_2S$	Dixyrazine, 3403
427.272259	$C_{26}H_{37}NO_4$	4-(Hexyloxy)benzilic Acid 2-Diethylaminoethyl Ester, 4587
427.272259	$[C_{52}H_{74}N_2O_8]^{2+}$	Laudexium (half mass), 5226

Molecular Weight	Empirical Formula	Compound Name, Monograph Number
427.574669	As_4S_4	Arsenic Disulfide, 822
427.974142	$C_{14}H_{12}N_4O_4S_4$	Thioaurin, 9050
427.998507	$C_{19}H_{15}Cl_3O_5$	Nornidulin, 6522
428.038832	$C_{21}H_{16}O_6S_2$	Methargen, Free acid, 5820
428.059961	$C_{18}H_{20}O_8S_2$	Diethylstilbestrol 4,4'-Disulfuric Ester, 3116
428.078995	$C_{18}H_{22}O_8P_2$	Fosfestrol, 4109
428.084053	$C_{18}H_{22}Cl_2N_4O_2S$	Asazol, 849
428.110733	$C_{22}H_{20}O_9$	Hesperetin, Triacetate, 4534
428.147118	$C_{23}H_{24}O_8$	Wortmannin, 9717
428.205970	$C_{22}H_{28}N_4O_5$	Nitroakridin 3582, 6402
428.231122	$C_{24}H_{32}N_2O_5$	Metoserpate, 6026
428.292660	$C_{27}H_{40}O_4$	Gentrogenin, 4233 17α-Hydroxyprogesterone Caproate, 4757 Lontanyl, 5403
428.329046	$C_{28}H_{44}O_3$	Anagestone, Hexanoate, 662 Nandrolone Decanoate, 6186
428.365431	$C_{29}H_{48}O_2$	Allocholesterol, Acetate, 246 Cholesterol, Acetate, 2192 Epicholesterol, Acetate, 3538
429.100108	$C_{28}H_{15}NO_4$	Anthrimide, 725
429.251524	$C_{25}H_{35}NO_5$	Mebeverine, 5594
429.324295	$C_{27}H_{43}NO_3$	Imperialine, 4818
430.025307	$C_{14}H_{14}N_4O_8S_2$	Dinsed, 3285
430.094645	$C_{18}H_{24}O_8P_2$	Hexestrol 4,4'-Diphosphoric Ester, 4573
430.126383	$C_{22}H_{22}O_9$	Formononetin, 7-Glucoside, 4102
430.162768	$C_{23}H_{26}O_8$	Sikkimotoxin, 8229
430.199154	$C_{24}H_{30}O_7$	Athamantin, 881
430.214410	$C_{28}H_{30}O_4$	Thymolphthalein, 9145
430.218935	$C_{18}H_{26}N_{10}O_3$	Congocidine, 2466
430.235539	$C_{25}H_{34}O_6$	4-Pregnene-20,21-diol-3,11-dione, Diacetate, 7529 Tanghinigenin, Acetate, 8826
430.262029	$C_{28}H_{34}N_2O_2$	Carbiphene, 1803
430.273262	$C_{27}H_{34}N_4O$	Piritramide, 7284

Molecular Weight	Empirical Formula	Compound Name, Monograph Number
430.308310	$C_{27}H_{42}O_4$	Convallamarogenin, 2481 Hecogenin, 4476 Pseudohecogenin, 7701 Ruscogenin, 8055
430.344696	$C_{28}H_{46}O_3$	Tocol, Acetate, 9195
430.355929	$C_{27}H_{46}N_2O_2$	Solanocapsine, 8480
430.381081	$C_{29}H_{50}O_2$	Cholestanol, Acetate, 2191 Epicholestanol, Acetate, 3537 Vitamin E, 9681
430.983499	$C_{15}H_{14}ClN_3O_4S_3$	Benzthiazide, 1134
431.113565	$C_{22}H_{22}ClNO_6$	Ochratoxin C, 6549
431.286765	$[C_{30}H_{40}P]^+$	Dodecyltriphenylphosphonium, 3412
431.339945	$C_{27}H_{45}NO_3$	Verticine, 9624
432.020685	$C_{11}H_{11}AuN_2O_2S$	Lopion, Free acid, 5406
432.031349	$C_{17}H_{18}Cl_2N_2O_5S$	Clometacillin, 2346
432.105647	$C_{21}H_{20}O_{10}$	Apigetrin, 765 Genistein, 7-D-Glucoside, 4221 Sophoricoside, 8493
432.142033	$C_{22}H_{24}O_9$	Methylgardenin, 4203 Homonataloin, 4620 Podophyllic Acids, 7323
432.157289	$C_{26}H_{24}O_6$	Phenolphthalol, Triacetate, 7043
432.194817	$C_{24}H_{29}FO_6$	Fluprednidene Acetate, 4074
432.233227	$C_{28}H_{33}ClN_2$	Buclizine, 1456
432.251189	$C_{25}H_{36}O_6$	Allopregnane-3β,21-diol-11,20-dione, 3,21-Diacetate, 255 Hydroxydione, 4724 3-O-Acetylperiplogenin, 6962 4-Pregnene-17α,20β,21-triol-3-one, 20,21-Diacetate, 7532
432.302831	$C_{30}H_{40}O_2$	β-Citraurin, 2305
432.323960	$C_{27}H_{44}O_4$	Chlorogenin, 2122 Digalogenin, 3129 Gitogenin, 4259 Acetyllithocholic Acid Methyl Ester, 5388
432.867197	$C_9H_9I_2NO_3$	β-Amino-4-hydroxy-3,5-diiodohydrocinnamic Acid, 452 3,5-Diiodotyrosine, 3176
433.113472	$[C_{21}H_{21}O_{10}]^+$	Pelargonidin, 3-Glucoside, 6865
433.133259	$C_{16}H_{23}N_3O_{11}$	Streptozocin, Tetraacetate, 6820
433.203529	$C_{22}H_{31}N_3O_4S$	Penethamate, 6877
433.282824	$C_{25}H_{39}NO_5$	Cassamine, 1885

433.

Molecular Weight	Empirical Formula	Compound Name, Monograph Number
433.721568	$AuBr_3$	Gold Tribromide, 4368
433.910628	Ir_2O_3	Iridium Sesquioxide, 4936
434.118228	$C_{21}H_{29}Cl_3O_3$	Testosterone 17-Chloral Hemiacetal, 8891
434.119770	$C_{19}H_{28}Cl_2N_2O_3S$	Phenamet, 6993
434.121297	$C_{21}H_{22}O_{10}$	Eriodictyol, 7-L-Rhamnoside, 3591
434.142427	$C_{18}H_{26}O_{12}$	Hexa-O-acetylgalactitol, 4177
434.149617	$C_{23}H_{27}ClO_6$	Chloroprednisone, 21-Acetate, 2139
434.163970	$C_{23}H_{25}F_3N_2OS$	Flupentixol, 4071
434.210467	$C_{24}H_{31}FO_6$	Betamethasone, 21-Acetate, 1211 Dexamethasone, 21-Acetate, 2899 Fluperolone Acetate, 4072 Paramethasone, 21-Acetate, 6834 Triamcinolone Acetonide, 9280
434.212492	$C_{27}H_{31}ClN_2O$	Chlorbenzoxamine, 2048
434.266839	$C_{25}H_{38}O_6$	Allopregnane-3β,11β,21-triol-20-one, 3,21-Diacetate, 265 Allopregnane-3β,17α,21-triol-20-one, 3,21-Diacetate, 266
434.366100	$C_{30}H_{46}N_2$	N-Benzilidenechonemorphine, 2214
435.065570	$C_{19}H_{18}ClN_3O_5S$	Cloxacillin, 2372
435.262088	$C_{24}H_{37}NO_6$	3-O-Lauroylpyridoxol Diacetate, 5234
435.575543	Br_4Sn	Stannic Bromide, 8551
436.061178	$C_{14}H_{10}HgN_4$	Barbak, 961
436.123029	$C_{18}H_{20}N_4O_9$	Hypoglycine B, N-2,4-Dinitrophenylhydrazone, 4788
436.136947	$C_{21}H_{24}O_{10}$	Phlorizin, 7137
436.226117	$C_{24}H_{33}FO_6$	Flurandrenolide, 4077
436.261360	$C_{28}H_{36}O_4$	Androstenediol, 3-Acetate-17-benzoate, 673 Oxymetholone, Enol benzoate, 6780
436.282489	$C_{25}H_{40}O_6$	Chenodeoxycholic Acid, Diformate, 2010
436.308979	$C_{28}H_{40}N_2O_2$	Bialamicol, 1223
436.318875	$C_{26}H_{44}O_5$	Cholic Acid, Ethyl ester, 2197
436.334131	$C_{30}H_{44}O_2$	Dehydroergosterol, Acetate, 2847
436.882317	C_4BaN_4Pt	Barium Platinous Cyanide, 997
436.887386	$C_{14}H_8Br_2F_3NO_2$	Fluorosalan, 4065
437.023910	$C_{18}H_{15}NO_8S_2$	Picosulfuric Acid, 7208
437.174869	$C_{22}H_{26}F_3N_3OS$	Fluphenazine, 4073

Use in conjunction with The Merck Index, Ninth Edition

Molecular Weight	Empirical Formula	Compound Name, Monograph Number
437.212157	$C_{27}H_{32}ClNO_2$	Triparanol, 9403
437.843352	$Cl_6H_8N_2Os$	Ammonium Osmium Chloride, 566
438.179087	$C_{24}H_{26}N_2O_6$	Suxibuzone, 8799
438.188983	$C_{22}H_{30}O_9$	Simarubin, 8281
438.204239	$C_{26}H_{30}O_6$	Erythrocentaurin, Dimedone deriv, 3599 3,6-Dimethylmangostin, 5573
438.225368	$C_{23}H_{34}O_8$	Ouabagenin, 6734
438.277010	$C_{28}H_{38}O_4$	Promethestrol, Dibutyrate, 7582
438.291615	$C_{24}H_{42}N_2O_3S$	Stearylsulfamide, 8584
438.313395	$C_{29}H_{42}O_3$	Estradiol, 17-Undecenoate, 3626
438.349781	$C_{30}H_{46}O_2$	Ergosterol, Acetate, 3580 Fungisterol, Acetate, 4142 Isopyrocalciferol, Acetate, 5081 Lumisterol, Acetate, 5418 Pyrocalciferol, Acetate, 7778
438.480067	$C_{30}H_{62}O$	1-Triacontanol, 9277
438.970884	$C_{11}H_{13}ClF_3N_3O_4S_3$	Polythiazide, 7362
438.999619	$C_{12}H_5O_{12}N_7$	Dipicrylamine, 3352
439.027676	$C_{11}H_{21}NO_{11}S_3$	Glucocheirolin, Free acid, 4282
439.047422	$C_{20}H_{13}N_3O_7S$	Eriochrome® Black T, Free acid, 3590
439.168100	$C_{23}H_{31}Cl_2NO_3$	Estramustine, 3632
439.345030	$C_{29}H_{45}NO_2$	Acetylsolanidine, 8478
440.155666	$C_{19}H_{20}N_8O_5$	Aminopterin, 484
440.160007	$C_{21}H_{29}O_8P$	Cortisone Phosphate, 2516 Prednisolone Phosphoric Acid, 7512
440.204633	$C_{22}H_{32}O_9$	Simarubidin, 8280
440.217444	$C_{24}H_{31}F_3O_4$	Flumedroxone Acetate, 4024
440.241018	$C_{23}H_{36}O_8$	Dihydroouabagenin, 6734
440.267508	$C_{26}H_{36}N_2O_4$	Bisobrin, 1322
440.292660	$C_{28}H_{40}O_4$	Azafrin, Methyl ester, 910 17α-Hydroxyprogesterone 3-Cyclopentyl Enol Ether, Acetate, 4758
440.329046	$C_{29}H_{44}O_3$	Estradiol, 17-Undecanoate, 3626
440.365431	$C_{30}H_{48}O_2$	22,23-Dihydroergosterol, Acetate, 3580
440.846650	$Cl_6H_8N_2Pt$	Ammonium Platinic Chloride, 578

Molecular Weight	Empirical Formula	Compound Name, Monograph Number
441.139682	$C_{19}H_{19}N_7O_6$	Folic Acid, 4085, 8369
441.189986	$C_{23}H_{27}N_3O_6$	Ergonovine Maleate, 3572
441.870601	O_5Ta_2	Tantalum Pentoxide, 8834
441.917864	Au_2O_3	Gold Trioxide, 4374
441.920892	STl_2	Thallium Sulfide, 8984
441.938071	F_2Hg_2	Mercurous Fluoride, 5737
442.014157	$C_{20}H_{17}Cl_3O_5$	Nidulin, 6351
442.095357	$C_{28}H_{14}N_2O_4$	Indanthrene®, 4821
442.137616	$C_{22}H_{22}N_2O_8$	Methacycline, 5798
442.175657	$C_{21}H_{31}O_8P$	Hydrocortisone Phosphate, 4676
442.199154	$C_{25}H_{30}O_7$	Picrolichenic Acid, 7212
442.225643	$C_{28}H_{30}N_2O_3$	Rhodamine B, Inner salt, 7973
442.283158	$C_{26}H_{38}N_2O_4$	Depersolon, Free base, 2871
442.381081	$C_{30}H_{50}O_2$	Betulin, 1220 Campesterol, Acetate, 1730 α-Ergostenol, Acetate, 3577 β-Ergostenol, Acetate, 3578 γ-Ergostenol, Acetate, 3579
443.136888	$C_{26}H_{21}NO_6$	Triacetyldiphenolisatin, 9276
443.170121	$C_{23}H_{29}N_3O_2S_2$	Thiothixene, 9105
443.176461	$C_{16}H_{25}N_7O_8$	Gougerotin, 4378
443.910620	$C_{13}H_{14}As_2N_2O_4S$	Neoarsphenamine, Free acid, 6269
444.069262	$C_{21}H_{16}O_{11}$	Anhydromethylenecitric Acid Disalicylic Acid Ester, 690
444.105647	$C_{22}H_{20}O_{10}$	Granaticin, 4382
444.112760	$C_{12}H_{36}O_6Si_6$	Dodecamethylcyclohexasiloxane, 3408
444.153266	$C_{22}H_{24}N_2O_8$	Doxycycline, 3429 Quatrimycin, 7816 Tetracycline, 8913
444.251189	$C_{26}H_{36}O_6$	Bufotalin, 1466 Prednisolone 21-Trimethylacetate, 7517 Prednival, 7519
444.396731	$C_{30}H_{52}O_2$	Ergostanol, Acetate, 3576
445.104934	$C_{22}H_{24}ClN_3OS_2$	Clospirazine, 2367
445.149386	$C_{22}H_{27}N_3O_3S_2$	Metopimazine, 6022
445.159077	$C_{23}H_{28}ClN_3O_2S$	Thiopropazate, 9094

Use in conjunction with The Merck Index, Ninth Edition

Molecular Weight	Empirical Formula	Compound Name, Monograph Number
445.173667	$C_{23}H_{27}NO_8$	Narceine, 6248
445.178376	$C_{21}H_{27}N_5O_4S$	Glipizide, 4273
445.579573	Br_4Te	Tellurium Tetrabromide, 8869
445.735396	$BiBr_3$	Bismuth Bromide, 1281
446.110750	$C_{16}H_{22}N_4O_9S$	Cephamycin C, 1940
446.121297	$C_{22}H_{22}O_{10}$	Prunetin, 4'-Glucoside, 7693
446.136553	$C_{26}H_{22}O_7$	D-2-Deoxyribose, 1,3,4-tribenzoate, 2869
446.142427	$C_{19}H_{26}O_{12}$	Gaultherin, 4210 Violutin, 9653
446.168916	$C_{22}H_{26}N_2O_8$	Podophyllic Acids, Hydrazides, 7323
446.181020	$C_{22}H_{30}N_4O_2S_2$	Thioproperazine, 9095
446.194068	$C_{24}H_{30}O_8$	Desaspidin, 2877 Flavaspidic Acid, 4004
446.230454	$C_{25}H_{34}O_7$	4-Pregnene-17α,20β,21-triol-3,11-dione, 20,21-Diacetate, 7531
446.247047	$C_{30}H_{30}N_4$	Bis(p-dimethylaminobenzylidene)benzidine, 1268
446.304562	$C_{28}H_{38}N_4O$	Carpipramine, 1859
446.339610	$C_{28}H_{46}O_4$	Acetyllithocholic Acid Ethyl Ester, 5388
446.363420	$[C_{28}H_{48}NO_3]^+$	Methylverticinium, 9624
446.810512	$C_{13}H_8Br_3NO_2$	Tribromsalan, 9296
446.882847	$C_{10}H_{11}I_2NO_3$	Propyliodone, 7652
447.269299	$C_{19}H_{37}N_5O_7$	Sisomicin, 8292
447.554369	Se_4S_4	Selenium Sulfides, 8187
447.594420	C_3HBr_5O	Pentabromoacetone, 6898
447.794527	$C_{13}H_7Br_3O_3$	Tribromophenyl Salicylate, 9293
448.092496	$C_{22}H_{21}ClO_8$	Monorden, Diacetate, 6089
448.100562	$C_{21}H_{20}O_{11}$	Kaempferol, 3-Glucoside, 5126 Luteolin, 5-Glucoside, 5428 Luteolin, 7-Glucoside, 5428 Quercitrin, 7824
448.136947	$C_{22}H_{24}O_{10}$	Sakuranetin, 5-Glucoside, 8078
448.209718	$C_{24}H_{32}O_8$	Amarolide, Diacetate, 381

448.

Molecular Weight	Empirical Formula	Compound Name, Monograph Number
448.246104	$C_{25}H_{36}O_7$	Allopregnane-3β,17α,21-triol-11,20-dione, 3,21-Diacetate, 264 Allotetrahydrocortisone, 3α,21-Diacetate, 273 4-Pregnene-11β,17α,20β,21-tetrol-3-one, 20,21-Diacetate, 7530
448.261360	$C_{29}H_{36}O_4$	Algestone Acetophenide, 228
448.270463	$C_{27}H_{45}Br$	Cholesteryl Bromide, 2193
448.282489	$C_{26}H_{40}O_6$	Ursodeoxycholic Acid, Diformate, 9551
448.318875	$C_{27}H_{44}O_5$	Digitogenin, 3136
448.773739	I_2Pt	Platinous Iodide, 7308
448.825726	$C_8H_5I_2NO_5$	Iodomethamic Acid, 8389
449.108387	$[C_{21}H_{21}O_{11}]^+$	Cyanidin, 3-Galactoside, 2695 Cyanidin, 3-Glucoside, 2695
449.173291	$C_{20}H_{27}N_5O_5S$	Glisoxepid, 4274
449.277738	$C_{25}H_{39}NO_6$	Erythrophlamine, 3608
449.284949	$C_{19}H_{39}N_5O_7$	Gentamicin C_{1a}, 4224
449.686565	$Cu_4H_6O_{10}S$	Copper Sulfate Tribasic, 2658
449.954198	$C_{16}H_{16}As_2N_2O_4$	Arsenophenylglycine, 837
449.993137	$C_{15}H_6N_4O_{13}$	Aloetic Acid, 302
450.022531	$C_{15}H_{18}N_2O_8S_3$	Noprylsulfamide, Free acid, 6492
450.046670	$C_{24}H_{19}BrO_4$	4,4'-(β-Bromostyrylidene)diphenol Diacetate, 1439
450.055545	$C_{19}H_{18}N_2O_7S_2$	Ponceau 3R, Free acid, 7366
450.116212	$C_{21}H_{22}O_{11}$	Carthamin, 1863
450.150280	$C_{24}H_{30}O_3Si_3$	2,4,6-Triethyl-2,4,6-triphenylcyclotrisiloxane, 9352
450.189853	$C_{24}H_{34}O_4S_2$	Thiomesterone, 9078
450.261754	$C_{25}H_{38}O_7$	Allopregnane-3α,11β,17α,21-tetrol-20-one, 3,21-Diacetate, 260 Allopregnane-3β,11β,17α,21-tetrol-20-one, 3,21-Diacetate, 261 Laserpitin, 5220
450.290923	$[C_{31}H_{36}N_3]^+$	Stilbazium, 8598
450.298139	$C_{26}H_{42}O_6$	Citrullol, Diacetate, 2313
450.349781	$C_{31}H_{46}O_2$	Vitamin K_1, 9684
450.757441	Cl_6Na_2Pt	Sodium Hexachloroplatinate(IV), 8374
451.190519	$C_{23}H_{28}F_3N_3OS$	Homofenazine, 4618

Use in conjunction with The Merck Index, Ninth Edition

Molecular Weight	Empirical Formula	Compound Name, Monograph Number
451.264214	$C_{18}H_{37}N_5O_8$	Dibekacin, 2969
451.608549	$As_2Ni_3O_8$	Nickel Arsenate, 6315
451.781741	$CoK_3N_6O_{12}$	Cobaltic Potassium Nitrite, 2384
451.989508	$C_6H_8N_6O_{18}$	Mannitol Hexanitrate, 5576
452.034809	$C_{18}H_{16}N_2O_8S_2$	Allura® Red AC, Free acid, 276
452.131862	$C_{21}H_{24}O_{11}$	Helicin, Tetraacetate, 4484
452.176580	$C_{24}H_{30}ClFO_5$	Clocortolone, 21-Acetate, 2337
452.201045	$C_{24}H_{30}F_2O_6$	Flumethasone, 21-Acetate, 4025 Fluocinolone Acetonide, 4030
452.246378	$C_{30}H_{32}N_2O_2$	Diphenoxylate, 3321
452.280084	$[C_{28}H_{38}NO_4]^+$	Butropium, 1518
452.329046	$C_{30}H_{44}O_3$	Boldenone, 10-Undecenoate, 1338
452.365431	$C_{31}H_{48}O_2$	Dihydrovitamin K_1, 3162
453.056149	$C_{19}H_{17}ClFN_3O_5S$	Floxacillin, 4017
453.251524	$C_{27}H_{35}NO_5$	Etorphine, 3-Acetate, 3832
453.272653	$C_{24}H_{39}NO_7$	Delcosine, 2852
453.324295	$C_{29}H_{43}NO_3$	Tomatillidine, O-Acetate, 9246
453.623719	$O_8P_2Sr_3$	Strontium Phosphate, Tribasic, 8641
453.933320	$C_{20}H_{10}Cl_4O_4$	Phenoltetrachlorophthalein, 7046
454.030017	$C_{14}H_{14}N_8O_4S_3$	Cefazolin, 1911
454.073711	$C_{13}H_{19}N_4O_{12}P$	SAICAR, 8077
454.089997	$C_{23}H_{18}O_{10}$	Tetraacetylbaptigenin, 959 Datiscetin, Tetraacetate, 2811 Scutellarein, Tetraacetate, 8165
454.119858	$C_{23}H_{22}N_2O_6S$	Carfecillin, 1839
454.131380	$C_{23}H_{28}Cl_2O_5$	Dichlorisone, 21-Acetate, 3019
454.171316	$C_{20}H_{22}N_8O_5$	A-Ninopterin, 695, 6978 Methotrexate, 5858
454.192230	$C_{24}H_{32}ClFO_5$	Halcinonide, 4445
454.227494	$C_{17}H_{34}N_4O_{10}$	Ribostamycin, 8000
454.283158	$C_{27}H_{38}N_2O_4$	Verapamil, 9604
454.344696	$C_{30}H_{46}O_3$	Anagestone, Cyclopentylpropionate, 662

454.

Molecular Weight	Empirical Formula	Compound Name, Monograph Number
454.381081	$C_{31}H_{50}O_2$	Chondrillasterol, Acetate, 2209 7-Dehydrositosterol, Acetate, 2848 Fucosterol, Acetate, 4131 Stigmasterol, Acetate, 8596
454.602102	$As_2Co_3O_8$	Cobaltous Arsenate, 2386
455.155332	$C_{20}H_{21}N_7O_6$	Methopterin, 5857 Ninopterin, 6374
455.189504	$C_{26}H_{31}Cl_2N_3$	Aminoquinol, 488
455.635027	AsI_3	Arsenic Triiodide, 831
455.779586	HgI_2	Mercuric Iodide, Red, 5716
455.907253	O_4PbW	Lead Tungstate(VI), 5285
456.008753	$C_{12}H_{26}O_6P_2S_4$	Dioxathion, 3302
456.105647	$C_{23}H_{20}O_{10}$	Fustin, Tetraacetate, 4173
456.214804	$C_{26}H_{32}O_7$	Methyl Picrolichenate, 7212
456.226037	$C_{25}H_{32}N_2O_6$	Vindoline, 9641
456.253869	$[C_{30}H_{34}NO_3]^+$	8-p-Phenylbenzylatropinium, 7076
456.323960	$C_{29}H_{44}O_4$	Diosgenin, Acetate, 3295 Lapinone, 5212
456.325298	$[C_{30}H_{40}N_4]^{2+}$	Dequalinium, 2873, 2874
456.360346	$C_{30}H_{48}O_3$	β-Boswellic Acid, 1366 Nandrolone, Dodecanoate, 6185 Oleanolic Acid, 6673 Ursolic Acid, 9552
456.396731	$C_{31}H_{52}O_2$	β-Sitosterol, Acetate, 8294 γ-Sitosterol, Acetate, 8295
457.152538	$C_{27}H_{23}NO_6$	Benzoylchelidonine, 2009
457.158411	$C_{20}H_{27}NO_{11}$	Amygdalin, 630
457.184901	$C_{23}H_{27}N_3O_7$	Minocycline, 6053
457.766942	$C_3La_2O_9$	Lanthanum Carbonate, 5209
457.933564	O_3Tl_2	Thallium Sesquioxide, 8982
458.164795	$C_{14}H_{42}O_5Si_6$	Tetradecamethylhexasiloxane, 8915
458.178812	$C_{21}H_{30}O_{11}$	Gein, 4214
458.191624	$C_{23}H_{29}F_3O_6$	Fluprostenol, 4076
458.230454	$C_{26}H_{34}O_7$	Cinobufotalin, 2297 Fumagillin, 4136
458.266839	$C_{27}H_{38}O_6$	Prednisolone Tebutate, 7516

Use in conjunction with The Merck Index, Ninth Edition

Molecular Weight	Empirical Formula	Compound Name, Monograph Number
458.339610	$C_{29}H_{46}O_4$	Sarsasapogenin, Acetate, 8134 Smilagenin, Acetate, 8301 Acetyltigogenin, 9163
458.375996	$C_{30}H_{50}O_3$	β-Tocopherol, Acetate, 9196 γ-Tocopherol, Acetate, 9197
458.412381	$C_{31}H_{54}O_2$	Stigmastanol, Acetate, 8595
459.121906	$C_{28}H_{17}N_3O_4$	Bis(4-amino-1-anthraquinonyl)amine, 1257
459.176741	$[C_{23}H_{27}N_2O_8]^+$	Methylquatrimycinium, 7816
459.630383	$Cr_3Fe_2O_{12}$	Ferric Chromate(VI), 3934
459.765116	$C_3Ce_2O_9$	Cerous Carbonate, 1952
459.959678	$C_{14}H_{18}As_2N_2O_6$	Diphetarsone, Free acid, 3351
460.148181	$C_{22}H_{24}N_2O_9$	Oxytetracycline, 6791
460.209718	$C_{25}H_{32}O_8$	Albaspidin, 198 Aspidin, 869 α-Kosin, 5168 Prednisolone 21-Succinate, 7514
460.224969	$[C_{28}H_{26}N_7]^+$	Isometamidium, 5037
460.282489	$C_{27}H_{40}O_6$	Hydrocortisone Tebutate, 4678
461.227869	$C_{28}H_{29}F_2N_3O$	Pimozide, 7235
461.304228	$C_{29}H_{39}N_3O_2$	Echinuline, 3468
461.530129	P_2Se_5	Phosphorus Pentaselenide, 7163
461.710124	$C_9H_6Br_4O_2$	3,4,5,6-Tetrabromo-o-cresol, Acetate, 8901
461.782801	Br_2O_6Pb	Lead Bromate, 5249
461.785595	I_2Pb	Lead Iodide, 5264
462.079826	$C_{21}H_{18}O_{12}$	Scutellarein, Glucuronide, 8165
462.091930	$C_{21}H_{22}N_2O_6S_2$	Dibenzoyldjenkolic acid, 3404
462.116212	$C_{22}H_{22}O_{11}$	Scoparin, 8154 Tectorigenin, 7-Glucoside, 8860
462.225368	$C_{25}H_{34}O_8$	Hydrocortisone Succinate, 4677
462.261754	$C_{26}H_{38}O_7$	Pleuromutilin, Diacetate, 7314
462.298139	$C_{27}H_{42}O_6$	Hydrallostane, 21-tert-Butylacetate, 4646
463.177708	$C_{22}H_{29}N_3O_6S$	Pivampicillin, 7296
463.300599	$C_{20}H_{41}N_5O_7$	Gentamicin C_2, 4224
463.938026	O_3Pb_2	Lead Sesquioxide, 5274

Molecular Weight	Empirical Formula	Compound Name, Monograph Number
463.938799	$C_{18}H_{12}Cl_4O_6$	Diploicin, Acetate, 3356
464.095476	$C_{21}H_{20}O_{12}$	Isoquercitrin, 5083 Myricetin, 3-Rhamnoside, 6158 Quercetin, 3-D-Galactoside, 7821 Quercimeritrin, 7822
464.098644	$C_{21}H_{21}ClN_2O_8$	Demeclocycline, 2856
464.267508	$C_{28}H_{36}N_2O_4$	Psychotrine, 7715
464.313789	$C_{27}H_{44}O_6$	α-Ecdysone, 3459
465.103301	$[C_{21}H_{21}O_{12}]^+$	Delphinidin, 3-Glucoside, 2850
465.309038	$C_{26}H_{43}NO_6$	Glycocholic Acid, 4330
465.666903	$O_{12}S_3Y_2$	Yttrium Sulfate, 9773
465.945520	Bi_2O_3	Bismuth Oxide, 1298
466.107595	$C_{19}H_{23}N_4O_6PS$	Benfotiamine, 1044
466.143177	$C_{16}H_{24}HgO_3$	Acetomeroctol, 51
466.181396	$C_{27}H_{30}O_5S$	Thymol Blue, 9142
466.192230	$C_{25}H_{32}ClFO_5$	Clobetasol, 17-Propionate, 2332
466.230783	$C_{22}H_{43}IO_2$	Iodobehenic Acid, 1675
466.234191	$C_{25}H_{26}N_{10}$	Homocongasine, 4614
466.283158	$C_{28}H_{38}N_2O_4$	Cephaeline, 1931
466.344696	$C_{31}H_{46}O_3$	Lithocholic Acid, Benzyl ester, 5388 Vitamin K_1 Oxide, 9685
466.403548	$C_{30}H_{50}N_4$	Sarothamnine, 8131
466.403548	$[C_{30}H_{50}N_4]^{2+}$	Isocurine, Cation, 5020
467.259128	$C_{18}H_{37}N_5O_9$	Tobramycin, 9193
467.288303	$C_{25}H_{41}NO_7$	Delsoline, 2852 Lycoctonine, 5435
467.314792	$C_{28}H_{41}N_3O_3$	Oxethazaine, 6758
467.979648	$C_{17}H_{18}Br_2N_4O_2$	Dibromopropamidine, 2992
468.004572	$C_{16}H_{12}N_4O_9S_2$	Tartrazine, Free acid, 8847
468.069262	$C_{23}H_{16}O_{11}$	Cromolyn, 2585
468.110644	$C_{23}H_{26}Cl_2O_6$	Simfibrate, 8286
468.186966	$C_{21}H_{24}N_8O_5$	A-Denopterin, 142
468.191307	$C_{23}H_{33}O_8P$	Cortisone Phosphate, Dimethyl ester, 2516

Molecular Weight	Empirical Formula	Compound Name, Monograph Number
468.243144	$C_{18}H_{36}N_4O_{10}$	Gentamicin A, 4224
468.346427	$C_{28}H_{44}N_4O_2$	Tomatillidine, Semicarbazone, 9246
468.371579	$C_{30}H_{48}N_2O_2$	Hexestrol Bis(β-diethylaminoethyl ether), 4572
468.396731	$C_{32}H_{52}O_2$	α-Amyrin, Acetate, 656 β-Amyrin, Acetate, 657 Lupeol, Acetate, 5424 Taraxasterol, Acetate, 8837 Taraxerol, Acetate, 8839
468.988587	$C_{16}H_{11}N_3O_{10}S_2$	p-Nitrobenzeneazochromotropic Acid, 6410
469.026598	$C_{19}H_{17}Cl_2N_3O_5S$	Dicloxacillin, 3060
469.170982	$C_{21}H_{23}N_7O_6$	Denopterin, 2861
469.257672	$C_{26}H_{35}N_3O_5$	Reserpilic Acid, Dimethylaminoethyl ester, 7936
469.303953	$C_{25}H_{43}NO_7$	Methymycin, 6008 Neomethymycin, 6277
469.589944	$As_2O_8Zn_3$	Zinc Ortho-arsenate, 9810
469.809110	$C_{10}H_{18}Br_4O$	Nerol, Tetrabromide, 6299
469.933564	CO_3Tl_2	Thallium Carbonate, 8973
470.048526	$C_{22}H_{14}O_{12}$	Ellagic Acid, Tetraacetate, 3497
470.142427	$C_{21}H_{26}O_{12}$	Plumieride, 7318
470.191383	$C_{22}H_{26}N_6O_6$	Oxotremorine, Picrolonate, 6767
470.194068	$C_{26}H_{30}O_8$	Limonin, 5334 Physodic Acid, 7187
470.230454	$C_{27}H_{34}O_7$	Methyl O-methylpicrolichenate, 7212
470.266839	$C_{28}H_{38}O_6$	Withaferin A, 9712
470.299202	$C_{24}H_{42}N_2O_7$	Fusaria Antibiotics, Sambucinin, 4167
470.303225	$C_{29}H_{42}O_5$	Antheridiol, 716 Gentrogenin, Acetate, 4233
470.339610	$C_{30}H_{46}O_4$	Enoxolone, 3526 Gypsogenin, 4438
470.375996	$C_{31}H_{50}O_3$	β-Boswellic Acid, Methyl ester, 1366 Oleanolic Acid, Methyl ester, 6673 Ursolic Acid, Methyl ester, 9552
470.964617	$C_{16}H_{15}CrN_6S_4$	Rhodanilic Acid, 7974
471.060569	$C_{12}H_{15}HgNO_6$	Mercuderamide, 5703
471.223017	$C_{22}H_{29}N_7O_5$	Puromycin, 7730
471.942327	$C_2Hg_2N_2O$	Mercuric Oxycyanide, 5721

Molecular Weight	Empirical Formula	Compound Name, Monograph Number
472.122152	$C_{27}H_{22}Cl_2N_4$	Clofazimine, 2338
472.166235	$C_{22}H_{30}FO_8P$	Dexamethasone, 21-Phosphate, 2899
472.246104	$C_{27}H_{36}O_7$	Fumagillin, Methyl ester, 4136
472.355260	$C_{30}H_{48}O_4$	Gratiogenin, 4385 Hederagenin, 4479
472.366494	$C_{29}H_{48}N_2O_3$	N'-Acetylsolanocapsine, 8480
472.391646	$C_{31}H_{52}O_3$	Vitamin E Acetate, 9682
472.402879	$C_{30}H_{52}N_2O_2$	N,N,N'-Trimethylsolanocapsine, 8480
473.048207	$C_{18}H_{20}ClN_3O_6S_2$	Sporidesmin A, 8543
473.143430	$C_{22}H_{23}N_3O_9$	Aluminon, 314
473.165896	$C_{20}H_{23}N_7O_7$	Folinic Acid-SF, 4086
473.204967	$C_{25}H_{31}NO_8$	Ethylnarceine, 6248
473.277738	$C_{27}H_{39}NO_6$	Prednisolone 21-Diethylaminoacetate, 7511
473.878969	Cl_2Hg_2	Mercurous Chloride, 5736
473.912890	$C_{14}H_{16}Cl_6O_5$	Toloxychlorinol, 9218
474.200216	$C_{24}H_{30}N_2O_8$	Podophyllinic Acid 2-Ethylhydrazide, 7324
474.225368	$C_{26}H_{34}O_8$	6α-Methylprednisolone, 21-Succinate, 5980
474.261754	$C_{27}H_{38}O_7$	Sarmentogenin, Diacetate, 8129
474.278153	$C_{28}H_{39}FO_5$	Fluocortolone, 21-Hexanoate, 4032
474.298139	$C_{28}H_{42}O_6$	Cynanchogenin, 2766
475.196336	$C_{24}H_{33}N_3O_3S_2$	Pipotiazine, 7277
475.243519	$C_{29}H_{31}F_2N_3O$	Fluspirilene, 4082
475.261020	$[C_{30}H_{31}N_6]^+$	Janus Green B, Cation, 5107
475.293388	$C_{27}H_{41}NO_6$	Hydrocortamate, 4673
476.128793	$C_{23}H_{31}Cl_3O_4$	Testosterone 17-Chloral Hemiacetal, Acetate, 8891
476.152991	$C_{20}H_{28}O_{13}$	Primulaverin, 7546
476.223056	$C_{29}H_{33}ClN_2O_2$	Loperamide, 5404
476.241018	$C_{26}H_{36}O_8$	Andrographolide, Triacetyl deriv, 668
476.257417	$C_{27}H_{37}FO_6$	Betamethasone, 17-Valerate, 1211
476.267508	$C_{29}H_{36}N_2O_4$	Emetamine, 3507
476.313789	$C_{28}H_{44}O_6$	Ursodeoxycholic Acid, Diacetate, 9551

Use in conjunction with The Merck Index, Ninth Edition

Molecular Weight	Empirical Formula	Compound Name, Monograph Number
476.440076	$C_{30}H_{57}BO_3$	Menthyl Borate, 5666
477.068633	$C_{15}H_{17}HgNO_2S$	N-(Ethylmercuri)-p-toluenesulfonanilide, 3765
477.156972	$C_{22}H_{27}N_3O_7S$	Griseoviridin, 4394
477.316249	$C_{21}H_{43}N_5O_7$	Gentamicin C_1, 4224
477.934294	$C_{16}H_{16}Cl_6N_2O_2$	Triclofenol Piperazine, 9335
478.114294	$C_{22}H_{23}ClN_2O_8$	Chlortetracycline, 2181
478.192230	$C_{26}H_{32}ClFO_5$	Clobetasone, 17-Butyrate, 2333
478.200297	$C_{25}H_{31}FO_8$	Triamcinolone, 16α,21-Diacetate, 9279
478.235539	$C_{29}H_{34}O_6$	Tribenoside, 9286
478.273068	$C_{27}H_{39}FO_6$	Fludrocortisone, 21-tert-Butylacetate, 4022
478.283158	$C_{29}H_{38}N_2O_4$	Dehydroemetine, 2845 O-Methylpsychotrine, 7715
478.293054	$C_{27}H_{42}O_7$	Allopregnane-3β,17α,20β,21-tetrol, 3,20,21-Triacetate, 259
478.309647	$C_{32}H_{38}N_4$	Etioporphyrin, 3825
478.377058	$C_{28}H_{50}N_2O_4$	Carpaine, 1857
479.267174	$C_{29}H_{37}NO_5$	Cytochalasin B, 2787
479.702019	Cl_6K_2Os	Potassium Hexachloroosmate(IV), 7416
479.865126	$Al_2F_{18}Si_3$	Aluminum Hexafluorosilicate, 338
479.957181	$C_{20}H_{18}Br_2O_4$	8,6'-Dibromootobain, 6733
479.989326	$N_4O_{12}Th$	Thorium Nitrate, 9119
480.049670	$C_{14}H_{14}HgO_6$	Mercumallylic Acid, 5704, 5705
480.090391	$C_{21}H_{20}O_{13}$	Quercetagetin, 7-Glucoside, 7820
480.102228	$C_{20}H_{25}IN_4O_2$	Iodohexamidine, 4898
480.265590	$C_{28}H_{37}ClN_4O$	Clocapramine, 2335
480.298808	$C_{29}H_{40}N_2O_4$	Emetine, 3508-3511
480.308704	$C_{27}H_{44}O_7$	β-Ecdysone, 3459
480.490632	$C_{32}H_{64}O_2$	Cetyl Palmitate, 1986
481.130758	$C_{24}H_{23}N_3O_6S$	Talampicillin, A14
481.218601	$C_{22}H_{27}N_9O_4$	Distamycin A, 3379
481.557209	Sb_2Se_3	Antimony Triselenide, 750
481.940435	Bi_2O_4	Bismuth Tetroxide, 1316

482.

Molecular Weight	Empirical Formula	Compound Name, Monograph Number
482.121297	$C_{25}H_{22}O_{10}$	Silymarin-Group, 8279
482.142427	$C_{22}H_{26}O_{12}$	Catalposide, 1897 Glucovanillin, Tetraacetate, 4297
482.366100	$[C_{34}H_{46}N_2]^{2+}$	Hedaquinium, 4478
482.705317	Cl_6K_2Pt	Potassium Hexachloroplatinate(IV), 7417
483.115753	$C_{24}H_{26}BrN_3O_3$	Nicergoline, 6310
483.179421	$C_{28}H_{25}N_3O_5$	β- or γ-Aminoorceimin, 6704
483.254043	$C_{18}H_{37}N_5O_{10}$	Kanamycin B, 5132
483.606284	I_3KZn	Potassium Triiodozincate, 7490
483.968533	$C_9H_{15}N_2O_{15}P_3$	Uridine 5'-Triphosphate, 9543
484.085306	$C_{20}H_{20}O_{14}$	Hamamelitannin, 4459
484.100562	$C_{24}H_{20}O_{11}$	Rhamnetin, Tetraacetate, 7962
484.158077	$C_{22}H_{28}O_{12}$	Oosporein, Tetraacetate, 6689
484.163437	$C_{28}H_{24}N_2O_6$	β- or γ-Aminoorcein, 6704
484.199822	$C_{29}H_{28}N_2O_5$	Monotrityl Thymidine, 9138
484.209718	$C_{27}H_{32}O_8$	Muconomycin B, 6127 Physodic Acid, Methyl ester, 7187
484.238058	$C_{18}H_{36}N_4O_{11}$	Kanamycin A, 5132 Kanamycin C, 5132
484.248784	$[C_{31}H_{34}NO_4]^+$	Fentonium, 3921
484.282489	$C_{29}H_{40}O_6$	Cortisone, 21β-Cyclopentanepropionate, 2515
484.314852	$C_{25}H_{44}N_2O_7$	Fusaria Antibiotics, Avenacein, 4167
484.355260	$C_{31}H_{48}O_4$	Gypsogenin, Methyl ester, 4438 Testosterone, 3-Oxododecanoate, 8890
485.076219	$C_{13}H_{17}HgNO_6$	Mersalyl, Free acid, 5749
485.147452	$C_{28}H_{23}NO_7$	β- or γ-Hydroxyorcein, 6704
485.201610	$C_{25}H_{32}ClN_5OS$	Chlorimpiphenine, 2062
485.277738	$C_{28}H_{39}NO_6$	Prednylidene 21-Diethylaminoacetate, 7521
485.683442	$C_{10}Cl_{10}O$	Chlordecone, 2053
485.875932	O_7Re_2	Rhenium Heptoxide, 7968
485.918899	$C_{14}H_{16}I_2O_3$	Monophen®, 6088
486.103247	$C_{26}H_{25}Cl_3N_2O$	Hetolin®, 4539
486.116212	$C_{24}H_{22}O_{11}$	Irigenin, Triacetate, 4939

Use in conjunction with The Merck Index, Ninth Edition

Molecular Weight	Empirical Formula	Compound Name, Monograph Number
486.137341	$C_{21}H_{26}O_{13}$	Fabiatrin, 3861 Leucoglycodrin, 5306
486.137608	$C_{24}H_{29}Cl_2FO_5$	Flucloronide, 4020
486.225368	$C_{27}H_{34}O_8$	Myricetin, Hexaethyl ether, 6158
486.261754	$C_{28}H_{38}O_7$	Bongkrekic Acid, 1347
486.309373	$C_{28}H_{42}N_2O_5$	Fenoxedil, 3910
486.334525	$C_{30}H_{46}O_5$	Ceanothic Acid, 1908 Quillaic Acid, 7826 Quinovic Acid, 7899
486.370910	$C_{31}H_{50}O_4$	Sulphurenic Acid, 8785
487.683678	BaI_2O_6	Barium Iodate, 984
487.763855	$C_6Ba_2FeN_6$	Barium Ferrocyanide, 978
487.991166	$C_{16}H_{18}CrN_7S_4$	Rhodanilic Acid, Ammonium salt, 7974
488.107335	$C_{14}H_{26}N_4O_{11}P_2$	Citicoline, 2300
488.204633	$C_{26}H_{32}O_9$	Estriol, 16,17-Bis(hemisuccinic acid), 3635
488.277404	$C_{28}H_{40}O_7$	Diginin, 3132
488.350175	$C_{30}H_{48}O_5$	Acacic Acid, 11 Cimigenol, 2269 Escigenin, 3617
488.365431	$C_{34}H_{48}O_2$	7-Dehydrocholesterol, Benzoate, 2842
489.143905	$C_{14}H_{25}HgNO_5$	Mercurin®, 5731
489.293782	$C_{24}H_{43}NO_9$	Pseudopederin, 7705
489.793644	$AuNa_3O_6S_4$	Gold Sodium Thiosulfate, 4366
489.849335	Au_2S_3	Gold Trisulfide, 4375
489.865341	$SeTl_2$	Thallium Selenide, 8981
490.119674	$C_{18}H_{18}N_8O_9$	Kethoxal, 2,4-Dinitrophenylosazone, 5149
490.256669	$C_{27}H_{38}O_8$	Tetrahydrocortisone, Triacetate, 8928
490.273068	$C_{28}H_{39}FO_6$	Dexamethasone, 21-(3,3-Dimethylbutyrate), 2899
490.381081	$C_{34}H_{50}O_2$	Cholesterol, Benzoate, 2192
491.242021	$C_{28}H_{33}N_3O_5$	O-Methylhaplophytine, 4464
491.274819	$C_{30}H_{35}F_2N_3O$	Lidoflazine, 5324
492.090391	$C_{22}H_{20}O_{13}$	Carminic Acid, 1844
492.252526	$C_{31}H_{32}N_4O_2$	Bezitramide, 1222

Molecular Weight	Empirical Formula	Compound Name, Monograph Number
492.272319	$C_{27}H_{40}O_8$	Cortolone, 3,20,21-Triacetate, 2519
492.396731	$C_{34}H_{52}O_2$	Epicholestanol, Benzoate, 3537
493.134602	$[C_{23}H_{25}O_{12}]^+$	Malvidin, 3β-Glucoside, 5543 Malvidin, 3-Galactoside, 5543
493.143821	$C_{23}H_{28}ClN_3O_5S$	Glyburide, 4311
493.231182	$C_{25}H_{35}NO_9$	Ryanodine, 8065
493.303953	$C_{27}H_{43}NO_7$	Zygadenine, 9854
493.887598	$C_{15}H_{12}I_2O_3$	Iodoalphionic Acid, 4887
494.151159	$C_{26}H_{26}N_2O_6S$	Carindacillin, 1840
494.211610	$C_{26}H_{32}F_2O_7$	Fluocinonide, 4031
494.223531	$C_{27}H_{36}ClFO_5$	Clocortolone, 21-Pivalate, 2337
494.247996	$C_{27}H_{36}F_2O_6$	Flumethasone, 21-Pivalate, 4025
494.258794	$C_{20}H_{38}N_4O_{10}$	Neamine, N-Acetyl deriv, 6261
494.266839	$C_{30}H_{38}O_6$	Periplogenin, Monobenzoate, 6962
494.278073	$C_{29}H_{38}N_2O_5$	Strophanthidin, Phenylhydrazone, 8645
494.287969	$C_{27}H_{42}O_8$	α-Cortol 3,20,21-Triacetate, 2518
494.967777	$KN_3O_{11}U$	Potassium Uranyl Nitrate, 7493
495.179421	$C_{29}H_{25}N_3O_5$	Nicomorphine, 6336
496.226117	$C_{29}H_{33}FO_6$	Betamethasone, 17-Benzoate, 1211
496.230848	$C_{25}H_{36}O_{10}$	Glaucarubin, 4266
496.246104	$C_{29}H_{36}O_7$	O-Cinnamoyltaxicin-I, 8853
496.256604	$C_{27}H_{45}I$	Cholesteryl Iodide, 2195
496.282489	$C_{30}H_{40}O_6$	Absinthin, 7 Anabsinthin, 659
496.314852	$C_{26}H_{44}N_2O_7$	Fusaria Antibiotics, Fructigenin, 4167
497.210841	$C_{20}H_{35}NO_{13}$	Validamycin A, 9570
497.221366	$C_{28}H_{32}FNO_6$	Dexamethasone, 21-Isonicotinate, 2899
498.225368	$C_{28}H_{34}O_8$	Uliginosin B, 9501
498.272987	$C_{28}H_{38}N_2O_6$	Simetride, 8284
498.277010	$C_{33}H_{38}O_4$	5-Androstene-3β,16α-diol, Dibenzoate, 674
498.330502	$C_{26}H_{46}N_2O_7$	Fusaria Antibiotics, Lateritiin-I, 4167 Fusaria Antibiotics, Lateritiin-II, 4167

Use in conjunction with The Merck Index, Ninth Edition

Molecular Weight	Empirical Formula	Compound Name, Monograph Number
498.370910	$C_{32}H_{50}O_4$	β-Boswellic Acid, Acetate, 1366 Oleanolic Acid, Acetate, 6673 Ursolic Acid, Acetate, 9552
498.944971	$C_8H_9AsBiNO_6$	Glycobiarsol, 4329
499.278132	$C_{25}H_{41}NO_9$	Aconine, 110
499.892994	Hg_2O_4S	Mercurous Sulfate, 5741
500.080319	$C_{20}H_{28}Cl_4N_2O_4$	Teclozan, 8858
500.095317	$C_{19}H_{25}N_4O_6PS_2$	Diuredosan, 3398
500.183504	$C_{30}H_{28}O_7$	Curvularin, Dibenzoate, 2685
500.215866	$C_{26}H_{32}N_2O_8$	Tritoqualine, 9428
500.241018	$C_{28}H_{36}O_8$	Uliginosin A, 9501
500.336256	$[C_{28}H_{44}N_4O_4]^{2+}$	Hexadistigmine, Cation, 4551
500.350175	$C_{31}H_{48}O_5$	Quillaic Acid, Methyl ester, 7826
500.365431	$C_{35}H_{48}O_2$	Ergosterol, Benzoate, 3580
500.386560	$C_{32}H_{52}O_4$	Sulphurenic Acid, Methyl ester, 8785
501.125584	$C_{22}H_{29}Cl_2N_3O_4S$	Chlorobutin Penicillin, 2106
501.617255	I_3Sb	Antimony Triiodide, 748
501.699149	$Al_2O_{16}S_4Zn$	Aluminum Zinc Sulfate, 372
502.125732	$C_{20}H_{26}N_2O_{11}S$	N^4-β-D-Glucosylsulfanilamide, 2,3,4,6-Tetraacetate, 4296
502.220283	$C_{27}H_{34}O_9$	Muconomycin A, 6127 Sarverogenin, Diacetate, 8136
502.249960	$C_{19}H_{34}N_8O_8$	Streptothricin F, 8617
502.334800	$[C_{36}H_{42}N_2]^{2+}$	Hexafluorenium, 4552
502.365825	$C_{31}H_{50}O_5$	Acacic Acid, Methyl ester, 11
502.377058	$C_{30}H_{50}N_2O_4$	β-Tocopherol, Allophanate, 9196 γ-Tocopherol, Allophanate, 9197
503.161389	$C_{25}H_{29}NO_8S$	Guaiacodeine, 4398
503.309432	$C_{25}H_{45}NO_9$	Pederin, 6863
504.003480	$C_{12}H_{19}N_4O_{10}P_3S$	Thiamine Triphosphoric Acid Ester, 9030
504.084518	$C_{30}H_{16}O_8$	Hypericin, 4785
504.169035	$C_{18}H_{32}O_{16}$	Melezitose, 5638 Raffinose, 7914
504.203197	$C_{22}H_{30}Cl_2N_{10}$	Chlorhexidine, 2060

504.

Molecular Weight	Empirical Formula	Compound Name, Monograph Number
504.222014	$C_{24}H_{32}N_4O_8$	Scillabiose, Phenylosazone, 8150
504.252332	$C_{28}H_{37}FO_7$	Betamethasone, 17,21-Dipropionate, 1211
504.317252	$C_{24}H_{40}N_8O_4$	Dipyridamole, 3366
504.396731	$C_{35}H_{52}O_2$	α-Ergostenol, Benzoate, 3577 β-Ergostenol, Benzoate, 3578 γ-Ergostenol, Benzoate, 3579
505.221286	$C_{28}H_{31}N_3O_6$	Hepronicate, 4512
505.340339	$C_{29}H_{47}NO_6$	Coumingine, 2551
505.900550	O_4STl_2	Thallium Sulfate, 8983
506.092122	$C_{20}H_{18}N_4O_{12}$	Hexamethylene Glycol, Bis(3,5-dinitrobenzoate), 4560
506.143764	$C_{25}H_{22}N_4O_8$	Streptonigrin, 8615
506.245710	$C_{34}H_{34}O_4$	Promethestrol, Dibenzoate, 7582
506.291792	$C_{28}H_{41}FNO_6$	Dexamethasone, 21-Diethylaminoacetate, 2899
506.339610	$C_{33}H_{46}O_4$	Nandrolone p-Hexyloxyphenylpropionate, 6187
506.412381	$C_{35}H_{54}O_2$	Ergostanol, Benzoate, 3576
506.995751	$C_{10}H_{16}N_5O_{13}P_3$	Adenosine Triphosphate, 145
507.216734	$C_{26}H_{32}F_3N_3O_2S$	Oxaflumazine, 6740
507.923615	$Na_2O_{10}S_2U$	Sodium Uranyl Sulfate, 8473
508.124858	$C_{23}H_{25}ClN_2O_9$	Clomocycline, 2351
508.246104	$C_{30}H_{36}O_7$	3-Benzoylstrophanthidin, 8645
508.468942	$C_{26}H_{56}N_{10}$	Alexidine, 224
509.298868	$C_{27}H_{43}NO_8$	Cevine, 1989 Germine, 4245
509.314124	$C_{31}H_{43}NO_5$	Diacetyljervine, 5115
509.335253	$C_{28}H_{47}NO_7$	Narbomycin, 6247
509.935349	CBi_2O_5	Bismuth Subcarbonate, 1308
510.463468	$[C_{30}H_{60}N_3O_3]^{3+}$	Triethylgallaminium, 4188
511.168976	$C_{23}H_{29}NO_{12}$	Hygromycin, 4774
511.189592	$C_{33}H_{25}N_3O_3$	Norbormide, 6496
511.437588	$[C_{32}H_{55}N_4O]^+$	Thonzonium, 9113
512.350175	$C_{32}H_{48}O_5$	Gypsogenin, Acetate, 4438
512.386560	$C_{33}H_{52}O_4$	β-Boswellic Acid, Acetate of methyl ester, 1366 Oleanolic Acid, Acetate of methyl ester, 6673

Use in conjunction with The Merck Index, Ninth Edition

Molecular Weight	Empirical Formula	Compound Name, Monograph Number
513.647729	$AuBr_4H$	Gold Tribromide, Acid, 4369
513.876991	Bi_2S_3	Bismuth Sulfide, 1313
514.293054	$C_{30}H_{42}O_7$	Cucurbitacin I, 2611
514.329440	$C_{31}H_{46}O_6$	Convallamarogenin, Diacetate, 2481 Pseudohecogenin, 3,26-Diacetate, 7701 Ruscogenin, Diacetate, 8055
514.365825	$C_{32}H_{50}O_5$	Ceanothic Acid, Dimethyl ester, 1908 Quinovic Acid, Dimethyl ester, 7899
514.377058	$C_{31}H_{50}N_2O_4$	N,N'-Diacetylsolanocapsine, 8480
515.291675	$C_{26}H_{45}NO_7S$	Taurocholic Acid, 8851
515.979486	$C_{21}H_{17}AsN_2O_5S_2$	Thiocarbamizine, 9054
516.126777	$C_{25}H_{24}O_{12}$	Cynarin(e), 2767
516.178418	$C_{30}H_{28}O_8$	Rottlerin, 8033
516.308704	$C_{30}H_{44}O_7$	Cucurbitacin D, 2610 Cynanchogenin, Monoacetate, 2766
516.309575	$C_{31}H_{48}O_2S_2$	Probucol, 7552
516.345090	$C_{31}H_{48}O_6$	Fusidic Acid, 4172 Gitogenin, Diacetate, 4259
516.392709	$C_{31}H_{52}N_2O_4$	Vitamin E, Allophanate, 9681
516.396731	$C_{36}H_{52}O_2$	Chondrillasterol, Benzoate, 2209 α-Spinasterol, Benzoate, 8521
516.651915	$C_2H_6Co_5O_{12}$	Cobaltous Carbonate Basic, 2388
517.662941	$In_2O_{12}S_3$	Indium Sulfate, 4835
517.887598	$C_{17}H_{12}I_2O_3$	Benziodarone, 1090
518.131553	$C_{14}H_{42}O_7Si_7$	Tetradecamethylcycloheptasiloxane, 8914
518.194068	$C_{30}H_{30}O_8$	Gossypol, 4377
518.324354	$C_{30}H_{46}O_7$	Cucurbitacin F, 2610
518.387229	$C_{34}H_{50}N_2O_2$	N-Benzylidenesolanocapsine, 8480
518.429488	$[C_{28}H_{58}N_2O_6]^{2+}$	Prodeconium, 7562
519.134432	$C_{23}H_{32}Cl_2NO_6P$	Estramustine, 17-Phosphate, 3632
519.617908	CI_4	Carbon Tetraiodide, 1823
520.148181	$C_{27}H_{24}N_2O_9$	Albofungin, 200
520.209718	$C_{30}H_{32}O_8$	Tetrahydrorottlerin, 8033

520.

Molecular Weight	Empirical Formula	Compound Name, Monograph Number
520.216929	$C_{24}H_{32}N_4O_9$	Cellobiose Phenylosazone, 1916 Turanose, Phenylosazone, 9472
520.222782	$C_{28}H_{37}ClO_7$	Beclomethasone, Dipropionate, 1025
520.242081	$C_{26}H_{36}N_2O_9$	Antimycin A_3, 753
520.355260	$C_{34}H_{48}O_4$	Estradiol 3,17β-Dicypionate, 3630
520.618873	CeI_3	Cerous Iodide, 1955
521.153326	$C_{24}H_{27}NO_{12}$	Dhurrin, Pentaacetate, 2915
521.195071	$C_{31}H_{27}N_3O_5$	Nalorphine Dinicotinate, 6181
521.901303	$C_7H_6BiIO_6$	Bismuth Iodosubgallate, 1293
522.093371	$C_{20}H_{20}Cl_2N_8O_5$	Dichloromethotrexate, 3043
522.132343	$C_3H_{16}N_4O_{11}U$	Ammonium Uranium Carbonate, 599
522.137341	$C_{24}H_{26}O_{13}$	Centaurein, 1924 Irigenin, 7-Glucoside, 4939 Scopolin, Tetraacetate, 8162
523.112435	$C_{20}H_{27}Cl_2N_3O_9$	Chloramphenicol Pantothenate, 2042
523.170134	$C_{28}H_{27}ClF_5NO$	Penfluridol, 6879
524.054105	$C_{11}H_{18}HgN_2O_7S$	Diurgin, 3399
524.152991	$C_{24}H_{28}O_{13}$	Gentiopicrin, Tetraacetate, 4229
524.459332	$C_{36}H_{60}O_2$	Vitamin A, Palmitate, 9666
524.893412	$C_{15}H_{13}I_2NO_4$	3,5-Diiodothyronine, 3175
525.199882	$C_{28}H_{31}NO_9$	Rhodomycin B, 7982
525.247501	$C_{28}H_{35}N_3O_7$	Virginiamycin M_1, 9658
525.293782	$C_{27}H_{43}NO_9$	Protoverine, 7688
525.330168	$C_{28}H_{47}NO_8$	Picromycin, 7214
525.478662	Na_2Se_6	Sodium Hexaselenide, 8376
526.311420	$C_{30}H_{50}O_4Cr$	Ledol, Chromate, 5288
526.402211	$C_{34}H_{54}O_4$	Betulin, Diacetate, 1220
527.049080	$C_{20}H_{21}N_3O_8S_3$	Solupyridine, Free acid, 8486
527.076734	$C_{18}H_{25}NO_{13}S_2$	Sinigrin, Free acid tetraacetate, 8289
527.179146	$C_{27}H_{29}NO_{10}$	Daunorubicin, 2815
527.209008	$C_{27}H_{33}N_3O_6S$	Gliquidone, A9
527.226765	$C_{27}H_{33}N_3O_8$	Rolitetracycline, 8020

Use in conjunction with The Merck Index, Ninth Edition

Molecular Weight	Empirical Formula	Compound Name, Monograph Number
527.232639	$C_{20}H_{37}N_3O_{13}$	Destomycin A, 2894 Hygromycin B, 4775
527.916900	$Hg_2N_2O_6$	Mercurous Nitrate, 5739
528.214153	$C_{24}H_{36}N_2O_9S$	Celesticetin, 1914
528.323960	$C_{35}H_{44}O_4$	Allopregnane-3β,20α-diol, Dibenzoate, 253 Allopregnane-3β,20β-diol, Dibenzoate, 254 Uranediol, Dibenzoate, 9510
528.345090	$C_{32}H_{48}O_6$	O-Acetylquinovic Acid, 7899
529.662838	$O_{12}S_3Sb_2$	Antimony Sulfate, 742
530.251583	$C_{29}H_{38}O_9$	Uscharidin, 9555
530.278073	$C_{32}H_{38}N_2O_5$	Cortivazol, 2517
530.287969	$C_{30}H_{42}O_8$	Proscillaridin, 7662
530.293842	$C_{23}H_{46}O_{13}$	Neomycin C, 6278
530.397125	$C_{33}H_{54}O_5$	α-Tocopherol Acid Succinate, 9203
530.412381	$C_{37}H_{54}O_2$	α-Amyrin, Benzoate, 656 Lupeol, Benzoate, 5424 Taraxerol, Benzoate, 8839
531.617908	C_2I_4	Tetraiodoethylene, 8935
532.267233	$C_{29}H_{40}O_9$	Calotropin, 1724
532.283632	$C_{30}H_{41}FO_7$	Triamcinolone Hexacetonide, 9281
532.303619	$C_{30}H_{44}O_8$	Cucurbitacin J, 2613 Cucurbitacin K, 2613
534.137341	$C_{25}H_{26}O_{13}$	Ruberythric Acid, 8038
534.194856	$C_{23}H_{34}O_{14}$	Strophanthobiose, Pentaacetate, 8647
534.210112	$C_{27}H_{34}O_{11}$	Phillyrin, 7133
534.226511	$C_{28}H_{35}FO_9$	Triamcinolone Acetonide, 21-Hemisuccinate, 9280
534.282883	$C_{29}H_{42}O_9$	Helveticoside, 4491 Neriantin, 6297
534.319269	$C_{30}H_{46}O_8$	Cucurbitacins G and H, 2612 Desacetyloleandrin, 6672 Periplocymarin, 6961
534.585201	$[C_{36}H_{74}N_2]^{2+}$	Triclobisonium, 9332
535.402545	$C_{35}H_{53}NO_3$	Vitamin E, α-Tocopheryl nicotinate, 9681
535.629674	$AuBr_4Na$	Sodium Tetrabromoaurate(III), 8457
536.095476	$C_{27}H_{20}O_{12}$	Collinomycin, 2444

Molecular Weight	Empirical Formula	Compound Name, Monograph Number
536.268482	$[C_{28}H_{42}Cl_2N_4O_2]^{2+}$	Ambenonium, 383
536.290467	$C_{30}H_{45}ClO_6$	Senegenin, 8199
536.438202	$C_{40}H_{56}$	α-Carotene, 1852 β-Carotene, 1853 γ-Carotene, 1854 δ-Carotene, 1855 Lycopene, 5438
537.330168	$C_{29}H_{47}NO_8$	Sabadine, 8068
537.908518	$C_{21}H_{16}Br_2O_5S$	Bromcresol Purple, 1387
537.981060	$C_{20}H_{14}N_2O_{10}S_3$	Amaranth, Free acid, 378
538.205027	$C_{26}H_{34}O_{12}$	Rhododendrin, Pentacetate, 7981
538.283158	$C_{34}H_{38}N_2O_4$	Nafiverine, 6175
538.304287	$C_{31}H_{42}N_2O_6$	Batrachotoxin, 1016
539.280258	$C_{21}H_{41}N_5O_{11}$	Apramycin, 781
539.626233	$C_{10}Cl_{12}$	Mirex, 6056
539.871492	$K_2O_{10}S_2U$	Potassium Uranyl Sulfate, 7494
540.184292	$C_{25}H_{32}O_{13}$	Oleuropein, 6677
540.222014	$C_{27}H_{32}N_4O_8$	Pyridomycin, 7760
540.235933	$C_{30}H_{36}O_9$	Nimbin, 6368
540.843875	$C_{15}H_{10}ClI_2NO_3$	Clioxanide, 2326
541.028345	$C_{20}H_{19}N_3O_9S_3$	Fuchsin(e) Acid, Free acid, 4127
542.163556	$C_{24}H_{30}O_{14}$	Swertiamarin, Tetraacetate, 8800
542.251583	$C_{30}H_{38}O_9$	Dihydronimbin, 6368
542.314458	$C_{34}H_{42}N_2O_4$	Belladonnine, 1032
543.174061	$C_{27}H_{29}NO_{11}$	Doxorubicin, 3428
543.749860	$C_6Ce_2O_{12}$	Cerous Oxalate, 1957
544.002858	$C_{18}H_{16}N_4O_{10}S_3$	Azosulfamide, Free acid, 929
544.137773	$C_{28}H_{32}O_4Si_4$	2,4,6,8-Tetramethyl-2,4,6,8-tetraphenylcyclotetrasiloxane, 8946
544.142821	$C_{23}H_{28}O_{15}$	Purpurogallin, Diglucoside, 7733
544.170181	$C_{27}H_{32}N_2O_6S_2$	Sulphan Blue, Free acid, 8782
544.176705	$C_{28}H_{32}O_9S$	Prednisolone 21-m-Sulfobenzoate, 7515
544.180543	$C_{25}H_{28}N_4O_{10}$	D-Araboflavin, Tetraacetate, 791

Use in conjunction with The Merck Index, Ninth Edition

Molecular Weight	Empirical Formula	Compound Name, Monograph Number
544.184786	$C_{23}H_{36}N_4O_5S_3$	Octotiamine, 6569
544.209718	$C_{32}H_{32}O_8$	Rottlerin, 5,7-Dimethyl ether, 8033
545.643763	HoI_3	Holmium Iodide, 4602
546.217323	$C_{22}H_{34}N_4O_{12}$	Oryzacidin A, 6722
546.393377	$[C_{34}H_{50}N_4O_2]^{2+}$	Benzoquinonium, 1116
547.205361	$C_{27}H_{33}NO_{11}$	Colchicum Glucoside Colchicoside, 2438
547.279470	$C_{30}H_{37}N_5O_5$	Ergoclavine, 3562 Ergosine, 3573 Ergosinine, 3574
547.901098	$CeH_8N_8O_{18}$	Ammonium Ceric Nitrate, 526
547.952343	$C_{12}H_{10}HgO_8S_2$	Mercuric p-Phenylsulfonic Acid, 5723
548.135920	$C_{19}H_{28}N_6O_9S_2$	Amicarbalide, Diisethionate, 397
548.231122	$C_{34}H_{32}N_2O_5$	Micranthine, 6044
548.273381	$C_{28}H_{40}N_2O_9$	Antimycin A_1, 752
548.298533	$C_{30}H_{44}O_9$	Cymarin, 2763 Peruvoside, 6968
548.532103	$C_{40}H_{68}$	Phytofluene, 7194
549.425794	As_2Se_5	Arsenic Pentaselenide, 825
549.986584	$C_{20}H_{24}I_2O_2$	Thymol Iodide, 9144
550.044099	$C_{18}H_{32}I_2O_3$	Periodyl, 6959
550.277798	$C_{29}H_{42}O_{10}$	Adonitoxin, 156 Convallatoxin, 2485
550.417467	$C_{40}H_{54}O$	Echinenone, 3464
550.981833	$C_{19}H_{23}I_2NO_2$	Bufeniode, 1461
551.309432	$C_{29}H_{45}NO_9$	Germine, 3-Acetate, 4245 Germine, 16-Acetate, 4245
551.603612	$AuBr_4K$	Potassium Tetrabromoaurate(III), 7471
551.749074	HgI_2O_6	Mercuric Iodate, 5715
552.293448	$C_{29}H_{44}O_{10}$	Tetramethoxyglaucarubin, 4266
552.433117	$C_{40}H_{56}O$	Cryptoxanthin, 2603 Lycoxanthin, 5445 Rubixanthin, 8049
553.844999	O_4SeTl_2	Thallium Selenate, 8980
553.878000	$C_{14}H_{16}As_2N_2O_8S_2$	Sulfarsphenamine, Free acid, 8736

Use in conjunction with The Merck Index, Ninth Edition

554.

Molecular Weight	Empirical Formula	Compound Name, Monograph Number
554.215198	$C_{30}H_{34}O_{10}$	Physodic Acid, Diacetate, 7187
554.244408	$C_{22}H_{42}N_4O_8S_2$	Pantethine, 6817
554.304368	$C_{33}H_{43}FO_6$	Betamethasone, 21-Adamantoate, 1211
554.360740	$C_{34}H_{50}O_6$	Acacic Acid, Diacetyl lactone, 11
554.371973	$C_{33}H_{50}N_2O_5$	Chimyl Alcohol, Bis(phenylurethan), 2016
554.448767	$C_{40}H_{58}O$	Rhodopin, 7983
555.275172	$C_{21}H_{41}N_5O_{12}$	Butirosin, 1510
556.159414	$C_{29}H_{24}N_4O_8$	Niceritrol, 6311
556.303619	$C_{32}H_{44}O_8$	Cucurbitacin E, 2611
556.398857	$[C_{32}H_{52}N_4O_4]^{2+}$	Demecarium, 2855
556.748199	$C_9H_6I_3NO_3$	Acetrizoic Acid, 70, 5627
557.768600	$C_{10}H_9I_3O_3$	Phenobutiodil, 7034
558.131468	$C_{34}H_{22}O_8$	3,3'-Ethylidenebis(4-hydroxycoumarin), Dibenzoate, 3749
558.137341	$C_{27}H_{26}O_{13}$	Leucocyanidin, Hexaacetate, 5304
558.268965	$C_{28}H_{38}N_4O_8$	Pactamycin, 6796
558.319269	$C_{32}H_{46}O_8$	Cucurbitacin B, 2610
560.241993	$C_{27}H_{42}Cl_2N_2O_6$	Chloramphenicol Palmitate, 2041
560.334919	$C_{32}H_{48}O_8$	Cucurbitacin C, 2609
560.353363	$C_{25}H_{48}N_6O_8$	Deferoxamine, 2837
561.178102	$C_{26}H_{31}N_3O_9S$	Griseoviridin, Diacetate, 4394
561.295120	$C_{31}H_{39}N_5O_5$	Ergocornine, 3563, 3586, 3587 Ergocorninine, 3564
561.381809	$C_{36}H_{51}NO_4$	Stigmasterol, p-Nitrobenzoate, 8596
561.777936	Br_2Hg_2	Mercurous Bromide, 5734
562.038371	$C_{23}H_{19}ClN_4O_7S_2$	Polar® Yellow, Free acid, 7334
562.168641	$C_{27}H_{30}O_{13}$	Methyl Tetra-O-methylcarminate, 1844 α-Peltatin-β-D-glucoside, 6870
562.214445	$C_{24}H_{34}N_8O_4S_2$	Thiamine Disulfide, 9023
562.246772	$C_{35}H_{34}N_2O_5$	Trilobine, 9367
562.258006	$C_{34}H_{34}N_4O_4$	Protoporphyrin IX, 7685
562.381081	$C_{40}H_{50}O_2$	Rhodoxanthin, 7987
563.126542	$C_{16}H_{27}HgNO_6S$	Mercaptomerin, 5700

Use in conjunction with The Merck Index, Ninth Edition

Molecular Weight	Empirical Formula	Compound Name, Monograph Number
563.182518	$C_{27}H_{33}NO_{10}S$	Thiocolchicine, 2-Glucoside analog, 9059
564.126777	$C_{29}H_{24}O_{12}$	Theaflavine, 8987
564.147906	$C_{26}H_{28}O_{14}$	Apiin, 766
564.247166	$C_{31}H_{36}N_2O_8$	Raunescine, 7920
564.293448	$C_{30}H_{44}O_{10}$	Vernadigin, 9618
564.396731	$C_{40}H_{52}O_2$	Canthaxanthin, 1753 Torularhodin, 9250
565.428366	$C_{40}H_{55}NO$	Echinenone, Oxime, 3464
565.667901	$La_2O_{12}S_3$	Lanthanum Sulfate, 5209
566.055025	$C_{15}H_{24}N_2O_{17}P_2$	Uridine Diphosphate Glucose, 9542
566.272713	$C_{29}H_{42}O_{11}$	α-Antiarin, 727
566.371973	$C_{34}H_{50}N_2O_5$	η-Tocopherol, p-Nitrophenylurethane, 9202
567.286406	$C_{21}H_{41}N_7O_{11}$	Deoxydihydrostreptomycin, 2865
567.666075	$Ce_2O_{12}S_3$	Cerous Sulfate, 1958
567.888122	$C_{18}O_{16}O_6$	3,3',4'-Trimethoxyfisetin, 4003
568.136947	$C_{32}H_{24}O_{10}$	Thermorubin A, 9013
568.179206	$C_{26}H_{32}O_{14}$	Hamamelitannin, Hexamethyl ether, 4459
568.223925	$C_{29}H_{38}ClFO_8$	Formocortal, 4101
568.257337	$C_{34}H_{36}N_2O_6$	Pseudomorphine, 7704
568.303619	$C_{33}H_{44}O_8$	Helvolic Acid, 4492
568.428031	$C_{40}H_{56}O_2$	Lycophyll, 5439 Xanthophyll, 9726 Zeaxanthin, 9776
568.784585	$C_{11}H_{10}I_3NO_2$	Cinamiodyl, 2271
569.119816	$C_{15}H_{21}HgN_5O_6$	Oradon, 6698
569.670505	$O_{12}Pr_2S_3$	Praseodymium Sulfate, 7506
569.848457	$Cl_2Hg_2O_6$	Mercurous Chlorate, 5735
570.100956	$C_{27}H_{22}O_{14}$	Myricetin, Hexaacetate, 6158 Quercetagetin, Hexaacetate, 7820
570.182275	$C_{24}H_{26}N_8O_9$	Diopterin, 3291
570.230078	$C_{32}H_{34}N_4O_4S$	Diathymosulfone, 2956
570.355654	$C_{34}H_{50}O_7$	Carbenoxolone, 1796 Quillaic Acid, Diacetate, 7826

Molecular Weight	Empirical Formula	Compound Name, Monograph Number
570.392040	$C_{35}H_{54}O_6$	Sulphurenic Acid, Diacetate, 8785
570.628807	C_4HI_4N	Iodol®, 4900
570.800235	$C_{11}H_{12}I_3NO_2$	Iopanoic Acid, 4916
571.670653	$Nd_2O_{12}S_3$	Neodymium Sulfate, 6273
571.784250	$C_{11}H_{11}I_3O_3$	Iophenoxic Acid, 4919
572.108248	$C_{16}H_{22}HgN_2O_8$	Meluginan, 5647
572.116606	$C_{27}H_{24}O_{14}$	Ampelopsin, Hexaacetate, 612
572.167597	$C_{24}H_{32}N_2O_{12}S$	p,p'-Sulfonyldianiline-N,N'-digalactoside, 8758
572.371304	$C_{34}H_{52}O_7$	Cimigenol, Diacetate, 2269
572.455309	$[C_{35}H_{60}N_2O_4]^{2+}$	Pancuronium, 6814
572.459332	$C_{40}H_{60}O_2$	Kitol, 5163
574.314183	$C_{32}H_{46}O_9$	Cucurbitacin A, 2609
574.350569	$C_{33}H_{50}O_8$	Cephalosporin P_1, 1938
575.079392	$C_{32}H_{16}CuN_8$	Copper Phthalocyanine, 2497
575.310770	$C_{32}H_{41}N_5O_5$	Ergocryptine, 3567, 3587 Ergocryptinine, 3568, 3586
576.184292	$C_{28}H_{32}O_{13}$	Nodakenin, Tetraacetate, 6484 β-Peltatin-β-D-glucoside, 6871 Picropodophyllin-β-D-glucoside, 7215 Podophyllotoxin-β-D-glucoside, 7325
576.262423	$C_{36}H_{36}N_2O_5$	Tiliacorine, 9166 Trilobine, N-Methyl deriv, 9367
576.329834	$C_{32}H_{48}O_9$	Cerberin, 1944 Oleandrin, 6672
577.859652	$C_{20}H_8Br_2N_2O_9$	Eosine I Bluish, Free acid, 3530
578.163556	$C_{27}H_{30}O_{14}$	Glucofrangulin, 4283 Kaempferol, 3,7-Dirhamnoside, 5126 Morindin, 6102 Sophorabioside, 8492
578.184685	$C_{24}H_{34}O_{16}$	Hexacetylscillabiose, 8150
578.262817	$C_{32}H_{38}N_2O_8$	Deserpidine, 2880
578.301032	$C_{32}H_{47}ClO_7$	Senegenin, Monoacetate, 8199
578.412381	$C_{41}H_{54}O_2$	Torularhodin, Methyl ester, 9250
579.152281	$C_{25}H_{29}N_3O_{11}S$	Cephamycin B, 1940
580.179206	$C_{27}H_{32}O_{14}$	Naringin, 6252

Use in conjunction with The Merck Index, Ninth Edition

Molecular Weight	Empirical Formula	Compound Name, Monograph Number
580.257337	$C_{35}H_{36}N_2O_6$	Daphnoline, 2807
580.428031	$C_{41}H_{56}O_2$	Vitamin $K_{2(30)}$, 9686
581.263820	$C_{33}H_{35}N_5O_5$	Ergotamine, 3582, 3583 Ergotaminine, 3584
581.265670	$C_{21}H_{39}N_7O_{12}$	Streptomycin, 8611, 8613
581.356383	$C_{31}H_{51}NO_9$	Rosamicin, 8023
582.131468	$C_{36}H_{22}O_8$	Emodin, Tribenzoate, 3512
582.247835	$C_{33}H_{34}N_4O_6$	Biliverdine, 1237
582.319269	$C_{34}H_{46}O_8$	Helvolic Acid, Methyl ester, 4492
583.279470	$C_{33}H_{37}N_5O_5$	Dihydroergotamine, 3151
583.281320	$C_{21}H_{41}N_7O_{12}$	Dihydrostreptomycin, 3157, 3158, 8618
583.861778	$C_{20}H_{10}I_2O_5$	4',5'-Diiodofluorescein, 3172
584.263485	$C_{33}H_{36}N_4O_6$	Bilirubin, 1236
584.283277	$C_{29}H_{44}O_{12}$	Acocanthin, 109 Ouabain, 6735
584.309767	$C_{32}H_{44}N_2O_8$	Lappaconitine, 5214
584.422946	$C_{40}H_{56}O_3$	Capsanthin, 1767 Chrysanthemaxanthin, 2247 Flavoxanthin, 4011
585.285737	$C_{22}H_{43}N_5O_{13}$	Amikacin, A1
585.381809	$C_{38}H_{51}NO_4$	Myrophine, 6163
586.146861	$C_{24}H_{30}N_2O_{13}S$	N^4-β-D-Glucosylsulfanilamide, N,N',2,3,4,6-Hexaacetate, 4296
586.156615	$C_{30}H_{35}BrO_7$	3-(p-Bromobenzoyl)strophanthidin, 8645
586.168641	$C_{29}H_{30}O_{13}$	Amarogentin, 380
586.256669	$C_{35}H_{38}O_8$	Rottlerin, Pentamethyl ether, 8033
586.263879	$C_{29}H_{38}N_4O_9$	Mepicycline, 5684, 5685, 6894
586.369013	$C_{27}H_{50}N_6O_8$	Deferoxamine, N-Acetyl deriv, 2837
587.345818	$C_{33}H_{49}NO_8$	Pseudojervine, 7703
589.693819	BiI_3	Bismuth Iodide, 1291
590.163556	$C_{28}H_{30}O_{14}$	Pseudobaptigenin, 7-Rhamnoglucoside, 7695
590.241687	$C_{36}H_{34}N_2O_6$	Menisarine, 5662
590.289306	$C_{36}H_{38}N_4O_4$	Protoporphyrin IX, Dimethyl ester, 7685

Molecular Weight	Empirical Formula	Compound Name, Monograph Number
590.309098	$C_{32}H_{46}O_{10}$	Monoacetylcymarin, 2763 Tanghinin, 8827
590.335588	$C_{35}H_{46}N_2O_6$	Pyrocalciferol, 3,5-Dinitrobenzoate, 7778
590.381869	$C_{34}H_{54}O_8$	Lasalocid, 5219
590.915375	$C_{15}H_{17}AsN_5O_6SSb$	Arsant, Free acid, 818
591.340733	$C_{32}H_{49}NO_9^+$	Cevadine, 1988
591.377118	$C_{33}H_{53}NO_8$	Imperialine, β-D-Glucopyranoside, 4818
591.694673	$O_{12}S_3Sm_2$	Samarium Sulfate, 8106
592.150346	$C_{16}H_{48}O_8Si_8$	Hexadecamethylcyclooctasiloxane, 4548
592.179206	$C_{28}H_{32}O_{14}$	Acacetin, 7-Rhamnoglucoside, 9 Linarin, 5338
592.288363	$C_{31}H_{44}O_{11}$	3-O-Acetyladonitoxin, 156
592.299596	$C_{30}H_{44}N_2O_{10}$	Hexobendine, 4578
592.355260	$C_{40}H_{48}O_4$	Astacin, 875
593.356383	$C_{32}H_{51}NO_9$	Protoveratridine, 7686
593.392768	$C_{33}H_{55}NO_8$	Verticine, 3-Glucoside, 9624
593.697677	$Eu_2O_{12}S_3$	Europic Sulfate, 3853
594.150280	$C_{36}H_{30}O_3Si_3$	Hexaphenylcyclotrisiloxane, 4566
594.158471	$C_{27}H_{30}O_{15}$	Datiscetin, Rutinoside, 2811
594.272987	$C_{36}H_{38}N_2O_6$	Bebeerine, 1021 d-Chondrocurine, 2210 Curine, 2682 Daphnandrine, 2804 Isochondodendrine, 5013 Protocuridine, 7681
594.403273	$C_{36}H_{54}N_2O_5$	Vitamin E, p-Nitrophenylurethane, 9681
594.443681	$C_{42}H_{58}O_2$	Cryptoxanthin, Monoacetate, 2603
594.466148	$[C_{40}H_{58}N_4]^{2+}$	Bisdequalinium, 1267
595.166296	$[C_{27}H_{31}O_{15}]^+$	Cyanidin, 3-Rhamnoglucoside, 2695 Pelargonidin, 3,5-Diglucoside, 6865
596.263485	$C_{34}H_{36}N_4O_6$	Phytochlorin, 7193
596.283277	$C_{30}H_{44}O_{12}$	Estradiol Benzoate, 17β-Maltoside, 3628
596.288637	$C_{36}H_{40}N_2O_6$	Magnoline, 5525
596.351512	$[C_{40}H_{44}N_4O]^{2+}$	C-Curarine I, 2679
596.386560	$C_{40}H_{52}O_4$	Astaxanthin, 877

Use in conjunction with The Merck Index, Ninth Edition

Molecular Weight	Empirical Formula	Compound Name, Monograph Number
596.459332	$C_{42}H_{60}O_2$	Rhodoviolascin, 7986
596.506950	$C_{42}H_{64}N_2$	Hydrabamine, 4640, 4641
597.206421	$C_{31}H_{36}ClN_3O_5S$	Methophenazine, 5855
597.260585	$C_{21}H_{39}N_7O_{13}$	Hydroxystreptomycin, 4766
597.268630	$C_{31}H_{39}N_3O_9$	Mikamycin A, 6047
597.522616	I_4Zr	Zirconium Iodide, 9844
597.811134	$C_{12}H_{13}I_3N_2O_2$	Ipodate, 4928
598.279135	$C_{34}H_{38}N_4O_6$	Hematoporphyrin, 4495
598.293054	$C_{37}H_{42}O_7$	Sarmentogenin, Dibenzoate, 8129
599.309432	$C_{33}H_{45}NO_9$	Delphinine, 2851
600.149243	$C_{30}H_{24}N_4O_{10}$	Nicofuranose, 6334
600.169035	$C_{26}H_{32}O_{16}$	Monotropein, Pentaacetate, 6092
600.304681	$C_{32}H_{44}N_2O_9$	Streptolydigin, 8610
600.348278	$C_{27}H_{48}N_6O_9$	Nocardamin, 6483
600.392709	$[C_{38}H_{52}N_2O_4]^{2+}$	Diethylbelladonninium, 1032
600.417861	$C_{40}H_{56}O_4$	Violaxanthin, 9650
602.258794	$C_{29}H_{38}N_4O_{10}$	Lymecycline, 5446
603.361862	$C_{30}H_{53}NO_{11}$	Benzonatate, 1107
603.703413	$Gd_2O_{12}S_3$	Gadolinium Sulfate, 4174
604.299596	$C_{31}H_{44}N_2O_{10}$	Dilazep, 3185
604.351238	$C_{36}H_{48}N_2O_6$	Tachysterol, 4-Methyl-3,5-dinitrobenzoate, 8811
606.257731	$C_{33}H_{38}N_2O_9$	Raunescine, Monoacetate, 7920
606.267627	$C_{31}H_{42}O_{12}$	Tetraacetylouabagenin, 6734
606.272987	$C_{37}H_{38}N_2O_6$	Cepharanthine, 1942
606.282883	$C_{35}H_{42}O_9$	O-Cinnamoyltaxicin-II Triacetate, 8853
606.294117	$C_{34}H_{42}N_2O_8$	Ethyserpine, 3818
607.069935	$C_{19}H_{29}NO_{15}S_3$	Glucocheirolin, Free acid tetraacetate, 4282
608.103558	$C_{28}H_{24}N_4O_8S_2$	Blancophor® R, 1332
608.174121	$C_{28}H_{32}O_{15}$	Diosmetin, 7-Rutinoside, 3296
608.273381	$C_{33}H_{40}N_2O_9$	Deserpidine, 10-Methoxy deriv, 2880 Reserpine, 7937

608.

Molecular Weight	Empirical Formula	Compound Name, Monograph Number
608.288637	$C_{37}H_{40}N_2O_6$	Berbamine, 1176 Chondrofoline, 2211 Oxyacanthine, 6769
609.295120	$C_{35}H_{39}N_5O_5$	Ergocristine, 3565, 3587 Ergocristinine, 3566, 3586
609.296462	$[C_{37}H_{41}N_2O_6]^+$	d-Methylchondrocurine, 2210 Methylcurinium, 2682
610.132256	$C_{30}H_{26}O_{14}$	Ergoflavin, 3569
610.153385	$C_{27}H_{30}O_{16}$	Rutin, 8062 Saponarin, 8122
610.189771	$C_{28}H_{34}O_{15}$	Hesperidin, 4535
610.279135	$C_{35}H_{38}N_4O_6$	Biliverdine, Dimethyl ester, 1237
610.304287	$[C_{37}H_{42}N_2O_6]^{2+}$	Tubocurarine, 9466
610.402211	$C_{41}H_{54}O_4$	24-Hydroxycholesterol, 24β-Epimer dibenzoate, 4721
610.413444	$C_{40}H_{54}N_2O_3$	η-Tocopherol, 4-Phenylazobenzoate, 9202
611.161210	$[C_{27}H_{31}O_{16}]^+$	Cyanidin, 3,5-Diglucoside, 2695 Cyanidin, 3-Sophoroside, 2695
611.357051	$C_{34}H_{49}N_3O_7$	η-Tocopherol, 3',5'-Dinitrophenylurethane, 9202
612.125629	$C_{16}H_{22}HgN_6O_7$	Meralluride, 5695
612.147906	$C_{30}H_{28}O_{14}$	Tetraacetylgranaticin, 4382
612.151965	$C_{20}H_{32}N_6O_{12}S_2$	Glutathione, Disulfide, 4307
612.205421	$C_{28}H_{36}O_{15}$	Neohesperidin Dihydrochalcone, 6276
612.231910	$C_{31}H_{36}N_2O_{11}$	Novobiocin, 6530
612.298096	$C_{31}H_{50}O_8P_2$	Dihydrovitamin K_1, Diphosphate, 3162
612.314577	$C_{31}H_{48}O_{12}$	Testosterone, β-Maltoside, 8890
613.769663	$C_{11}H_9I_3N_2O_4$	Diatrizoic Acid, 2959, 5628 Iothalamic Acid, 4922
613.806048	$C_{12}H_{13}I_3N_2O_3$	Iocetamic Acid, 4870 Iomeglamic Acid, 4914
614.184685	$C_{27}H_{34}O_{16}$	Monotropein, Pentaacetate methyl ester, 6092
614.312286	$C_{23}H_{46}N_6O_{13}$	Neomycin B, 6278
614.362077	$[C_{40}H_{46}N_4O_2]^{2+}$	C-Toxiferine I, 9253
614.381869	$C_{36}H_{54}O_8$	Cimigenol, Triacetate, 2269
615.296302	$C_{23}H_{45}N_5O_{14}$	Paromomycin, 6844
615.713557	$Dy_2O_{12}S_3$	Dysprosium Sulfate, 3457

Use in conjunction with The Merck Index, Ninth Edition

Molecular Weight	Empirical Formula	Compound Name, Monograph Number
616.177295	$C_{34}H_{32}FeN_4O_4$	Heme, 4498
616.377727	$[C_{40}H_{48}N_4O_2]^{2+}$	C-Calebassine, 1719
616.418649	$C_{33}H_{60}O_{10}$	Nonoxynol 9, 6487
618.143085	$C_{24}H_{44}I_2O_2$	Iodobrassid, 4892
618.301327	$C_{29}H_{42}N_6O_9$	Amicetin, 398
618.304013	$C_{33}H_{46}O_{11}$	Helveticoside, 3′,4′-Diacetate, 4491
618.382144	$C_{41}H_{50}N_2O_3$	ε-Tocopherol, 4-Phenylazobenzoate, 9199
619.526437	In_2Te_3	Indium Telluride, 4836
619.715801	$Er_2O_{12}S_3$	Erbium Sulfate, 3557
620.283277	$C_{32}H_{44}O_{12}$	Scilliroside, 8153
620.288637	$C_{38}H_{40}N_2O_6$	Insularine, 4858
620.311597	$C_{34}H_{49}ClO_8$	Senegenin, Diacetate, 8199
620.743076	$C_6K_2N_6PtS_6$	Potassium Hexathiocyanatoplatinate(IV), 7422
621.763515	$C_{14}H_9I_3O_4$	3,3′,5-Triiodothyroacetic Acid, 9363
622.002418	$C_{27}H_{28}Br_2O_5S$	Bromthymol Blue, 1448
622.007714	$C_{23}H_{28}I_2O_4$	Hinderin, 4590
622.189771	$C_{29}H_{34}O_{15}$	Pectolinarigenin, 7-Rutinoside, 6862
622.277798	$C_{35}H_{42}O_{10}$	O-Cinnamoyltaxicin-I Triacetate, 8853
622.304287	$C_{38}H_{42}N_2O_6$	Berbamine, Methyl ether, 1176 Tetrandrine, 8950
624.134722	$H_8N_2O_7U_2$	Ammonium Uranate(VI), 598
624.147906	$C_{31}H_{28}O_{14}$	Ergochrysins, 3561
624.319937	$C_{38}H_{44}N_2O_6$	Dauricine, 2816
624.319937	$[C_{38}H_{44}N_2O_6]^{2+}$	d-Dimethylchondrocurine, 2210
624.409815	$C_{32}H_{56}N_4O_8$	Sporidesmolide III, 8544
624.429094	$C_{41}H_{56}N_2O_3$	ζ₂-Tocopherol, 4-Phenylazobenzoate, 9201
624.820553	$C_{19}H_{11}Cl_2I_2NO_3$	Rafoxanide, 7915
625.176860	$[C_{28}H_{33}O_{16}]^+$	Peonin, 6939
625.372701	$C_{35}H_{51}N_3O_7$	ζ₂-Tocopherol, 3′,5′-Dinitrophenylurethane, 9201
625.723641	$O_{12}S_3Tm_2$	Thulium Sulfate, 9135
625.842434	$C_{14}H_{17}I_3N_2O_2$	Ipodate, Ethyl ester, 4928

626.

Molecular Weight	Empirical Formula	Compound Name, Monograph Number
626.251583	$C_{37}H_{38}O_9$	Sarverogenin, Dibenzoate, 8136
626.281261	$C_{29}H_{38}N_8O_8$	Guamecycline, 4409
627.156125	$[C_{27}H_{31}O_{17}]^+$	Delphinidin, 3,5-Diglucoside, 2850
627.520107	I_4Sn	Stannic Iodide, 8555
627.785313	$C_{12}H_{11}I_3N_2O_4$	Iodamide, 4871 Metrizoic Acid, 6028
627.984179	$C_{22}H_{26}As_2N_2O_8S$	Spirotrypan, Free acid, 8539
628.251977	$C_{33}H_{40}O_{12}$	Kosin, Tetraacetate, 5168
628.278467	$C_{36}H_{40}N_2O_8$	Hydroxycodeine, 4722
628.303619α	$C_{38}H_{44}O_8$	Gambogic Acid, 4199
628.335982	$C_{34}H_{48}N_2O_9$	Ajacine, 182
629.023260	$C_{26}H_{19}N_3O_{10}S_3$	Anazolene, 666
629.319997	$C_{34}H_{47}NO_{10}$	Indaconitine, 4819
629.566461	$K_4O_{16}S_4Zr$	Potassium Zirconium Sulfate, 7497
629.731253	$C_{20}H_{10}Br_4O_4$	3',3'',5',5''-Tetrabromophenolphthalein, 8903
630.253708	$C_{30}H_{38}N_4O_{11}$	Apicycline, 763
632.372642	$C_{40}H_{48}N_4O_3$	Geissospermine, 4216
632.382660	$C_{32}H_{64}O_4Sn$	Dibutyltin Dilaurate, 3012
632.392434	$C_{36}H_{56}O_9$	Quinovin, Glycoside A, 7900 Quinovin, Glycoside B, 7900
633.180035	$C_{34}H_{33}FeN_4O_5$	Hematin, 4494
633.682682	Au_2Se_3	Gold Selenide, 4364
634.045513	$Na_2O_7U_2$	Sodium Uranate(VI), 8472
634.132256	$C_{32}H_{26}O_{14}$	Actinorhodine, 138
634.289031	$C_{35}H_{42}N_2O_9$	Rescinnamine, 7934
635.722382	$Na_4O_9PbS_6$	Lead Sodium Thiosulfate, 5275
635.732937	$O_{12}S_3Yb_2$	Ytterbium Sulfate, 9772
635.779165	$C_{15}H_{11}I_3O_4$	3,3',5-Triiodothyropropionic Acid, 9365
636.096264	$C_{27}H_{24}O_{18}$	Corilagin, 8828
637.524137	I_4Te	Tellurium Tetraiodide, 8871
637.736761	$Lu_2O_{12}S_3$	Lutetium Sulfate, 5429
638.107454	$C_{38}H_{30}NiO_2P_2$	Bis(triphenylphosphine)dicarbonylnickel, 1327

Use in conjunction with The Merck Index, Ninth Edition

Molecular Weight	Empirical Formula	Compound Name, Monograph Number
638.132081	$C_{15}H_{32}HgO_{12}S$	Diglucomethoxane, 3141
638.163556	$C_{32}H_{30}O_{14}$	Chrysergonic Acid, 2253 Secalonic Acids, 8170
638.335588	$C_{39}H_{46}N_2O_6$	Magnoline, Trimethyl ether, 5525
638.335588	$[C_{39}H_{46}N_2O_6]^{2+}$	Tubocurarine, Dimethyl ether, 9466
638.425465	$C_{33}H_{58}N_4O_8$	Sporidesmolide I, 8544
638.826450	$C_{15}H_{16}I_3NO_3$	Buniodyl, Free acid, 1476
639.409481	$C_{33}H_{57}N_3O_9$	Enniatin B, 3525
640.179206	$C_{32}H_{32}O_{14}$	Chartreusin, 1997
640.216929	$C_{34}H_{32}N_4O_9$	Nicomol, 6335
640.251977	$C_{34}H_{40}O_{12}$	Filixic Acids, PBP, 3999
640.805714	$C_{14}H_{14}I_3NO_4$	Propyl Docetrizoate, 7637
640.842100	$C_{15}H_{18}I_3NO_3$	Tyropanoic Acid, 9491
641.153796	$C_{20}H_{26}HgN_5NaO_5$	Tachydrol, 8810
641.171775	$[C_{28}H_{33}O_{17}]^+$	Petunidin, 3,5-Diglucoside, 6975
641.800963	$C_{13}H_{13}I_3N_2O_4$	Diprotrizoic Acid, 8359
642.351632	$C_{35}H_{50}N_2O_9$	Batyl Alcohol, Bis(p-nitrobenzoate), 1018
643.392286	$C_{39}H_{59}Cl_2NO_2$	Phenesterine, 7011
643.710518	$C_{20}H_8Br_4O_5$	Eosine Yellowish—(YS), 3531
645.023698	$C_{25}H_{29}I_2NO_3$	Amiodarone, 501
645.314912	$C_{34}H_{47}NO_{11}$	Aconitine, 113
645.646301	$C_6HBiBr_4O_3$	Bibrocathol, 1226
645.920693	$C_{24}H_{21}Cl_6O_6P$	Falone®, 3873
648.387348	$C_{36}H_{56}O_{10}$	Quinovin, Glycoside C, 7900
648.490632	$C_{46}H_{64}O_2$	Vitamin $K_{2(35)}$, 9686
650.790064	$C_{15}H_{12}I_3NO_4$	3,5,3'-Triiodothyronine, 9364
651.037471	$C_{28}H_{21}N_5O_6S_4$	Clayton® Yellow, Free acid, 2318
651.146148	$C_{34}H_{32}ClFeN_4O_4$	Hemin, 4501
652.119876	$C_{32}H_{24}N_6O_6S_2$	Congo Red, Free acid, 2467
652.289700	$C_{37}H_{40}N_4O_7$	Haplophytine, 4464
652.351238	$[C_{40}H_{48}N_2O_6]^{2+}$	Metocurine, 6019

Molecular Weight	Empirical Formula	Compound Name, Monograph Number
652.351740	$C_{25}H_{44}N_{14}O_7$	Capreomycin IB, 1756
652.441115	$C_{34}H_{60}N_4O_8$	Sporidesmolide II, 8544 Sporidesmolide IV, 8544
653.221281	$C_{32}H_{40}BrN_5O_5$	Bromocriptine, A2
654.267627	$C_{35}H_{42}O_{12}$	Filixic Acids, PBB, 3999
654.366888	$[C_{40}H_{50}N_2O_6]^{2+}$	Dimethyldauricinium, 2816
654.397913	$C_{35}H_{58}O_{11}$	Filipin III, 3998
655.187425	$[C_{29}H_{35}O_{17}]^+$	Malvidin, 3,5-Diglucoside, 5543
656.264823	$[C_{40}H_{32}N_8O_2]^{2+}$	Blue Tetrazolium, Cation, 1335
656.305744	$C_{33}H_{44}N_4O_{10}$	Riboflavine, Tetrabutyrate, 7993
656.334919	$C_{40}H_{48}O_8$	Gambogic Acid, Methyl ester monomethyl ether, 4199
656.392434	$C_{38}H_{56}O_9$	Escigenin, Tetraacetate, 3617
657.322122	$C_{29}H_{47}N_5O_{12}$	Sisomicin, Penta-N-acetyl deriv, 8292
657.710338	Bi_2Se_3	Bismuth Selenide, 1303
657.750218	Hg_2I_2	Mercurous Iodide, 5738
657.781170	$C_{12}H_9O_{16}S_4Sb$	Stibophen, Free acid, 8594
658.423340	$C_{42}H_{58}O_6$	Fucoxanthin, 4132
659.109097	$C_{25}H_{29}N_3O_{14}S_2$	CephamycinA, 1940
659.861382	$C_6H_{18}O_{24}P_6$	Phytic Acid, 7191
660.314577	$C_{35}H_{48}O_{12}$	Deoxycorticosterone, Tetraacetyl-β-D-glucoside, 2863
661.806048	$C_{16}H_{13}I_3N_2O_3$	Iobenzamic Acid, 4868
663.109126	$C_{21}H_{27}N_7O_{14}P_2$	Nadide, 6170
664.615832	$C_{46}H_{80}O_2$	β-Amyrin, Palmitate, 657
665.304741	$C_{33}H_{47}NO_{13}$	Pimaricin, 7230
665.698240	$C_{19}H_{10}Br_4O_5S$	Bromphenol Blue, 1447
665.993390	$K_2O_7U_2$	Potassium Uranate(VI), 7492
666.169139	$C_{18}H_{54}O_9Si_9$	Octadecamethylcyclononasiloxane, 6553
666.278861	$C_{35}H_{42}N_2O_{11}$	Syrosingopine, 8807
666.393377	$[C_{44}H_{50}N_4O_2]^{2+}$	Alcuronium, 214
666.464811	$C_{45}H_{62}O_4$	Vitamin(s) K_2, Dihydrodiacetate, 9686
667.206699	$C_{38}H_{29}N_5O_7$	Cordycepin, Tetrabenzoate, 2506

Use in conjunction with The Merck Index, Ninth Edition

Molecular Weight	Empirical Formula	Compound Name, Monograph Number
668.201481	$C_{37}H_{36}N_2O_6S_2$	Guinea Green B, Free acid, 4426
668.283277	$C_{36}H_{44}O_{12}$	Filixic Acids, BBB, 3999
668.330896	$C_{36}H_{48}N_2O_{10}$	Lycaconitine, 5433
668.346655	$C_{25}H_{44}N_{14}O_8$	Capreomycin IA, 1756
668.444075	$C_{44}H_{60}O_5$	Capsanthin, Diacetate, 1767
670.392828	$C_{35}H_{58}O_{12}$	Fungichromin, 4140 Lagosin, 5198
670.429213	$C_{36}H_{62}O_{11}$	Monensin, 6081
673.346212	$C_{36}H_{51}NO_{11}$	Bikhaconitine, 1235 Veratridine, 9608
675.325477	$C_{35}H_{49}NO_{12}$	Jesaconitine, 5116
675.821698	$C_{17}H_{15}I_3N_2O_3$	Iobenzamic Acid, Methyl ester, 4868
676.309492	$C_{35}H_{48}O_{13}$	Tri-O-acetyl-convallatoxin, 2485
676.376390	$C_{44}H_{52}O_6$	Astacin, Diacetate, 875
678.200730	$C_{28}H_{38}O_{19}$	Octaacetyl-aldehydro-cellobiose, 1916 Octaacetyl-α-cellobiose, 1916 Octa-O-acetyl-β-sophorose, 8494 Sucrose Octaacetate, 8682
679.177448	$C_{36}H_{36}ClFeN_4O_4$	Hemin, Dimethyl ester, 4501
680.151177	$C_{34}H_{28}N_6O_6S_2$	Benzopurpurine 4B, Free acid, 1111
680.221175	$C_{20}H_{60}O_8Si_9$	Eicosamethylnonasiloxane, 3482
680.407690	$C_{44}H_{56}O_6$	Astaxanthin, Diacetate, 877
681.336041	$C_{34}H_{51}NO_{13}$	Tetrin A, 8960
681.456431	$C_{36}H_{63}N_3O_9$	Enniatin A, 3525
683.212380	$C_{37}H_{37}N_3O_6S_2$	Acid Violet 7B, Free acid, 107
684.460119	$C_{41}H_{64}O_8$	Prednisolone 21-Stearoylglycolate, 7513
685.325585	$C_{25}H_{43}N_{13}O_{10}$	Viomycin, 9654, 9655
685.462579	$C_{34}H_{63}N_5O_9$	Pepstatin, 6943
687.419377	$C_{35}H_{61}NO_{12}$	Oleandomycin, 6671
687.909582	O_4Pb_3	Lead Tetroxide, 5283
689.341127	$C_{36}H_{51}NO_{12}$	Pseudoaconitine, 7694
689.925232	$H_2O_4Pb_3$	Lead Hydroxide, 5262
691.287173	$C_{34}H_{46}ClN_3O_{10}$	Maytansine, 5590

Molecular Weight	Empirical Formula	Compound Name, Monograph Number
692.340792	$C_{36}H_{52}O_{13}$	Scillaren A, 8151
692.411158	$C_{36}H_{61}NaO_{11}$	Monensin, Sodium salt, 6081
693.372427	$C_{36}H_{55}NO_{12}$	Desatrine, 2878
693.523329	$C_{46}H_{67}N_3O_2$	Pilocereine, Methyl ether, 7227
693.729540	$C_{21}H_{14}Br_4O_5S$	Bromcresol Green, 1386
695.294176	$C_{37}H_{45}NO_{12}$	Rifamycin S, 8008
696.095320	$C_{30}H_{28}N_6O_6S_4$	Verticillin A, 9623
696.341067	$C_{41}H_{48}N_2O_8$	Thalicarpine, 8967
696.372092	$C_{36}H_{56}O_{13}$	Periplocin, 6960
697.309826	$C_{37}H_{47}NO_{12}$	Rifamycin SV, 8009
697.330956	$C_{34}H_{51}NO_{14}$	Tetrin B, 8960
698.351357	$C_{35}H_{54}O_{14}$	Uzarin, 9561
699.224868	$C_{29}H_{33}N_9O_{12}$	Pteropterin, 7719
699.448685	CdI_4K_2	Potassium Tetraiodocadmate, 7480
700.314017	$C_{27}H_{44}N_{10}O_{12}$	Streptonicozid, 8614
700.320725	$C_{36}H_{48}N_2O_{12}$	Rhodomycin A, 7982
700.367007	$C_{35}H_{56}O_{14}$	Chalcomycin, 1990
702.152597	$C_{43}H_{26}O_{10}$	Tetrabenzoylbaptigenin, 959
702.298175	$C_{32}H_{46}N_8O_6S_2$	Thiamine Disulfide, O,O-Diisobutyrate, 9023
702.582693	I_4Pt	Platinic Iodide, 7305
704.393771	$C_{43}H_{52}N_4O_5$	Voacamine, 9696
705.815967	$Bi_2O_{12}S_3$	Bismuth Sulfate, 1312
705.855042	$C_{20}H_{10}Br_2HgO_6$	Merbromin, 5697
706.283671	$C_{35}H_{46}O_{15}$	Glaucarubin, Pentaacetate, 4266
706.465599	$C_{40}H_{66}O_{10}$	Venturicidin B, 9602
707.208748	$C_{38}H_{40}ClFeN_4O_4$	Hemin, Diethyl ester, 4501
707.305410	$C_{37}H_{45}N_3O_{11}$	Rifamycin X, 8008
707.351691	$C_{36}H_{53}NO_{13}$	Lucensomycin, 5411
707.378181	$C_{39}H_{53}N_3O_9$	Bietaserpine, 1231
712.090235	$C_{30}H_{28}N_6O_7S_4$	Verticillin B, 9623
712.367007	$C_{36}H_{56}O_{14}$	Digitalin, 3133

Use in conjunction with The Merck Index, Ninth Edition

Molecular Weight	Empirical Formula	Compound Name, Monograph Number
713.240518	$C_{30}H_{35}N_9O_{12}$	Pteropterin, Methyl ester, 7719
713.426002	$C_{40}H_{63}N_3O_4S_2$	Pipotiazine, Palmitic ester, 7277 Pipotiazine, Undecylenic ester, 7277
714.109037	$C_{34}H_{26}N_4O_{10}S_2$	Benzo Azurine G, Free acid, 1092
715.247620	$C_{35}H_{41}NO_{15}$	Colchicum Glucoside Colchicoside, Tetraacetate, 2438
718.356442	$C_{38}H_{54}O_{13}$	Colocynthin, 2453
719.351691	$C_{37}H_{53}NO_{13}$	Germine, 3,4,7,15,16-Pentaacetate, 4245
719.710750	$C_{13}H_{16}Cl_{12}O_8$	Pentaerythritol Chloral, 6905
720.444863	$C_{40}H_{64}O_{11}$	Dermostatin A, 2875
722.642441	$C_{45}H_{86}O_6$	Trimyristin, 9392
724.330622	$C_{36}H_{52}O_{15}$	Hellebrin, 4487
725.971210	BaO_7U_2	Barium Uranium Oxide, 1009
726.231242	$C_{40}H_{38}O_{13}$	Rottlerin, Pentaacetate, 8033
726.379972	$C_{33}H_{54}N_6O_{12}$	Nocardamin, Triacetate, 6483
733.461242	$C_{37}H_{67}NO_{13}$	Erythromycin, 3602, 3603-3605, 3607
733.853456	Hg_3O_6S	Mercuric Subsulfate, 5725
734.202663	$C_{30}H_{42}N_2O_{15}S_2$	Sinalbin, 8287
734.205815	$C_{34}H_{38}O_{18}$	Catalposide, Hexaacetate, 1897
734.460513	$C_{41}H_{66}O_{11}$	Dermostatin B, 2875
734.527411	$C_{50}H_{70}O_4$	Vitamin(s) K_2, Dihydrodiacetate, 9686
736.112528	$C_{24}H_{36}N_2O_{18}S_3$	Glucosulfone, 4295
736.439778	$C_{40}H_{64}O_{12}$	Caincin, 1630 Nonactin, 6486
738.200730	$C_{33}H_{38}O_{19}$	Fabiatrin, Hexaacetate, 3861
738.331240	$Fe_5O_{12}Y_3$	Yig, 9764
739.655962	I_4Th	Thorium Iodide, 9118
740.216380	$C_{33}H_{40}O_{19}$	Robinin, 8017
740.230233	$C_{27}H_{42}FeN_9O_{12}$	Ferrichrome, 3941
741.203075	$[C_{36}H_{37}O_{17}]^+$	Monardein, 6078
742.232030	$C_{33}H_{42}O_{19}$	Troxerutin, 9450
743.075458	$C_{21}H_{28}N_7O_{17}P_3$	NADP, 6172
743.318494	$C_{27}H_{49}N_7O_{17}$	Streptomycin B, 8612

Molecular Weight	Empirical Formula	Compound Name, Monograph Number
743.487337	$C_{45}H_{65}N_3O_6$	Pilocereine, 7227
743.548176	AuI_4K	Potassium Tetraiodoaurate(III), 7479
744.062307	$C_{30}H_{28}N_6O_7S_5$	Verticillin C, 9623
744.605662	$C_{50}H_{80}O_4$	Diethylstilbestrol Dipalmitate, 3114
745.475545	$Fe_4O_{21}P_6$	Ferric Pyrophosphate, 3949
748.158297	$C_{37}H_{36}N_2O_9S_3$	Light Green SF Yellowish, Free acid, 5325
749.471413	$C_{41}H_{67}NO_{11}$	Venturicidin A, 9602
749.857108	$O_4P_2Pb_3$	Lead Phosphate, 5271
750.455428	$C_{41}H_{66}O_{12}$	α-Hederin, 4480
752.277404	$C_{50}H_{40}O_7$	Quercetin, Pentabenzyl ether, 7821
752.295381	$C_{43}H_{48}N_2O_6S_2$	Indocyanine Green, Free acid, 4839
753.299656	$C_{39}H_{47}NO_{14}$	Rifamycin O, 8008 Streptovaricin F, 8619
753.336041	$C_{40}H_{51}NO_{13}$	Streptovaricin D, 8619
754.377572	$C_{38}H_{58}O_{15}$	16-Acetyldigitalinum verum, 3133
755.106595	$C_{37}H_{29}N_3O_9S_3$	Methyl Blue, Free acid, 5903
755.315306	$C_{39}H_{49}NO_{14}$	Rifamycin B, 8008
756.190165	$C_{36}H_{36}O_{18}$	Scoparin, Heptaacetate, 8154
756.430945	$C_{40}H_{60}N_4O_{10}$	Bufotoxin, 1468
760.107992	$C_{34}H_{28}N_6O_9S_3$	Vital Red, Free acid, 9665
764.434693	$C_{41}H_{64}O_{13}$	Digitoxin, 3139
767.115216	$C_{21}H_{36}N_7O_{16}P_3S$	Coenzyme A, 2431
767.315306	$C_{40}H_{49}NO_{14}$	Streptovaricin E, 8619
767.331248	$C_{37}H_{49}N_7O_9S$	Pentagastrin, 6909
769.330956	$C_{40}H_{51}NO_{14}$	Streptovaricin C, 8619
770.226944	$C_{34}H_{42}O_{20}$	Xanthorhamnin, 9728
770.266875	$C_{38}H_{42}N_8O_6S_2$	Bisbentiamine, 1261
771.518637	$C_{47}H_{69}N_3O_6$	Pilocereine, Ethyl ether, 7227
771.788999	$Al_6Bi_2O_{12}$	Bismuth Aluminate, 1280
772.212439	$C_{40}H_{40}N_2O_{10}S_2$	Quinaphthol, 7838
774.652612	$C_{52}H_{86}O_4$	Promethestrol, Dipalmitate, 7582

Use in conjunction with The Merck Index, Ninth Edition

Molecular Weight	Empirical Formula	Compound Name, Monograph Number
776.686716	$C_{15}H_{11}I_4NO_4$	Thyroxine, 9155, 2912, 8394
780.429607	$C_{41}H_{64}O_{14}$	Digoxin, 3144 Gitoxin, 4262
781.672954	$C_{12}H_{12}O_{12}S_6Sb_2$	Stibocaptate, Free acid, 8593
785.157139	$C_{27}H_{33}N_9O_{15}P_2$	Flavine-Adenine Dinucleotide, 4007
785.325870	$C_{40}H_{51}NO_{15}$	Streptovaricin G, 8619
785.456156	$C_{40}H_{67}NO_{14}$	Leucomycin A_1, 5307
785.497902	$C_{47}H_{67}N_3O_7$	Pilocereine, Acetate, 7227
785.887676	$Cl_6H_{42}N_{14}O_2Ru_3$	Ruthenium Red, 8059
787.515956	HgI_4K_2	Potassium Tetraiodomercurate(II), 7481
788.316327	$C_{35}H_{48}N_8O_{11}S$	Phalloidine, 6979
788.325536	$C_{40}H_{52}O_{16}$	Tetraacetyl Scilliroside, 8153
789.487457	$C_{40}H_{71}NO_{14}$	Erythromycin Propionate, 3606, 3603
789.644885	$C_{20}H_{10}Br_4O_{10}S_2$	Sulfobromophthalein, 8751
792.200374	$C_{48}H_{40}O_4Si_4$	Octaphenylcyclotetrasiloxane, 6565
792.678433	$C_{56}H_{88}O_2$	Vitamin D_1, 9677
793.424856	$C_{41}H_{63}NO_{14}$	Protoveratrine A, 7687
794.445257	$C_{42}H_{66}O_{14}$	β-Methyldigoxin, 3144
796.424522	$C_{41}H_{64}O_{15}$	Diginatin, 3131
796.460907	$C_{42}H_{68}O_{14}$	Gratioside, 4386
796.483374	$C_{40}H_{68}N_4O_{12}$	Amidomycin, 404
798.234117	$C_{34}H_{36}HgN_4O_6$	Merphyrin, Free acid, 5748
802.486728	$C_{45}H_{70}O_{12}$	Oligomycin B, 6680
804.045713	$C_{30}H_{32}N_2O_{14}S_5$	Solasulfone, Free acid, 8485
804.377966	$C_{38}H_{60}O_{18}$	Stevioside, 8589
806.445257	$C_{43}H_{66}O_{14}$	Acetyldigitoxins, 81
806.736341	$C_{51}H_{98}O_6$	Tripalmitin, 9402
807.679463	Bi_2Te_3	Bismuth Telluride, 1315
808.470290	$C_{54}H_{64}O_6$	Violaxanthin, Dibenzoate, 9650
809.419771	$C_{41}H_{63}NO_{15}$	Protoveratrine B, 7687
810.192171	$C_{42}H_{30}N_6O_{12}$	Inositol Niacinate, 4856

Use in conjunction with The Merck Index, Ninth Edition

Molecular Weight	Empirical Formula	Compound Name, Monograph Number
810.393890	$C_{43}H_{58}N_2O_{13}$	Rifamide, 8006
810.420380	$C_{46}H_{58}N_4O_9$	Vinblastine, 9637
811.341521	$C_{42}H_{53}NO_{15}$	Streptovaricin B, 8619 Streptovaricin J, 8619
813.451071	$C_{41}H_{67}NO_{15}$	Troleandomycin, 9432
821.302001	$C_{35}H_{53}FeN_6O_{13}$	Coprogen, 2499
821.675817	$C_{20}H_{10}I_4O_4$	Iodophthalein, 4904, 4905 Phentetiothalein, 7063
822.403787	$C_{42}H_{62}O_{16}$	Glycyrrhizic Acid, 4342
822.405124	$C_{43}H_{58}N_4O_{12}$	Rifampin, 8007
822.440172	$C_{43}H_{66}O_{15}$	α- and β-Acetyldigoxin, 3144 α- and β-Acetylgitoxin, 4262
822.446218	$C_{42}H_{62}N_8O_7S$	Bottromycin A_2 (revised structure, 2nd printing), 1368
823.354091	$C_{43}H_{49}N_7O_{10}$	Virginiamycin S, 9658
824.399645	$C_{46}H_{56}N_4O_{10}$	Vincristine, 9640
825.621657	$Au_2O_{12}Se_3$	Gold Selenate, 4363
827.336435	$C_{42}H_{53}NO_{16}$	Streptovaricin A, 8619
827.466721	$C_{42}H_{69}NO_{15}$	Leucomycin A_3, 5307
827.685034	$C_{48}H_{93}NO_9$	Phrenosin, 7175
828.258761	$C_{24}H_{72}O_{10}Si_{11}$	Tetracosamethylhendecasiloxane, 8911
828.274683	$C_{30}H_{52}O_{26}$	Verbascose, 9613
829.992216	$C_8N_8Pt_2Th$	Thorium Tetracyanoplatinate(II), 9122
832.482037	$C_{42}H_{72}O_{16}$	Lankamycin, 5206
834.761768	$C_{60}H_{98}O$	Di-β-amyrin Ether, 657
835.447738	Cu_2HgI_4	Cuprous Mercuric Iodide, 2669
835.655081	$C_{20}H_8I_4O_5$	Erythrosine, Free acid, 3615
837.794311	$C_{19}H_{10}HgI_2O_7S$	Meralein, 5694
838.364990	$C_{43}H_{50}N_8O_{10}$	Vernamycin B_δ, 9619
838.435087	$C_{43}H_{66}O_{16}$	Acetyldiginatin-α, 3131
841.445986	$C_{42}H_{67}NO_{16}$	Carbomycin A, 1812
842.430001	$C_{42}H_{66}O_{17}$	Echujin, 3471
842.514006	$C_{43}H_{74}N_2O_{14}$	Spiramycin I, 8525

Use in conjunction with The Merck Index, Ninth Edition

Molecular Weight	Empirical Formula	Compound Name, Monograph Number
847.493776	BaHgI$_4$	Barium Mercuric Iodide, 987
847.684246	C$_{58}$H$_{89}$NO$_3$	Rhodoquinone, 7985
852.380640	C$_{44}$H$_{52}$N$_8$O$_{10}$	Vernamycin B$_\beta$, 9619, 7549 Vernamycin B$_\gamma$, 9619, 7549
854.544518	[C$_{52}$H$_{74}$N$_2$O$_8$]$^{2+}$	Laudexium, 5226
858.424916	C$_{42}$H$_{66}$O$_{18}$	Cerberoside, 1945
859.599907	C$_{18}$Fe$_7$N$_{18}$	Ferric Ferrocyanide, 3936
861.508586	C$_{43}$H$_{75}$NO$_{16}$	Erythromycin, Ethyl succinate, 3602
862.195644	C$_{42}$H$_{38}$O$_{20}$	Ergoflavin, Hexaacetate, 3569 Sennoside A & B, 8201
862.683912	C$_{59}$H$_{90}$O$_4$	Ubiquinone (10), 9496
864.414351	C$_{44}$H$_{64}$O$_{17}$	Tetraacetylperiplocin, 6960
866.396290	C$_{45}$H$_{54}$N$_8$O$_{10}$	Mikamycin B, 6047
867.498021	C$_{45}$H$_{73}$NO$_{15}$	Solanine, 8479
868.482037	C$_{45}$H$_{72}$O$_{16}$	Dioscin, 3292
872.054637	C$_{34}$H$_{28}$N$_6$O$_{14}$S$_4$	Evan's Blue, Free acid, 3855 Trypan Blue, Free acid, 9452
872.404181	C$_{42}$H$_{64}$O$_{19}$	Thevetin A, 9015
878.351357	C$_{50}$H$_{54}$O$_{14}$	α-Antiarin, Tribenzoate, 727
878.453805	C$_{44}$H$_{62}$N$_8$O$_{11}$	Etamycin®, 3640
879.515152	C$_{45}$H$_{74}$BNO$_{15}$	Boromycin, 1356
880.575937	C$_{45}$H$_{84}$O$_{16}$	Nonoxynol 15, 6487
883.492936	C$_{45}$H$_{73}$NO$_{16}$	Solasonine, 8484
884.524570	C$_{45}$H$_{76}$N$_2$O$_{15}$	Spiramycin II, 8525
884.783292	C$_{57}$H$_{104}$O$_6$	Petroselinic Acid, Glyceryl triester, 6974 Triolein, 9398
889.482312	BiI$_5$Na$_2$	Bismuth Sodium Iodide, 1304
890.830242	C$_{57}$H$_{110}$O$_6$	Tristearin, 9422
891.990324	C$_{32}$H$_{24}$N$_6$O$_{15}$S$_5$	Trypan Red, Free acid, 9453
892.535317	C$_{55}$H$_{72}$MgN$_4$O$_5$	Chlorophyll a, 2137
893.370728	C$_{49}$H$_{62}$ClFeN$_4$O$_6$	Cytohemin, 2790
894.514581	C$_{54}$H$_{70}$MgN$_4$O$_6$	Chlorophyll d, 2137
898.440447	C$_{54}$H$_{62}$N$_2$O$_{10}$	Violaxanthin, Di-p-nitrobenzoate, 9650

Molecular Weight	Empirical Formula	Compound Name, Monograph Number
898.540221	$C_{46}H_{78}N_2O_{15}$	Spiramycin III, 8525
902.359254	$C_{39}H_{54}N_{10}O_{13}S$	α-Amanitin, 375
903.343270	$C_{39}H_{53}N_9O_{14}S$	β-Amanitin, 375
906.514581	$C_{55}H_{70}MgN_4O_6$	Chlorophyll b, 2137
911.487851	$C_{46}H_{73}NO_{17}$	Triacetylleucomycin A_1, 5307
911.497736	$C_{42}H_{65}N_{13}O_{10}$	Saralasin, 8125
914.248074	$C_{43}H_{46}O_{22}$	Glucofrangulin A Octaacetate, 4283
915.519151	$C_{46}H_{77}NO_{17}$	Tylosin, 9486
918.181660	$C_{55}H_{46}OP_3Rh$	Hydridocarbonyltris(triphenylphosphine)rhodium(I), 4661
921.472201	$C_{47}H_{71}NO_{17}$	Candidin, 1745
923.398730	Ag_2HgI_4	Silver Tetraiodomercurate(II), 8277
923.487851	$C_{47}H_{73}NO_{17}$	Amphotericin B, 623
924.147773	$C_{54}H_{45}ClP_3Rh$	Chlorotris(triphenylphosphine)rhodium, 2163
925.503501	$C_{47}H_{75}NO_{17}$	Nystatin, 6547
942.221859	$C_{47}H_{42}O_{21}$	Theaflavine, Nonaacetate, 8987
942.482431	$C_{47}H_{74}O_{19}$	Deslanoside, 2882
946.274289	$C_{44}H_{50}O_{23}$	Hesperidin, Octaacetate, 4535
950.257457	$C_{57}H_{42}O_{14}$	Ampelopsin, Hexabenzoate, 612
954.380464	$C_{50}H_{63}ClO_{16}$	Chlorothricin, 2156
958.513731	$C_{48}H_{78}O_{19}$	Asiaticoside, 859
968.498081	$C_{49}H_{76}O_{19}$	Lanatoside A, 5202
971.499192	$C_{20}H_4Cl_4I_4O_5$	Rose Bengal, Free acid, 8025
976.378754	$C_{44}H_{64}O_{24}$	Crocin, 2584
984.042696	$C_{33}H_{24}Hg_2O_6S_2$	Hydrargaphen, 4649
984.492996	$C_{49}H_{76}O_{20}$	Lanatoside B, 5203 Lanatoside C, 5204
987.758593	$C_{55}H_{105}NO_{13}$	Cytolipin H, 2791
990.482431	$C_{51}H_{74}O_{19}$	Gitoxin, Pentaacetate, 4262
991.425561	$C_{43}H_{65}N_{11}O_{12}S_2$	Deaminooxytocin, 2823
994.383445	$C_{51}H_{62}O_{20}$	Estradiol Benzoate, 17β-Maltoside heptaacetate, 3628
1000.487910	$C_{49}H_{76}O_{21}$	Lanatoside D, 5205

Use in conjunction with The Merck Index, Ninth Edition

Molecular Weight	Empirical Formula	Compound Name, Monograph Number
1002.539946	$C_{50}H_{82}O_{20}$	Vincetoxin, 9639
1006.436460	$C_{43}H_{66}N_{12}O_{12}S_2$	Oxytocin, 6793
1030.534850	$C_{49}H_{70}N_{14}O_{11}$	Angiotensin II Amide 5-Valine, 684
1033.545760	$C_{50}H_{83}NO_{21}$	Tomatine, 9247
1039.436794	$C_{46}H_{65}N_{13}O_{11}S_2$	Felypressin, 3885
1041.495880	$C_{50}H_{72}CoN_{13}O_8$	Etiocobalamin, 3824
1044.887363	$C_{72}H_{116}O_4$	Xanthophyll, Dipalmitate, 9726 Zeaxanthin, Dipalmitate, 9776
1050.524690	$C_{50}H_{82}O_{23}$	F-Gitonin, 4260
1052.314750	$C_{41}H_{58}FeN_9O_{20}$	Ferrichrome A, 3941
1055.431709	$C_{46}H_{65}N_{13}O_{12}S_2$	Lypressin, 5449
1059.561399	$C_{50}H_{73}N_{15}O_{11}$	Bradykinin, 1369
1060.882277	$C_{72}H_{116}O_5$	Capsanthin, Dipalmitate, 1767
1068.426958	$C_{46}H_{64}N_{14}O_{12}S_2$	Desmopressin, 2883
1068.814592	$C_{72}H_{108}O_6$	Astacin, Dipalmitate (astacein), 875
1077.728750	$C_{55}H_{103}N_3O_{17}$	Primycin, Free base, 7547
1083.437857	$C_{46}H_{65}N_{15}O_{12}S_2$	Arginine Vasopressin, 9596
1084.472654	$C_{52}H_{76}O_{24}$	Aureolic Acid, 898
1100.420810	$C_{51}H_{64}N_{12}O_{12}S_2$	Echinomycin, 3466
1110.631161	$C_{54}H_{90}N_6O_{18}$	Valinomycin, 9572
1125.689679	$C_{54}H_{95}N_9O_{16}$	Viscosin, 9662
1127.475663	$C_{18}H_{10}I_6N_2O_7$	Ioglycamic Acid, 4913
1129.681908	$C_{49}H_{91}N_{15}O_{15}$	Polymyxin D_2, 7354
1139.512048	$C_{20}H_{14}I_6N_2O_6$	Iodipamide, 4883
1140.705938	$C_{60}H_{92}N_{12}O_{10}$	Gramicidin S, 4380
1142.467760	$C_{54}H_{70}N_{12}O_{12}S_2$	Triostin C, 9399
1143.697558	$C_{50}H_{93}N_{15}O_{15}$	Polymyxin D_1, 7354
1154.749928	$C_{52}H_{98}N_{16}O_{13}$	Circulin B, 2299 Polymyxin E_2, 2441
1168.765578	$C_{53}H_{100}N_{16}O_{13}$	Circulin A, 2299 Colistin A, 2441
1177.422031	$H_{24}Mo_7N_6O_{24}$	Ammonium Molybdate(VI), 561

1181.

Molecular Weight	Empirical Formula	Compound Name, Monograph Number
1181.573026	$C_{55}H_{75}N_{17}O_{13}$	LH-RH, 5318
1182.509434	$C_{57}H_{82}O_{26}$	Chromomycin A_3, 2234
1187.600882	$C_{54}H_{85}N_{13}O_{15}S$	Eledoisin, 3491
1187.656362	$C_{56}H_{85}N_{17}O_{12}$	Kallidin, 5129
1188.734278	$C_{55}H_{96}N_{16}O_{13}$	Polymyxin B_2, 7354
1189.277107	$C_{18}H_6BiBr_9O_3$	Bismuth Tribromophenate, 1317
1194.509434	$C_{58}H_{82}O_{26}$	Cerberoside, Acetyl deriv, 1945
1196.525084	$C_{58}H_{84}O_{26}$	Olivomycin A, 6686
1196.582599	$C_{56}H_{92}O_{27}$	Tigonin, 9164
1202.749928	$C_{56}H_{98}N_{16}O_{13}$	Polymyxin B_1, 7354
1208.488698	$C_{58}H_{80}O_{27}$	Thevetin A, Acetyl deriv, 9015
1215.395241	$C_{49}H_{61}N_{13}O_{24}$	Pteroylhexaglutamylglutamic Acid, 7720
1228.572428	$C_{56}H_{92}O_{29}$	Digitonin, 3137
1239.000067	$C_{36}H_{50}N_3O_{22}Sb_3$	Stibamine Glucoside, Free acid, 8590
1253.166559	BiI_7K_4	Bismuth Potassium Iodide, 1301
1253.554976	$C_{24}H_{20}I_6N_4O_8$	Iocarmic Acid, 4869
1254.628476	$C_{62}H_{86}N_{12}O_{16}$	Dactinomycin, 2794
1264.591045	$C_{58}H_{84}N_{14}O_{16}S$	Physalaemin, 7186
1268.644126	$C_{63}H_{88}N_{12}O_{16}$	Actinomycin C_2, 1595
1269.654631	$C_{66}H_{87}N_{13}O_{13}$	Tyrocidine A, 9490
1271.545868	$C_{59}H_{77}N_{13}O_{19}$	Telomycin, 8873
1277.217561	$Fe_4Na_8O_{35}P_{10}$	Ferric Sodium Pyrophosphate, 3951
1282.659776	$C_{64}H_{90}N_{12}O_{16}$	Actinomycin C_3, 1595
1289.650333	$C_{58}H_{91}N_{13}O_{20}$	Amphomycin, 621
1296.046913	$C_{51}H_{40}N_6O_{23}S_6$	Suramin, 8796
1308.665530	$C_{68}H_{88}N_{14}O_{13}$	Tyrocidine B, 9490
1329.572157	$C_{62}H_{89}CoN_{13}O_{14}P$	Vitamin B_{12s}, 9675
1345.567071	$C_{62}H_{89}CoN_{13}O_{15}P$	Hydroxocobalamin, 4706
1346.728148	$C_{63}H_{98}N_{18}O_{13}S$	Substance P, 8663
1347.676429	$C_{70}H_{89}N_{15}O_{13}$	Tyrocidine C, 9490
1354.567406	$C_{63}H_{88}CoN_{14}O_{14}P$	Vitamin B_{12}, 9670

Use in conjunction with The Merck Index, Ninth Edition

Molecular Weight	Empirical Formula	Compound Name, Monograph Number
1387.577636	$C_{64}H_{91}CoN_{13}O_{16}P$	Hydroxocobalamin, Acetate, 4706
1408.521147	$C_{62}H_{88}CoN_{13}O_{17}PS$	Sulfitocobalamin, 8749
1527.552529	$C_{65}H_{85}N_{13}O_{30}$	Mycobacillin, 6149
1540.969159	$C_{75}H_{144}O_{31}$	Nonoxynol 30, 6487
1545.876656	$C_{73}H_{131}N_3O_{31}$	Ganglioside G_I, 4201
1578.658346	$C_{72}H_{100}CoN_{18}O_{17}P$	Cobamamide, 2404
1579.697511	$C_{62}H_{97}N_{23}O_{26}$	Scotophobin, 8164
1636.716658	$C_{76}H_{104}N_{18}O_{19}S_2$	Somatostatin, 8490
1663.492357	$C_{72}H_{85}N_{19}O_{18}S_5$	Thiostrepton, 9103
1836.972072	$C_{84}H_{148}N_4O_{39}$	Ganglioside G_{II}, 4201
1899.738331	$H_{12}Mo_{12}N_3O_{40}P$	Ammonium Phosphomolybdate, 575
1914.604518	$Mo_{12}Na_3O_{40}P$	Sodium Phosphomolybdate, 8426
2025.886613	$C_{79}H_{131}N_{31}O_{24}S_4$	Apamin, 758
2352.939839	$C_{107}H_{138}Cl_2N_{26}O_{31}$	Enduracidin A, 3523
2366.955489	$C_{108}H_{140}Cl_2N_{26}O_{31}$	Enduracidin B, 3523
2844.754174	$C_{131}H_{229}N_{39}O_{31}$	Melittin, 5644
2879.216250	$H_4O_{40}SiW_{12}$	Silicotungstic Acid, 8243
2913.594188	$H_{40}H_{10}O_{41}W_{12}$	Ammonium Tungstate(VI), 597
2931.580644	$C_{136}H_{210}N_{40}O_{31}S$	Cosyntropin, 2532
2932.284907	$H_{12}N_3O_{40}PW_{12}$	Ammonium Phosphotungstate, 576
3053.648265	$C_{130}H_{220}N_{44}O_{41}$	Porcine Secretin, 7173

APPENDIX

Atomic Masses and Abundances of Naturally Occurring Isotopes

Atomic No.	Element	Mass No.	Mass[a]	Interim Isotopic Composition for Average Properties[b] %
1	H	1	1.007825	99.985
		2	2.014102	0.015
2	He	3	3.016029	0.00013
		4	4.002603	99.99987
3	Li	6	6.015123	7.5
		7	7.016005	92.5
4	Be	9	9.012183	100.
5	B	10	10.012938	20.
		11	11.009305	80.
6	C	12	12.000000	98.89
		13	13.003355	1.11
7	N	14	14.003074	99.64
		15	15.000109	0.36
8	O	16	15.994915	99.76
		17	16.999131	0.04
		18	17.999159	0.20
9	F	19	18.998403	100.
10	Ne	20	19.992439	90.51
		21	20.993845	0.27
		22	21.991384	9.22
11	Na	23	22.989770	100.
12	Mg	24	23.985045	78.99
		25	24.985839	10.00
		26	25.982595	11.01
13	Al	27	26.981541	100.
14	Si	28	27.976928	92.23
		29	28.976496	4.67
		30	29.973772	3.10
15	P	31	30.973763	100.
16	S	32	31.972072	95.00
		33	32.971459	0.76
		34	33.967868	4.22
		36	35.967079	0.02
17	Cl	35	34.968853	75.77
		37	36.965903	24.23
18	Ar	36	35.967546	0.34
		38	37.962732	0.07
		40	39.962383	99.59
19	K	39	38.963708	93.26
		40	39.963999	0.01
		41	40.961825	6.73
20	Ca	40	39.962591	96.941
		42	41.958622	0.647
		43	42.958770	0.135
		44	43.955485	2.086
		46	45.953689	0.004
		48	47.952532	0.187
21	Sc	45	44.955914	100.

Atomic Masses and Abundances of Naturally
Occurring Isotopes (Continued)

Atomic No.	Element	Mass No.	Mass[a]	Interim Isotopic Composition for Average Properties[b] %
22	Ti	46	45.952633	8.0
		47	46.951765	7.5
		48	47.947947	73.7
		49	48.947871	5.5
		50	49.944786	5.3
23	V	50	49.947161	0.25
		51	50.943963	99.75
24	Cr	50	49.946046	4.35
		52	51.940510	83.79
		53	52.940651	9.50
		54	53.938882	2.36
25	Mn	55	54.938046	100.
26	Fe	54	53.939612	5.8
		56	55.934939	91.8
		57	56.935396	2.1
		58	57.933278	0.3
27	Co	59	58.933198	100.
28	Ni	58	57.935347	68.27
		60	59.930789	26.10
		61	60.931059	1.13
		62	61.928346	3.59
		64	63.927968	0.91
29	Cu	63	62.929599	69.2
		65	64.927792	30.8
30	Zn	64	63.929145	48.6
		66	65.926035	27.9
		67	66.927129	4.1
		68	67.924846	18.8
		70	69.925325	0.6
31	Ga	69	68.925581	60.
		71	70.924701	40.
32	Ge	70	69.924250	20.5
		72	71.922080	27.4
		73	72.923464	7.8
		74	73.921179	36.5
		76	75.921403	7.8
33	As	75	74.921596	100.
34	Se	74	73.922477	0.9
		76	75.919207	9.0
		77	76.919908	7.6
		78	77.917304	23.5
		80	79.916521	49.8
		82	81.916709	9.2
35	Br	79	78.918336	50.69
		81	80.916290	49.31
36	Kr	78	77.920397	0.35
		80	79.916375	2.25
		82	81.913483	11.6
		83	82.914134	11.5
		84	83.911506	57.0
		86	85.910614	17.3
37	Rb	85	84.911800	72.17
		87	86.909184	27.83

Atomic Masses and Abundances of Naturally Occurring Isotopes (Continued)

Atomic No.	Element	Mass No.	Mass[a]	Interim Isotopic Composition for Average Properties[b] %
38	Sr	84	83.913428	0.5
		86	85.909273	9.9
		87	86.908890	7.0
		88	87.905625	82.6
39	Y	89	88.905856	100.
40	Zr	90	89.904708	51.4
		91	90.905644	11.2
		92	91.905039	17.1
		94	93.906319	17.5
		96	95.908272	2.8
41	Nb	93	92.906378	100.
42	Mo	92	91.906809	14.8
		94	93.905086	9.3
		95	94.905838	15.9
		96	95.904676	16.7
		97	96.906018	9.6
		98	97.905405	24.1
		100	99.907473	9.6
43	Tc	—	—	—
44	Ru	96	95.907596	5.5
		98	97.905287	1.9
		99	98.905937	12.7
		100	99.904218	12.6
		101	100.905581	17.1
		102	101.904348	31.6
		104	103.905422	18.6
45	Rh	103	102.905503	100.
46	Pd	102	101.905609	1.0
		104	103.904026	11.0
		105	104.905075	22.2
		106	105.903475	27.3
		108	107.903894	26.7
		110	109.905169	11.8
47	Ag	107	106.905095	51.83
		109	108.904754	48.17
48	Cd	106	105.906461	1.2
		108	107.904186	0.9
		110	109.903007	12.4
		111	110.904182	12.8
		112	111.902761	24.0
		113	112.904401	12.3
		114	113.903361	28.8
		116	115.904758	7.6
49	In	113	112.904056	4.3
		115	114.903875	95.7
50	Sn	112	111.904823	1.0
		114	113.902781	0.7
		115	114.903344	0.4
		116	115.901744	14.7
		117	116.902954	7.7
		118	117.901607	24.3
		119	118.903310	8.6
		120	119.902199	32.4
		122	121.903440	4.6
		124	123.905271	5.6
51	Sb	121	120.903824	57.3
		123	122.904222	42.7

Atomic Masses and Abundances of Naturally
Occurring Isotopes (Continued)

Atomic No.	Element	Mass No.	Mass[a]	Interim Isotopic Composition for Average Properties[b] %
52	Te	120	119.904021	0.1
		122	121.903055	2.5
		123	122.904278	0.9
		124	123.902825	4.6
		125	124.904435	7.0
		126	125.903310	18.7
		128	127.904464	31.7
		130	129.906229	34.5
53	I	127	126.904477	100.
54	Xe	124	123.90612	0.1
		126	125.904281	0.1
		128	127.903531	1.9
		129	128.904780	26.4
		130	129.903510	4.1
		131	130.905076	21.2
		132	131.904148	26.9
		134	133.905395	10.4
		136	135.907219	8.9
55	Cs	133	132.905433	100.
56	Ba	130	129.906277	0.1
		132	131.905042	0.1
		134	133.904490	2.4
		135	134.905668	6.6
		136	135.904556	7.9
		137	136.905816	11.2
		138	137.905236	71.7
57	La	138	137.907114	0.09
		139	138.906355	99.91
58	Ce	136	135.90714	0.2
		138	137.905996	0.3
		140	139.905442	88.4
		142	141.909249	11.1
59	Pr	141	140.907657	100.
60	Nd	142	141.907731	27.2
		143	142.909823	12.2
		144	143.910096	23.8
		145	144.912582	8.3
		146	145.913126	17.2
		148	147.916901	5.7
		150	149.920900	5.6
61	Pm	—	—	—
62	Sm	144	143.912009	3.1
		147	146.914907	15.1
		148	147.914832	11.3
		149	148.917193	13.9
		150	149.917285	7.4
		152	151.919741	26.6
		154	153.922218	22.6
63	Eu	151	150.919860	47.8
		153	152.921243	52.2
64	Gd	152	151.919803	0.2
		154	153.920876	2.2
		155	154.922629	14.8
		156	155.922130	20.5
		157	156.923967	15.7
		158	157.924111	24.8
		160	159.927061	21.8
65	Tb	159	158.925350	100.

Atomic Masses and Abundances of Naturally
Occurring Isotopes (Continued)

Atomic No.	Element	Mass No.	Mass[a]	Interim Isotopic Composition for Average Properties[b] %
66	Dy	156	155.924287	0.06
		158	157.924412	0.1
		160	159.925203	2.34
		161	160.926939	18.9
		162	161.926805	25.5
		163	162.928737	24.9
		164	163.929183	28.2
67	Ho	165	164.930332	100.
68	Er	162	161.928787	0.1
		164	163.929211	1.6
		166	165.930305	33.4
		167	166.932061	22.9
		168	167.932383	27.0
		170	169.935476	15.0
69	Tm	169	168.934225	100.
70	Yb	168	167.933908	0.1
		170	169.934774	3.1
		171	170.936338	14.3
		172	171.936393	21.9
		173	172.938222	16.2
		174	173.938873	31.7
		176	175.942576	12.7
71	Lu	175	174.940785	97.4
		176	175.942694	2.6
72	Hf	174	173.940065	0.2
		176	175.941420	5.2
		177	176.943233	18.5
		178	177.943710	27.1
		179	178.945827	13.8
		180	179.946561	35.2
73	Ta	180	179.947489	0.012
		181	180.948014	99.988
74	W	180	179.946727	0.1
		182	181.948225	26.3
		183	182.950245	14.3
		184	183.950953	30.7
		186	185.954377	28.6
75	Re	185	184.952977	37.40
		187	186.955765	62.60
76	Os	184	183.952514	0.02
		186	185.953852	1.58
		187	186.955762	1.6
		188	187.955850	13.3
		189	188.958156	16.1
		190	189.958455	26.4
		192	191.961487	41.0
77	Ir	191	190.960603	37.3
		193	192.962942	62.7
78	Pt	190	189.959937	0.01
		192	191.961049	0.79
		194	193.962679	32.9
		195	194.964785	33.8
		196	195.964947	25.3
		198	197.967879	7.2
79	Au	197	196.966560	100.

Atomic Masses and Abundances of Naturally
Occurring Isotopes (Continued)

Atomic No.	Element	Mass No.	Mass[a]	Interim Isotopic Composition for Average Properties[b] %
80	Hg	196	195.965812	0.2
		198	197.966760	10.1
		199	198.968269	16.9
		200	199.968316	23.1
		201	200.970293	13.2
		202	201.970632	29.7
		204	203.973481	6.8
81	Tl	203	202.972336	29.5
		205	204.974410	70.5
82	Pb	204	203.973037	1.4
		206	205.974455	24.1
		207	206.975885	22.1
		208	207.976641	52.4
83	Bi	209	208.980388	100.
84	Po	—	—	—
85	At	—	—	—
86	Rn	—	—	—
87	Fr	—	—	—
88	Ra	—	—	—
89	Ac	—	—	—
90	Th	232	232.038054	100.
91	Pa	—	—	—
92	U	234	234.040947	0.005
		235	235.043925	0.720
		238	238.050786	99.275
93	Np	237	237.048169	—

[a] From data in A. H. Wapstra, K. Bos, *Atomic Data and Nuclear Data Tables* **19**(3), 177-214 (1977).
[b] From 1975 IUPAC in *Pure and Applied Chemistry* **47**(1), 77-95 (1976).

DATE LOANED

JUL 2 7 1978			